ELIZABETHAN
INSTRUMENT
MAKERS

To the Museum of the History of Science:
locus mirabilis

Frontispiece of Gemini's *Anatomy*, 1545. BL Oxford.

ELIZABETHAN INSTRUMENT MAKERS

The Origins of the London Trade in Precision Instrument Making

Gerard L'E Turner

Senior Research Associate
Museum of the History of Science
University of Oxford

OXFORD
UNIVERSITY PRESS

OXFORD
UNIVERSITY PRESS

Great Clarendon Street, Oxford OX2 6DP

Oxford University Press is a department of the University of Oxford.
It furthers the University's objective of excellence in research, scholarship,
and education by publishing worldwide in

Oxford New York

Athens Auckland Bangkok Bogotá Buenos Aires Calcutta Cape Town
Chennai Dar es Salaam Delhi Florence Hong Kong Istanbul Karachi
Kuala Lumpur Madrid Melbourne Mexico City Mumbai Nairobi Paris
São Paulo Singapore Taipei Tokyo Toronto Warsaw

with associated companies in Berlin Ibadan

Oxford is a registered trade mark of Oxford University Press in the UK and
in certain other countries

Published in the United States by Oxford University Press Inc., New York

British Library Cataloguing in Publication Data Data available

Library of Congress Cataloguing in Publication Data
Data applied for

ISBN 0 19 856566 6

Typeset in Bembo by EXPO HOLDINGS, Malaysia
Printed in Belgium on acid-free paper
by Snoeck Ducaju & Zoon N.V.

FOREWORD

R. G. W. Anderson
Director, The British Museum

In his essay *Of the True Greatnesse of Kingdomes and Estates*, Francis Bacon (1561–1626) makes the remark: 'But thus much is certaine; That hee that Commands the *Sea*, is at great liberty, and may take as much, and as little of the Warre, as he will.'

In the Elizabethan period, command of the sea, conduct of war, and control of the land were increasingly facilitated by the use of scientific instruments such as the nocturnal, the gunner's rule, and the theodolite. Instruments were not unknown in England prior to this time. Geoffrey Chaucer wrote a treatise on the astrolabe at the end of the fourteenth century, to instruct his son in its use. The few, tantalizing pieces of early evidence that do survive are insufficient for it to be certain that instruments were being constructed in England rather than being made abroad for an English market. The research of Professor Turner makes the position clear for the second half of the sixteenth century: that high-quality scientific instruments were being made by a number of craftsmen, mainly in London, during a period of ambition, confidence, and increasing wealth.

The use of material culture has been well established for studies in a number of disciplines, where museums are an early port of call for the scholars involved. This has not been the case in the history of science until quite recently, and the potential which scientific instruments hold has only been significantly demonstrated over the past thirty years or so. This is in part because the objects concerned have not been well documented, there are relatively few of them, and they are scattered in many locations. Undoubtedly there are other, cultural reasons as well.

The work of the Scientific Instrument Commission of the International Union for the History and Philosophy of Science, of which Professor Turner was secretary for twenty years, has helped to bring about the change, by the production of national inventories, and by providing a forum for scholarly discussion. It would seem clear that with work such as that which follows, the object as evidence will become even better established in this field. Indeed, it is clear that instruments are not an alternative source of knowledge to manuscripts and printed books; they are complementary, and without them our understanding of important areas of cultural history could not be illuminated.

PREFACE

The mathematical instrument makers, or 'mechanicians', as they were often called, working in London between 1540 and 1610 are the subject of this book. The purpose of the study is to present the work of these men in its entirety, since up till now they have been considered, if at all, for a single aspect of their achievement; to emphasize the importance of practical technology in the fields of navigation, surveying, and gunnery in the Elizabethan period; and to relate the individual craftsmen of the sixteenth century to the rapid growth of the scientific instrument trade in London in the following three hundred years.

Scientific instruments, insofar as they have been studied at all, have generally been examined individually, and in relation to a scientific figure, or publication. My contention has always been that it is essential for their proper evaluation to try to establish the existence of workshops, and craft centres, and to relate the making of scientific artefacts to economic factors. In attempting this, I worked first on optical instruments, from their invention in the early seventeenth century. The examination of 110 surviving telescopes by James Short (1710–68), all marked with serial and model numbers, enabled me to work out his production rate, and therefore his earnings. The detailed examination of the tooled decoration on the leather and vellum tubes of 70 microscopes and telescopes made between 1650 and 1750 established that they were made by specialist tube makers, not by the instrument makers who sold them, and provided a dating guide. Nor is there any lack of scientific content in the study of instruments, provided that a representative group can be examined together, for these artefacts are, in a real sense, ideas made brass.

I have built on the work of admirable, but few, predecessors. R.T. Gunther and Ernst Zinner amassed valuable and extensive information on astrolabes and Germanic instrument making, respectively. D.W. Waters' work on the history of navigation, and E.G.R. Taylor's numerous books on geography, and the practical applications of mathematics, are landmarks in this type of study. More recently, Joyce Brown most tellingly discovered a basic link between the new instrument trade and the London guild structure, and this was extended by the research of M.A. Crawforth and Gloria Clifton, resulting in the first *Directory* of scientific instrument makers.

The method used for identifying unsigned and undated instruments is best described as forensic. In the fine art world it is called connoisseurship. This amounts to a close examination and collation of small details that may be compared with signed and dated (or datable) artefacts. One is looking for characteristics of engraving, nuances in decoration, ingenuity, choice of data, sources of data, spelling, and also errors. To further this approach, the descriptions of instruments in

Part II include apparently slight details such as the alternate shading of degree intervals in the divisions of a circle, which may be recognized as the style of a particular craftsman. Decoration or the lack of it may suggest a period of time. Similarly, the spelling of personal names and towns tends to have settled down during the early seventeenth century.

It is sometimes argued that instruments that have survived from as long ago as the sixteenth century will inevitably be luxury objects, and can therefore provide little accurate information on the range and quantity of instruments being produced. It is true, of course, that the survival of any artefact over centuries is bound to be random. However, of the more than 100 instruments described in this book, 18 different types can be distinguished, and 46 were made for professional use. Of the rest, the 24 compendia, pocket-sized personal possessions, form the largest single group, and these were not luxuries any more than a watch is today. Only some of the astrolabes are expensive luxury items.

It was the discovery of a significant group of sixteenth-century English instruments in the Museo di Storia della Scienza in Florence that began my investigation of the London trade at that time. Not all these instruments were signed, but I found that detailed examination of the engraving and other craft features, usually with the aid of a microscope, enabled me to identify the hand of certain makers with sufficient confidence for attribution. The technique was first used to identify a superb unsigned Flemish astrolabe in the Florence collection as being the work of Gerard Mercator. Two more astrolabes in European collections were also attributed to him, the third of which bore the monogram of Mercator, not previously recognized. Thus the monogram verified the process of attribution.

This forensic method has made it possible to record and study over a hundred London-made instruments of the Elizabethan period. The five main makers were pioneers, whose lead was followed by the establishment of an increasing number of individual workshops, and eventually developed into a retail trade on a considerable scale. What can be found out about these men, and those of their products that have survived, is a fresh source of information for the study of the intellectual and economic aspects of the Elizabethan period.

Our knowledge of Thomas Gemini, Humfrey Cole, and Augustine Ryther comes from their work as map or book engravers. The study of engraving is well established, and Gemini's edition of Vesalius, and the maps of Cole and particularly of Ryther, are duly recorded. But their instruments receive no mention. All that has been written about Gemini is the work of historians of medicine, because he published Vesalius's text. James Kynvyn, who is not known to have engraved maps, receives virtually no attention. Charles Whitwell fares better, since he engraved both maps and instrument illustrations in textbooks. The fact is that all these men, whose signed instruments have survived, and others whom we know to have been producing instruments, were making a living in response to an urgent need for more accurate measurement demanded by the professions of navigation, surveying, and astronomy. I have tried to deal with the whole man, considering his maps and illustrations as well as his instruments, and relating him to his patrons or customers.

The later international importance of London as a centre for precision instrument making had its origin in the Tudor period, and put down strong roots because of the mathematical knowledge and manual skill of these first practitioners.

The debt to Continental learning, and the craft of copperplate engraving for printing maps perfected in Flanders is clear. But Elizabethan England, threatened by war, competing for overseas trade, and facing population growth and huge social change at home, provided numerous and eager patrons.

Oxford
November 1999

G.L'E.T.

CONTENTS

NOTE TO THE READER

Each instrument in Part Two has a number printed in bold (e.g. **87**), and these are used for reference throughout the book. All measurements are in millimetres, except where otherwise stated. Futher information is on pp. 93–4.

ACKNOWLEDGEMENTS

I am deeply indebted to the Leverhulme Trustees, who, in March 1996, granted me an Emeritus Fellowship for two years to work on mathematical instrument making in Elizabethan England. The Trust's generous grant enabled me to travel, and to finance the acquisition of photographs of instruments that were essential to my research.

I also acknowledge with gratitude the support I have received from the Renaissance Trust throughout my research into the history of scientific instruments.

This book has been a family affair. I am grateful to my daughter, Jane Bigos, a graphic designer, for her advice on calligraphy, and to my son, Mark Turner, an information technology specialist, for creating images of letter and number forms, and for advice on all computing matters. My wife, Helen Turner, has been a close colleague in all my work on the history of scientific instruments for the past 25 years, and shares in whatever I have achieved.

A great many people have given me their help and support over the years, of whom I single out Robert Anderson, David Bryden, Jeremy Collins, Howard Dawes, Elly Dekker, David King, Gerry Martin, Tony Simcock, and William Ryan. To these and other colleagues and friends who share my enthusiasm for instruments, I express my warmest thanks.

For specific help with aspects of this book, I wish to thank the following: S. Ackermann, J. Baddeley, H. Beez, J.A. Bennett, C.N. Brown, Joyce Brown, R. Bud, K. Van Cleempoel, G.C. Clifton, J. Van Damme, P. Galluzzi, M. Gandy, J.A. Goodall, W.D. Hackmann, F. Herbert, H. Higton, G. Hudson, S.J. Johnston, M. Miniati, A. Morrison-Low, F. Principe, R.W. Shirley, A.D.C. Simpson, B. Stephenson, L. Taub, D.R. Thompson, S. Vanden Broecke, S. Waterman, T.P. Waterman. Without the willing co-operation of all these, and others whom I may have failed to name, a project as complex as this book could never have reached publication.

ILLUSTRATION CREDITS

The illustrations fall into three categories. There are colour pictures, described as Plates, and identified by the number of the instrument illustrated, in bold; photographs in Part I, described as Figures, identified by the chapter number followed by the sequential number; and photographs of the instruments described in Part II, which are identified by the instrument number in bold.

The following museums, libraries, institutions, and individuals have kindly supplied photographs used for this book. Permission for use has been granted, and their help and support are warmly acknowledged. Each name is followed by the abbreviation used in the captions, and by the figure, or object, numbers credited.

Adler Planetarium, Chicago (AP Chicago): Figs 5.6, 5.7, 5.8.

Bodleian Library, Oxford (BL Oxford): Frontispiece (Vet. A1 b.1); Fig. 3.1 (Douce G subt. 45); Fig. 3.4 (Mason Q 233 map 19–20); Fig. 4.1 (4° Z 58(9) Med Sig. Gii r/v); Fig. 4.2 (Ash. 162(5) fol. 36); Fig. 4.3 (4° Z 58(9) Med Sig. Eiv r); Fig. 4.5 (4° c. 13(2) Art before p. 1); **30** (H 4.20) Art(3). fol. 87 r); **35** (Mason BB 184 (no.92) opp. p. 18), **53** (Mason BB 184 (no.4) opp. p. 4).

British Museum (BM London). © The British Museum: **2, 10, 12, 16, 21, 22, 24, 36, 44, 53, 55, 57, 66, 85**.

Christie's South Kensington, London: **26, 50, 59, 83**.

Collection of Historical Scientific Instruments, Harvard (CHSI Harvard): **93**.

Eton College Library (ECL Eton): Fig. 3.2.

Historisches Museum, Basel (HM Basel). Photo Historisches Museum Basel: Peter Portner: Fig. 2.4; **1, 65**.

Horniman Museum (HM London): **33, 99**.

Istituto e Museo di Storia della Scienza (IMSS Florence): Figs 5.2, 5.3, 5.4, 5.5; **3, 7, 20, 35, 42, 43, 47, 48, 49, 51, 52, 67, 68, 70, 71, 72, 73, 74, 75**.

Kunstgewerbemuseum, Berlin (KM Berlin): **14**.

Musée de la Vie Wallonne, Liège (MVW Liège): **27**.

Musée du Louvre, Paris (ML Paris): **88**.

Musées Royaux d'Art et d'Histoire, Brussels (MRAH Brussels): **45**.

Museum of Artillery, Woolwich (MA Woolwich): **11**.

Museum of the History of Science, Oxford (MHS Oxford): Figs 2.11, 3.5, 4.4; **6, 8, 28, 31, 32, 37, 41, 56, 61, 69, 79, 86, 95, 98, 100**.

National Maritime Museum (NMM London): Figs 2.3, 2.6; **1, 9, 15, 29, 40, 58, 81**.

LIST OF PLATES

Plates fall between pages 210 and 211

PART ONE

THE LONDON TRADE

CHAPTER ONE

THE LONDON TRADE IN THE SIXTEENTH CENTURY

The London trade in precision scientific instrument making, which achieved world-wide status in the eighteenth century, can be said to have come into being from the 1540s. It arose to meet the pressing need of the new professionals, surveyors, navigators, gunners. They in turn were responding to economic demands created by a changing and expanding society in England, and by the discovery of the American continent.

The changes in England were sufficiently startling. Between 1525 and 1601, the population rose from 2.26 million to 4.1 million; there was considerable migration from the country to the towns, and particularly to London; and for much of the Tudor period, there was high inflation (Burnett 1969; Patten 1978; Guy 1988). At the beginning of the sixteenth century, England was still largely medieval and life on the land was only slowly ceasing to be purely at the subsistence level. But land ownership was changing, and this change accelerated with the acquisition by Henry VIII of the vast monastic property. Gradually, and starting even before the Tudor period, open fields were being enclosed to make farms. Produce of all kinds was needed to feed a growing population, and throughout the Tudor period, England remained self-supporting for food. Church lands passed into individual ownership, and private estates were created, each with its new house. The great medieval forests that had covered large areas of the country were shrinking under the demands of agriculture, house-building, and ship-building. By the end of the century, there were plans to drain the fens to make more land for agriculture and housing, on the Dutch pattern.

Thus, during the first half of the sixteenth century, the scene was being set for the development of surveying into a profession requiring increasing skill, speed, and accuracy, and therefore the use of instruments devised to make angular measurements. Surveying became a science alongside, and to a certain extent as a result of, the development of deep-sea navigation, and the growing military skills of fortification and siege associated with the use of artillery. The accurate determination of position, both at sea and on land, increased in importance as ships left the security of coastal waters, as new tracts of unexplored territory were discovered, and as the reclaiming of land from the sea by means of waterways, dikes, and locks became necessary.

The success of ocean navigation and the growth of overseas trade depended increasingly on improved navigational techniques, which in turn relied upon more accurate astronomical observation. When out of sight of land, the latitude could be measured by taking the elevation above the horizon of the Pole Star, or by the meridian altitude of the Sun. The longitude is more difficult. A clock taken on board ship could in principle keep the time at the port of departure, and give the longitude by the difference between the time it told and local time, estimated from the Sun or stars, because one hour equals 15° of longitude. But no clock existed that could withstand the movement of a ship at sea until the marine chronometer was devised in the mid-eighteenth century. The Moon, therefore, had to be used as a clock, which is possible if the Moon's position against the background of the stars can be set out in tables; this is known as the lunar distance method. The most complex of the instruments taken by Sir Robert Dudley to Florence was a lunar computer devised by him for this very purpose (see Chapter 5 and **51**). As the safety of men and ships depended on the accuracy of such tables, the refinement of astronomical observation became of pressing importance, and this depended on ever greater accuracy of measurement.

England was at war for most of the Tudor period (Guy 1988). Elizabeth's reign began with war against both France and Scotland, and later the threat of the Spanish Armada had to be faced. Therefore the arts of fortification and ordnance were in demand, both depending on accurate measurement and the use of instruments. In 1558, for example, the town of Berwick in Northumberland, crucial in Border defence, had new walls built to resist artillery (Guy 1988, p. 249). At the other end of the country, the Thames was defended by gun batteries. William Bourne (*c.* 1535–82), who spent his life at Gravesend, was a volun-

teer gunner with the garrison manning the town's defensive bulwark, and wrote a textbook for gunners, *The Arte of Shooting in Great Ordnaunce* (1578), as well as his famous navigation manual, *A Regiment for the Sea* (1574). Important products of the instrument trade were measuring devices for gunnery.

The use of angular measurement by professional surveyors and navigators, and the provision of accurate maps, charts, and surveys depended on two things: knowledge of mathematics, and the new techniques of printing, first invented in the mid-fifteenth century. Throughout the sixteenth century, there was a massive expansion in the practice of the mathematical arts (Feingold 1984). The foundations of this were laid by Continental scholars, some of the most notable being Regiomontanus (1436–76) in Nuremberg (Zinner 1990), Martin Waldseemüller (1470–*c.* 1518) of Lorraine (*DSB*), and Gemma Frisius (1508–55) of Louvain (Ortroy 1920). This university city of Flanders was of particular significance in the development of the skills demanded by the age of exploration, since it was a centre for engraving and printing (Cleempoel 1998), where the greatest of all map-makers, Gerard Mercator (1512–94), was educated. It was also one of the European centres visited between 1547 and 1550 by John Dee (1527–1608), a founder Fellow of Trinity College, Cambridge, who pioneered the provision of mathematical textbooks in English. Highly influential was Dee's *Mathematicall Præface* (1570), written for the first English translation of Euclid by Henry Billingsley, where he discussed and tabulated the practical applications of arithmetic and geometry. At this point, instruments come into their own. As pointed out by Bryden (1992, p. 302):

Frequently the application of mathematics to the practical arts involved the use of instruments in the making of primary measurements. At the same time, the subsequent mathematical manipulation of those observations could also be assisted, or undertaken with greater rapidity or surety, by instrumental means.

Before the new angular measuring techniques could be imported from the Continent, the ability to do simple arithmetic was a prerequisite. Arabic had to replace roman numerals, which did not disappear from the English Exchequer records until well into the seventeenth century. Teaching the use of pen reckoning and arabic numerals to a wide public was the achievement of Robert Recorde (1510?–58), a fellow of All Souls College, Oxford, whose influential book, *The Ground of Artes Teachyng the Worke and practise of Arithmetike*, was published in London in 1543 and ran through 28 editions to 1699. His heirs were the mathematical practitioners, some but by no means all university educated, who provided the growing class of professional surveyors and seamen with the practical knowledge they needed.

The skills of printing, and the engraving of book illustrations, maps, charts, and brass scientific instruments came to England from Continental Europe, and more particularly from the Low Countries. It is significant that *The Ground of Artes* was printed by Reynor Wolfe, who left Drenthe in The Netherlands to settle in London in 1533. Thomas Gemini followed him a few years later from Flanders, establishing himself with his edition of Vesalius's *Anatomy*. Even as late as the end of the sixteenth century, it was said to be difficult to find an English printer to set a mathematical work correctly. For the *Atlas of the Counties of England and Wales*, published in 1579 by Christopher Saxton (d.1596), 23 of the 34 maps are signed, and only seven of these are by Englishmen. Throughout the century, there was steady immigration of craftsmen to London from the turmoil of the Spanish Netherlands. The style of engraving practised by the London instrument makers was strongly influenced by the calligraphic teaching and practice of Gerard Mercator.

Surveying textbooks at the beginning of the sixteenth century did not deal with geometrical surveying. *The Boke of Surueyeng and Improuementes* published in 1523 by John Fitzherbert was concerned with carrying out audits and valuations. Even as late as 1577, Valentine Leigh covered much the same ground in *The Moste Profitable and commendable science, of Surueying of Landes, Tenementes, and Hereditamentes*, but did provide tables for estimating unequal-sided areas, by the use of simple diagrams. The traditional equipment of the surveyor was a plane table with alidade, a chain, a square, and a pair of compasses. But the plane table was condemned by Thomas Digges as 'an Instrument only for the ignorante and unlearned, that haue no knowledge of Noumbers'. What the mathematicians offered, and what the textbooks recommended, was the instrument that eventually became the theodolite. It first appeared in recognizable form in the 1512 edition of *Margarita Philosophica* by Gregor Reisch, in a section on architecture and perspective by Waldseemüller. It was illustrated, and called the 'Polimetrum'. Leonard Digges, of University College, Oxford, and his son, Thomas, a Cambridge graduate, wrote two influential texts on mathematical surveying: *A Boke named Tectonicon*, published in 1556, and

A Geometrical Practise named Pantometria, written by Leonard, but completed and published in 1571, after his death, by Thomas. This described three instruments for the surveyor that could be combined into 'a topographicall instrument'. One of the three, called 'theodelitus', was a simple theodolite, while the composite instrument was an altazimuth theodolite. These were the new instruments which, once mastered, could increase the speed and accuracy of surveying, and which, increasingly, were in demand by surveyors.

In much the same way, seamen found that, though they still used the traditional lead, log-line, and magnetic compass, for dead reckoning while sailing in sight of land, or running along the latitude, transatlantic voyages required more sophisticated methods of determining position. A quadrant and a forestaff, among other instruments, were essential to make observations and calculations on ocean voyages. Navigators, too, were influenced by textbooks on the technique of their profession. Two of the most influential were those published by William Barlow (1544–1625), of Balliol College, Oxford, later chaplain to Prince Henry, eldest son of James I, and by Edward Wright (1558?–1615), a Cambridge mathematician who, unlike Barlow, had practical experience of ships. Barlow published *The Navigator's Supply* in 1597, and was an inventor of instruments, improving the variation compass, the pantometer, the hemisphere, and the traverse board, and inventing what he called 'the Traveller's Jewel', a combination of variation compass and equinoctial sundial. This was engraved for his book's title page by Charles Whitwell, as were eight other instruments illustrated in the book, and Whitwell also made and sold the instruments.

Wright's *Certaine Errors in Navigation, arising either of the ordinarie erroneous making or using of the sea Chart, Compasses, Crosse staffe, and Table of declination of the Sunne, and fixed starres detected and corrected* appeared in 1599, its title making clear the sort of problems the deep-sea navigator had to overcome. It has been said of Wright (Waters 1958, p. 359) that *Certaine Errors* is 'one of the three most important English books ever published for the improvement of navigation'.

The textbooks referred to above are all by university-educated men, and indeed their knowledge was crucial in teaching the mathematical arts. But by no means all of the mathematical practitioners, as they were called, were university graduates. Barlow, for example, gave an advertisement on the title page of his book for 'Iohn Goodwin, dwellinge in Bucklersburye, teacher of the grownds of these artes', a respected teacher of mathematics, who knew no Latin, and was therefore 'unlettered'. Another practical man who contributed to the development of navigation was Robert Norman, a seaman turned instrument maker of Ratcliff, East London. He wrote, in 1581, the earliest text in English on the variation of the magnetic compass, entitled *The New attractive*, referring to himself in the dedication as 'an unlearned mechanician'. The age of Elizabeth was remarkable for the interaction between men of different backgrounds when they were engaged by a common purpose. The practice of navigation provided such a purpose, and linked men as disparate as Sir Robert Dudley, Matthew Baker, the royal shipwright, Abram Kendal, ship master, and William Bourne, author of *A Regiment for the Sea*. The skilled technician began during this period to be accepted at his true worth.

Gabriel Harvey (*c.* 1550–1630), Cambridge graduate, poet, and satirist, was also a keen student of astronomy, and an experimentalist, who demanded: 'Give me ocular and rooted demonstration of every principle' (Stern 1979, p. 165). In *Pierces supererogation* (1593, p. 190), he wrote:

> He that remembreth Humfrey Cole, a Mathematicall Mechanician, Matthew Baker, a ship wright, Iohn Shute, an architect, Robert Norman, a Nauigatour, William Bourne a gunner, John Hester a Chimist, or any like cunning, and subtile empirique (Cole, Baker, Shute, Norman, Bourne, Hester will be remembered, when greater Clarkes shall be forgotten), is a prowd man if he contemne expert artisans, or any sensible industrious practitioner, howsoever unlectured in Schooles or vnlettered in bookes.

How successful were they in worldly terms, these instrument makers? That the London merchants grew in wealth and influence during the sixteenth century is certainly true, but while some grew rich, the majority made only a reasonable living, and all traders suffered the adverse effects of high inflation. Gemini secured a royal stipend of £10, giving him some security. Double that amount was Cole's annual salary from the Mint. But both men needed to augment their incomes by casual earnings, and in this way they became instrument makers. Again, £10 was the price paid by the Earl of Leicester for an instrument made by Gemini, almost certainly an astrolabe (see Chapter 2, 'Gemini', and **6**). It is difficult, and may be misleading, to try to give a modern equivalent of this sum, but perhaps £5000 would be a possible figure.

The practitioners of the scientific instrument trade in the Elizabethan period themselves reveal the various influences that led to its existence. These include the important input from Continental Europe, the demand for the copperplate engraving of maps, the close links with mathematical practitioners,

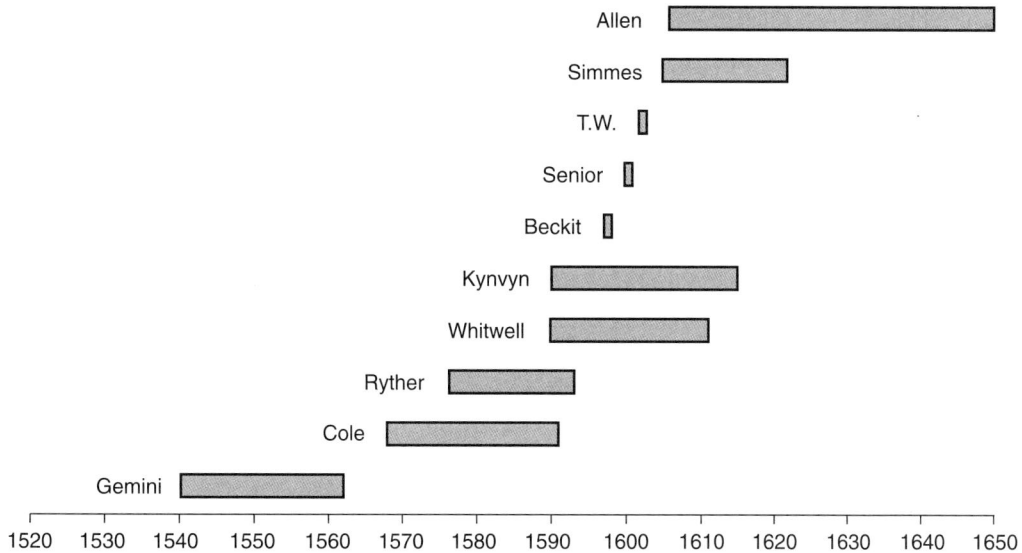

FIG. 1.1 Working dates of Elizabethan scientific instrument makers.

and the need to acquire and work for individual patrons. Those makers whose instruments have survived, and whom we therefore know something about, as well as others whose names only we know, or who are unknown, were finding and serving a new market, but one which grew with remarkable speed. A major reason for this growth was the size and opulence of London, which, by the 1550s, had a population estimated at 80,000, and astonished contemporary observers with its fine shops, new houses, and evidence of growing wealth. Though England as a whole had been poorly developed compared with the Continent, the changes that occurred during the Tudor period, and began in the capital, were truly astonishing. This is reflected in the growth of the new instrument trade. Only 80 or so years divide the arrival of Thomas Gemini in London and the establishment in business of Elias Allen, prolific and incomparable maker of the first half of the seventeenth century (Fig. 1.1).

A crucial factor in the establishment of the new trade in London was the way in which it became grafted onto the guild structure. How it did so was discovered and analysed by Joyce Brown (J. Brown 1979). She found that one of the Twelve Great Livery Companies, the Grocers, had provided a craft 'home' for a surprisingly large number of instrument makers, starting in the Elizabethan period with Augustine Ryther. Between 1688 and 1800, in particular, she found in the company records the names of 65 masters and over 200 apprentices. It was normal practice for an apprentice, at the end of his training, to become free of the same company as his master. There were, of course, instrument makers who belonged to other companies, for example, the Goldsmiths and the Clockmakers, but the Grocers' Company records yielded important material about the growth of this precision industry. In the eighteenth century, Thomas Heath, the Adams family over three generations, and the Troughtons, another dynasty, were all Grocers. See also Crawforth (1987).

The trade certainly did grow, for it depended upon a demand that steadily increased. The professional requirement for mathematical instruments at sea, for surveying, and other practical purposes continued to increase as more instruments were invented and trade expanded. To this was added, in the course of the seventeenth century, the making of the new optical instruments, the telescope and the microscope, both of which caught the public imagination. The telescope became a professional instrument comparatively early, used by astronomers increasingly as observatories were built all over Europe through the eighteenth century. A huge new market was added from the 1750s with the development of the teaching of experimental philosophy. The lecture demonstration of physical phenomena began in the universities, but became a popular form of general entertainment, promoted by the instrument makers themselves. The demonstrations required much apparatus, notably the air-pump, the electrical machine, and optical devices.

London in the eighteenth century was famous all over Europe for its shops, and not least for those selling precision instruments (G. Turner 1990, Chapter IX). During the seventeenth century, the individual workshop producing instruments to order was supplemented, and in the course of time almost replaced by the instrument retailer, with a range of goods to offer customers. A pioneer of instrument retailing was Benjamin Martin (1705–82), who was chosen to supply Harvard College, Massachusetts with replacement teaching apparatus after a fire. Visitors to London from all over Europe, some of whom have left diaries recording their travels, were enthusiastic over the range and quality of the instruments they found, and the knowledge of those who sold them. The instrument makers were, increasingly, men of intelligence and education, capable of writing about their products, willing and able to innovate, and recognized as distinguished members of the scientific community. The heirs of Humfrey Cole, Charles Whitwell, and Elias Allen were both numerous and gifted. The scientific instrument trade had grown to yield a rich harvest.

CHAPTER TWO
THE CRAFTSMEN

INTRODUCTION

The mathematical instrument makers who are the subject of this chapter are known to us from the survival of instruments signed by them (Table 2.1). All were primarily engravers, and all, with the exception of James Kynvyn, also engraved maps and other pictorial artefacts. Augustine Ryther, indeed, can most properly be described as a map engraver, though two instruments signed by him exist today. What occurred in the Elizabethan period was the response of skilled engravers to a demand for accurate instruments for time-telling, surveying, and navigation. We know that there were other engravers working in London in the second half of the sixteenth century; their names are recorded in Hind

TABLE 2.1 Elizabethan makers and their extant instruments

Name	Dates	Instruments in Part Two
Thomas Gemini	*c.* 1515–62†	7
Humfrey Cole	*c.* 1525–91†	26
James Kynvyn	*c.* 1550–1615†	12
Augustine Ryther	*c.* 1550–93†	2
Charles Whitwell	*c.* 1568–1611†	27
Robert Beckit	1597★	1
T.W.	1602★	1
Robert Grinkin	*c.* 1575–1626†	1
William Senior	*c.* 1575–1641†	1
Isaac Simmes	*c.* 1580–1622†	3
Elias Allen§	*c.* 1588–1653†	8¶
Anonymous	16th century	12
Total		101

Birth dates are estimates from a consideration of the career.

† Year of death.

★ Date from only known instrument.

§ Apprenticed to Whitwell *c.* 1602.

¶ Allen had a very considerable output, and many of his instruments survive. The 8 included here, made between 1606 and 1617, are chosen to show continuity of design from earlier years.

(1952, pp. 314–15). No signed instruments made by them have so far been discovered, but they may yet come to light.

The names of other London instrument makers of the time are known to us through contemporary accounts by the scholars and professional men who employed them. Edward Worsop published, in 1582, *A Discoverie of sundrie errours and faults committed by Landemeaters*, intended to provide a simple, understandable textbook for the average surveyor. In it he lists some names of makers who can provide instruments, including Humfrey Cole and John Bull for metal instruments, and John Reade, James Lockerson, and John Reynolds for wooden instruments. There are two important points to consider here. One is that, though we have a fine group of Cole's instruments, none are known by John Bull of Exchange Gate, the other maker of metal instruments mentioned. The second point is that, alongside the more expensive and durable metal instruments, many wooden instruments were being made for the surveyor and the navigator. Few, if any, of these have survived; the earliest known to the present author is a nocturnal dated 1637. For a thorough listing of instrument makers known from all sources, see the *Directory of British Scientific Instrument Makers 1550–1851* by Clifton (1995).

John Reade, who had a workshop in Hosier Lane, and specialized in making plane tables, is also recommended by Cyprian Lucar (1544–after 1590), an Oxford graduate who published, in 1590, *A Treatise named Lucarsolace*, again intended for the practical surveyor. So is John Reynolds 'dwelling right against the southeast end of Barking churchyard in tower

streete', and another name is added, that of Christopher Paine, also of Hosier Lane. We know of two more makers because of their association with mathematical writers or lecturers. Thomas Osborn was a kinsman of Thomas Fale, who wrote a book on dialling published in 1593, and had the instrument he devised made by Osborn. Francis Cooke, whose workshop was in Mark Lane, made instruments for Thomas Hood (c. 1557–1620), who devised a new form of sector. Such men are not known to us from surviving instruments, but they, too, were making a living in response to the professional market. These are only some of the names that can be collected from contemporary technical literature, and they indicate the scale on which the demand for scientific instruments grew during the Elizabethan period.

THOMAS GEMINI
c. 1515–62

In the heading to his entry on 'Thomas Geminus', Karrow (1993, p. 250) has: Geminus, Gemynous, Geminous, Geminie, Gemyny, Gemyne, Gemini, Lambert, Lambrit, Lambrechts, Lamberd, Thomas ♊. If the Latin versions, and the era's variations in spelling (see the variations of Humfrey Cole's name in Table 2.2) are ignored, one is left with the names Gemini and Lambrit. The dedication to his book *Compendiosa totius anatomie delineatio* is signed: Thomas Geminus Lysiensis; his Will is signed, in his elegant hand: Thomas Gemini. His executor was his brother, Jasper Lambrit. So we know that Gemini was the pseudonym of an immigrant to London from the Low Countries, whose given name was Lambrit.

There continues to be a mystery over where Gemini was born. The *DNB* entry for him, published in 1889, says the word 'Lysiensis' has baffled the most learned investigation, and adds that Gemini is generally supposed to have been an Italian, giving unacceptable 'evidence' for the assertion. In his paper 'Medicine and the Crown', Underwood (1953), commenting on the *Anatomy*, writes: 'Who was this man Geminus? He was certainly a foreigner, and he probably belonged to the village of Lys-les-Lanoy, which lies fifteen miles from Lille, in French Flanders.' This suggestion rests on Gemini's description of himself as 'Lysiensis'. The river Lys rises in Artois, and flows through Aire-sur-la-Lys, four other towns with 'sur la Lys' names, and Armentières, near Lille. At Kortrijk, just over the linguistic border, it is the Leie, and it reaches the sea at Knokke-Heist. Gemini's Will bequeathed to his brother, Jasper, living in London, his property 'as lyeing and beinge within Leighe nighe unto Marke Wessett within the bishopryke of Leuke [Liège] in the partes beyond the sea'. Michel (1961) assumes that this is the family property, and therefore Gemini's birthplace, and interprets Leighe as Lixhe, near Liège. Until the Belgian records have been exhaustively examined, there remains a choice between these alternative locations for Gemini's birthplace.

As to Gemini's profession, this seems to reflect the calling of those who write about him. Underwood (1953) wrote: 'Despite doubts which have been cast on his profession, Geminus was almost certainly a surgeon, and he was probably a maker of surgical instruments.' Mitchell (1953), attempting to answer Underwood's question, published a letter that draws on the *DNB* for the fact that Gemini was described as a surgeon in the list of Edward VI's disbursements (1547–52), and in the *Annals of the Royal College of Physicians* (1555). To medical historians, he was a surgeon or physician on the strength of his *Anatomy*. (See Figs 2.1, 2.2.) He became a Freeman of the Stationers' Company, setting up as a printer at Blackfriars. His imprint is on Leonard Digges, *A Prognostication* (1555), and on the 1559 edition of his *Anatomy*. In the *Register of the Stationers' Company* (Arber 1875, pp. 44, 48), he is described as a stranger (alien), and is recorded on 21 July 1555 as fined 12 pence for breaking the Company's rules, and as a benefactor, in the sum of 20 pence, to Bridewell. This was the name of a great house built by Henry VIII between Fleet Street and the Thames, made over by Edward VI in 1553 to the City of London as a penitentiary, and formally taken over by the Lord Mayor and Corporation in 1555. So Gemini can also be correctly described as a printer. His Will is headed: 'Thomas Lambrit als Geminy of the precincte of the late blacke friers nighe Ludgate of London, booke printer'. Finally, Henri Michel (1961), historian of scientific instruments, describes Gemini in the *Biographie Nationale* as: 'imprimeur, graveur et constructeur d'instruments de mathématique'.

In considering Gemini's background and early training, the most important element is the link that can be traced with one of the great figures of the Renaissance, Gerard Mercator (1512–94). A graduate of the University of Louvain, Mercator forsook the study of philosophy to make large globes for navigation, maps, and charts. His cylindrical projection of 1569 allowed seamen to plot a course

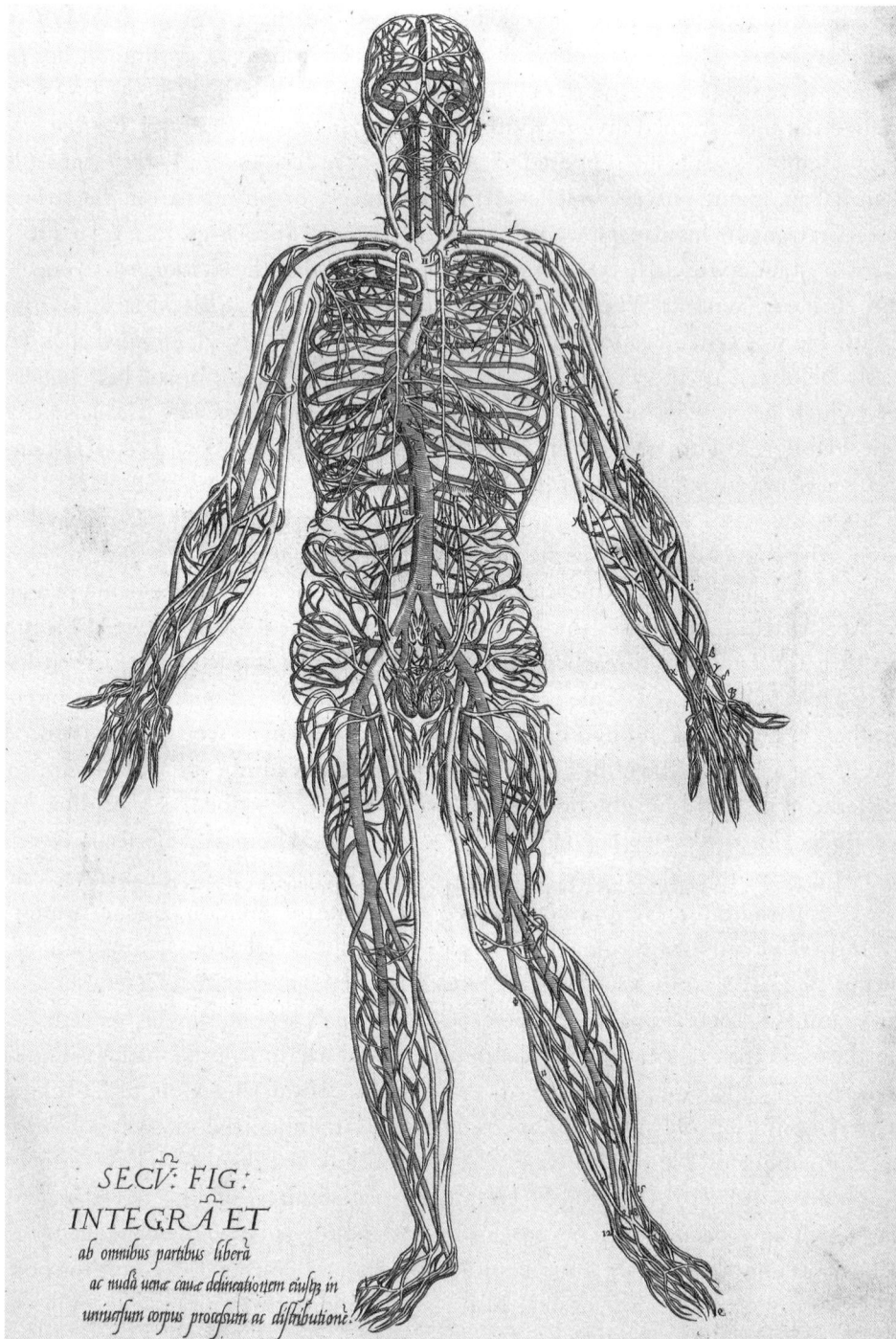

SECV̄: FIG:
INTEGRÂ ET
ab omnibus partibus liberâ
ac nudâ uena cauæ delineationem eiusꝫ in
uniuerſum corpus proceſſum ac diſtributionè

FIG. 2.1 The system of blood vessels in a man. Gemini, *Anatomy*, 1545, vol. I, plate between sigs. Eiii and Eiiii. WIL London.

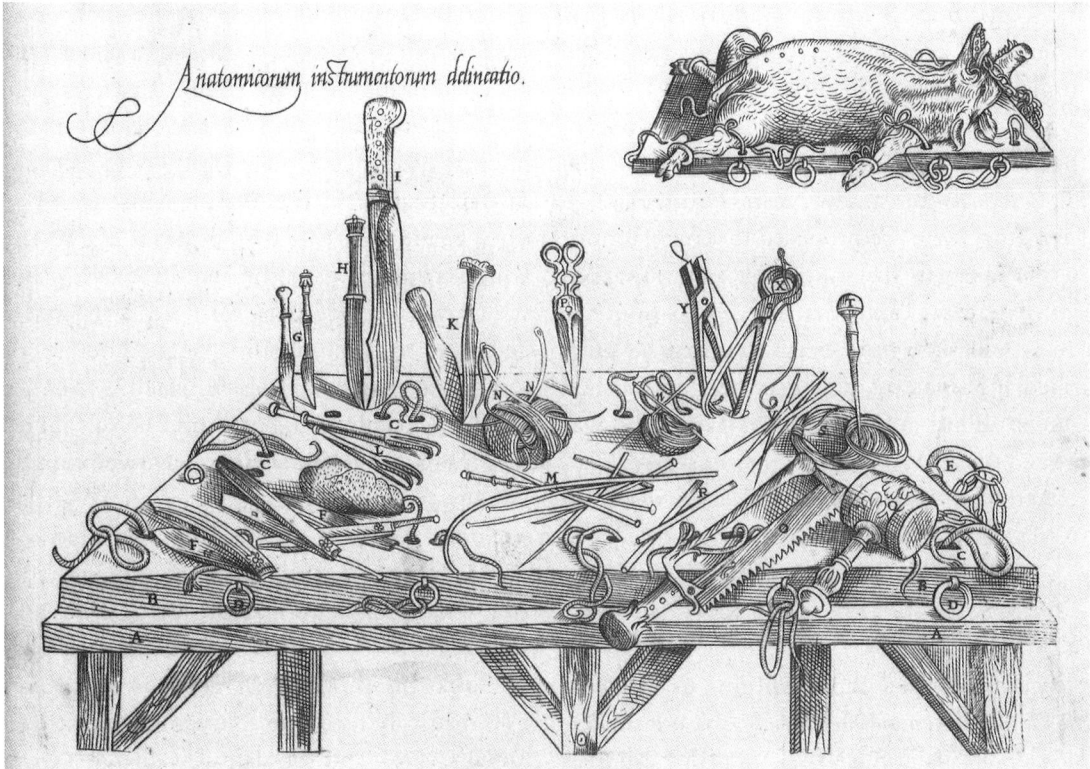

Anatomicorum instrumentorum delineatio.

FIG. 2.2 Anatomical, or surgical, instruments. Gemini, *Anatomy*, 1545, vol. II, lower half of plate after sig. Hii. WIL London.

much more easily and quickly (Watelet 1994). He was known to have made instruments in brass, but none was thought to have survived until 1992, when the present author identified an unsigned astrolabe in the Florence museum as by Mercator (G. Turner and Dekker 1993). This led to the discovery of two more of his astrolabes, in Augsburg and in Brno, Czech Republic (G. Turner 1994). The Brno astrolabe has a small convoluted monogram on the outer rim that had never been explained; it represents the letters GMR, Gerard Mercator Rupelmondanus. This early astrolabe, a product of Mercator's Louvain period, before 1552, has a rete that bears a

striking resemblance to those made by Gemini. This, and other features described below, suggest that Gemini may well have trained alongside Mercator, and his engraving is certainly of comparable quality.

Gemini moved to London in the early 1540s. Mercator set up his own workshop in 1540, but prior to that, as Walter Ghim tells us in his biography of Mercator (published in Osley 1969), he took his Master's degree at Louvain University in 1532, and then studied astronomy and mathematics, 'with such diligence … that, within a very few years, he was teaching their elements to a number of private students and from time to time was fashioning and constructing scientific instruments [i.e. spheres and astrolabes], astronomer's rings, and similar apparatus in bronze' (Osley 1969, p. 185). It is possible that Gemini was one of these private students, and he could be expected to be a few years younger than Mercator. It can be proposed that he was born in 1515, began his studies by 1535, worked with Mercator, and then moved to London. He would then have been 30 when he published his *Anatomy*, and 44 when he finished Queen Elizabeth's astrolabe. His Will was proved on 27 May 1562, so he must have died earlier that year. The Will is printed in O'Malley (1959, pp. 34–5).

Gemini's Search for Patronage

Gemini produced three editions of *Compendiosa totius anatomie delineatio* based on *De humani corporis fabrica* (Basel, 1543) by Andreas Vesalius (1514–64), a native of Brabant, who studied medicine at the universities of Louvain and Paris (Cushing 1962). These editions are described in Chapter 3. The first edition of *Anatomy* (1545) was dedicated to Henry VIII, who died on 28 January 1547. It earned for Gemini a royal pension of £10 a year, which continued until his death (Hind 1952, p. 41). The second edition (1553), was dedicated to Edward VI. It is a translation from Latin into English by Nicholas Udall, whose epistle 'To the jentill readers and Surgeons of Englande' is dated at Windsor 20 July 1552. The king died on 6 July 1553, aged 16. The third edition (1559), was dedicated to Queen Elizabeth.

It has been pointed out that the portrait on the title page of the third edition is that of Queen Mary, and not of Queen Elizabeth as is generally assumed. In the Bodleian Library copy (Douce G subt. 45), there is a paper note attached inside the front cover. It is written in ink in a late eighteenth-century hand, probably that of Francis Douce (1757–1834), keeper of manuscripts in the British Museum. The note reads: 'In the first edition H. 8. the King's arms were in the middle of ye print, which were replaced by ye head of Mary, & dedicated to Q Elizabeth. QM died before, or just as ye p[rinting] was finished, when ye book was coming out, & all printed off; therefore there was no time to engrave ye reigning Princess, & all yt cd be done was to dedicate ye book to her.' Queen Mary died on 17 November 1558, and Elizabeth immediately became Queen. The colophon reads: Anno salutis. 1559. Mense Septemb. It is possible that the month referred to is September 1558, just before the November death of Queen Mary, and the date was changed from 1558 to 1559. This construction would be necessary if the note is to be believed. This engraving resembles some paintings of Mary.

A separate portrait of Queen Elizabeth was published by Gemini a year or so later, a fine engraving, and a traditional early likeness (Strong 1987, p. 60). Its date of 156[] has lost the final number through damage, and not through some oversight (see Chapter 3).

Four of the seven known instruments made by Gemini have royal connections, and are fully described in Part Two. Astrolabe **1**, *c.* 1551, has the royal arms and cypher for Edward Rex. Quadrant **2**, dated 1551, is inscribed *Edwardus Rex*, and has the initials of members of the court, Sir John Cheke and William Buckley. Astrolabe **5**, dated 1552, has the royal arms and cypher for Edward Rex, and the arms and cyphers for the Duke of Northumberland, and Sir John Cheke. Astrolabe **6**, dated 1559, has the royal arms and the inscription: *Elizabeth Dei Gratia Anglia Franciæ & Hiberniæ Regina*. This is identical to the inscription Gemini used on his engraved portrait of the Queen preserved at Eton College (Fig. 3.2).

It is probably safe to assume that the instruments for King Edward were commissioned by Cheke and Buckley, who were court tutors. Whether payment for the instruments was made by these men or the royal purse we do not know. We are on safer ground for Elizabeth's astrolabe. The household accounts of Robert Dudley, Earl of Leicester (1532–88), record two payments to Gemini (Adams 1995, pp. 66, 138). Leicester, a favourite of the Queen, was appointed a Knight of the Garter on 23 April 1559, and he immediately ordered a case for the Garter, which cost 2s 6d. Towards the end of May, the accounts note:

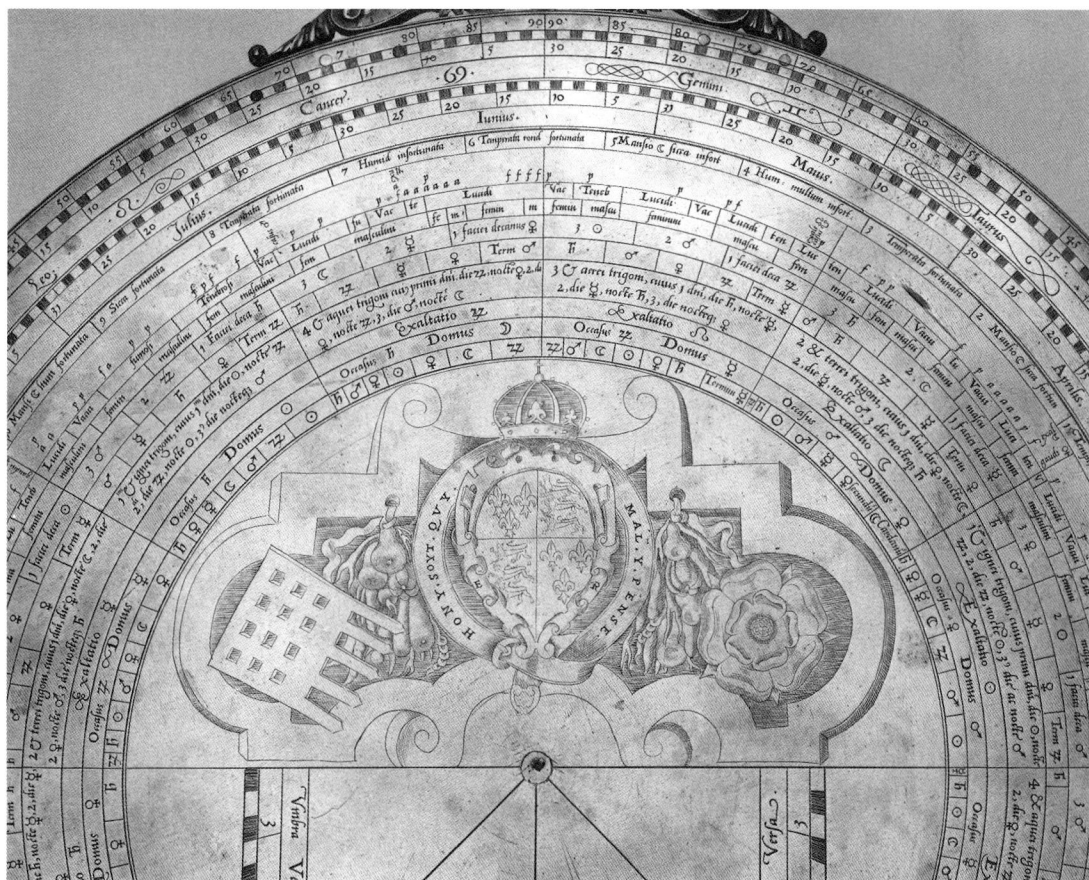

FIG. 2.3 Detail of Gemini astrolabe 1, *c.* 1550. NMM Greenwich.

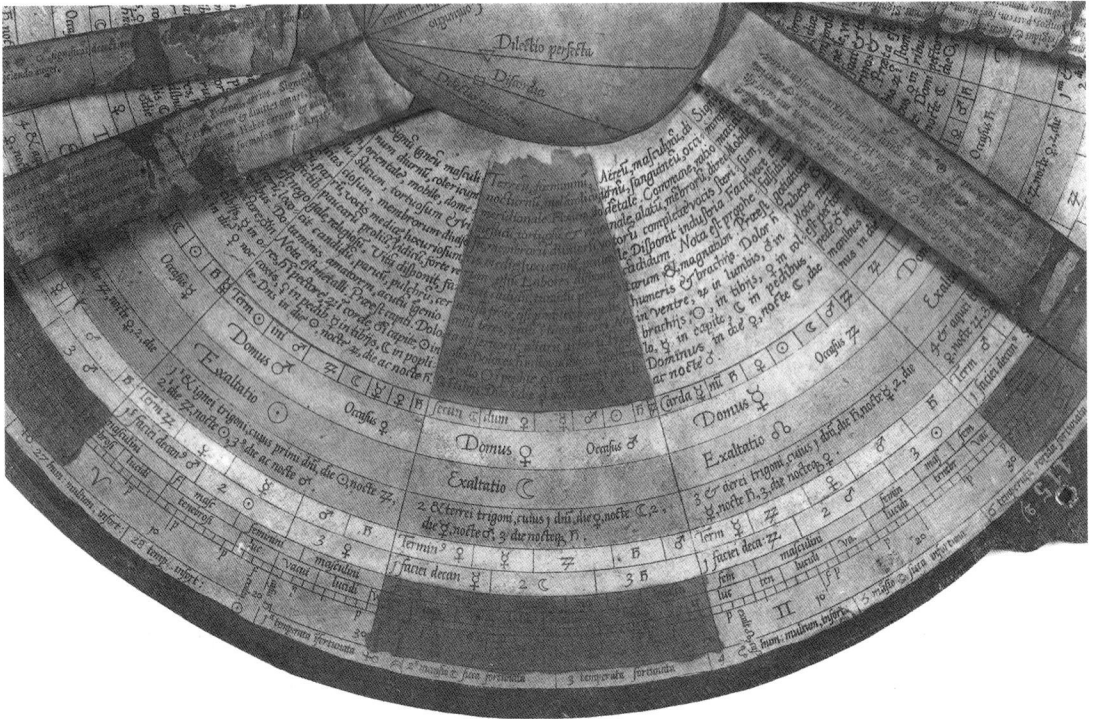

FIG. 2.4 Detail of Mercator astrological disc, 1551. HM Basel.

'Item to Jemynie the Frencheman for an instrment of astronymye *xli*'. There follow various expenses in going to Windsor for the Garter installation on 6 June 1559. One can scarcely doubt that the astrolabe (**6**) now on exhibition at the Museum of the History of Science, Oxford, is the 'instrment of astronymye' made by Gemini for the large sum of £10. In April 1560 we find another relevant payment: 'Item paid unto Gemyny of the Blacke Fryers for compasses xxxs.' No quantity is mentioned, so we do not know how many compasses could be had for 30s. They may have been similar to the gunner's compasses made by Humfrey Cole, now in the British Museum (**10, 22**). This is the only known reference to Gemini having produced this kind of instrument.

Gemini's Connection with Mercator

Gerard Mercator began his commercial life as a calligrapher, but not limited to handwriting. He must be regarded as a technical man who used lettering in his products, producing his first terrestrial globe in about 1536, a celestial in 1537, a map of the Holy Land in 1537, and a double-cordiform map of the world in 1538. In 1540 he published his manual on calligraphy. It is striking how alike are the engraving hands of Mercator and Gemini. Figs 2.3 and 2.4 compare similar parts of the astrological disc of Mercator, dated at the bottom of the explanatory text Louvain May 1551, with the back of Gemini's now incomplete astrolabe **1** (G. Turner and Cleempoel 2000). The layout, wording, and sigils are the same; they differ

FIG. 2.5 Comparison between the rete patterns of a Gemini astrolabe and two Mercator astrolabes. From the left: Gemini, 1559; Mercator, *c.* 1545; Mercator, *c.* 1570.

only in the reading position, Gemini from centre to rim, Mercator from rim to centre. Mercator's disc was supplied with his celestial globe of 1551, which has astrological information on the horizon ring (Dunn 1994; Broecke 2001). Mercator would have been working on the gores and the disc, among his other tasks, for some considerable time, so could have been engaged in their production from about 1548. Gemini's instrument is undated, but was most likely to have been made in the first years of King Edward's short reign. Whichever of the two men first drafted it, the common astrological material is firm evidence for a link between the two Flemings.

The rete design of Gemini's two astrolabes described under **6** and **7**, matches that of Mercator more closely than any other maker. So far three astrolabes of Mercator have been identified. The earliest of the three is in the Moravská Galerie, Brno, is datable to about 1545, and was definitely made in Louvain (G. Turner 1994, p. 350). The other two were made *c.* 1570 in Duisburg, where Mercator

went after leaving Louvain in 1552. Silhouettes of the retes are displayed in Fig. 2.5.

The sides of two quadrants by Gemini, **2**, dated 1551, and **3**, are engraved with an unusual sinical quadrant. A very full description of the manner of use is in a Mercator holograph of 1550 preserved in the Archives of the Royal Monastery of San Lorenzo el Escorial, Spain (MS &-III-8, ff. 237–64). I am grateful to Dr K. Van Cleempoel for drawing my attention to his discovery. Here is yet another example of the continuing connection between Gemini and Mercator.

All the known instruments produced by Gemini are described in Part Two. Of these, four are astrolabes, two are quadrants, and one is a horizontal sundial. The three conventional astrolabes (**1**, **6**, **7**) are remarkably similar in size, with diameters of 14 inches in English measure, consistent to less than 1 per cent. The smaller astrolabe (**5**) and the sundial (**4**) have diameters of 12 inches, with a difference of only 1.5 per cent. The radii of the two quadrants at $10^3/_4$ inches differ by only 1 per cent. The measurements show that Gemini was using the English inch, and had the idea of adhering to consistent sizing. Mercator used the length standard of the place where he was working, thus the Mechelen standard at Louvain, and the Rhineland standard at Duisburg (G. Turner 1994, p. 350).

Gemini's Calligraphy

The cursive italic hand was established in Italy and spread through Europe. With the use of copperplate engraving, italic lettering soon became the standard, and could be cut small, suitable for maps and instruments. Following Mercator's style (Osley 1969), Gemini's capitals are large, and his minuscules are open and regular, with rather short ascenders and descenders; *ct* and *st* ligatures are used, and flourishes are common. The numeral *1* has an angled serif at top left, and a curve at bottom left. The *3* is long, with two clear curves. The loop of the *6* and *9* is not closed, and the riser or descender is curved and not over-long. Gemini was a highly skilled engraver, and his calligraphy shows clearly the teaching of Mercator. The fine anatomical figures in his *Anatomy*, and the wording on each plate, show this graphically. When finishing the degree scales on instruments, alternate degree intervals are shaded by hatching in a characteristic manner.

HUMFREY COLE
c. 1525–91

What can be set down about the career of Humfrey Cole must start from the words he engraved in a cartouche on his Map of the Holy Land:

GRAVEN BI HUMFRAY COLE GOLDSMITH A ENGLISH MAN BORN IN YE NORTH AND PERTAYNING TO YE MINT IN THE TOWER · 1572

Cole's claim to be a goldsmith cannot be substantiated from the guild records, as the relevant documents have been destroyed. There is no reason, however, to doubt his assertion. An apprenticeship to a goldsmith might well have led to a position at the Mint. No date or location has been found for Cole's birth, beyond his claim that he was born in the north. There was an important Cole family in Newcastle at the end of the sixteenth century, but registers do not exist before 1560, too late for Cole's birth, and the name Humfrey does not occur. Cole is, however, a common name in the north-east (Ruth Wallis, personal communication). The various spellings of the name on Humfrey Cole's instruments emphasize the non-standardization of spelling in the period (see Table 2.2). It is a matter of regret that Henri Michel (1961, col. 392) tried to make out that Cole was a Fleming, and related to Abraham Ortelius (1527–98). There are some who have repeated this fable.

Cole wrote in December 1578 a letter to Lord Burghley (transcribed in Gunther 1927, p. 311) explaining that he had worked at the Tower Mint for 20 years. He could have worked first as a moneyer, before taking the comparatively senior position of sinker of the irons, also known as die-sinker or undergraver, which he is recorded as holding from 1563. He succeeded John Lawrence, who was appointed in 1545 at a salary of £10 a year, raised in 1553 to £20 a year, which was the payment Cole received throughout his tenure of the post (Challis 1975). After 1578, his name no longer appears in the Tower records.

TABLE 2.2 Instruments by Humfrey Cole

Date	Signature	Instrument	Number
1568	HVMFRAY COOLE	compendium	8
1569	Humfray Colle	compendium	9
1569★	Humfray Coolle	gunner's folding rule	10
1570	HC	gunner's folding rule	11
1570★	H. Cole	nocturnal	12
1574★	humfray Colle	altitude dial	13
1574	Humfridus Côle Anglus	altitude dial	14
1574	H. Cole	theodolite	15
1574	Humfrey Cole	astrolabe	16
1574	Humfrey Côle	surveyor's folding rule	17
1574	Humfrey Côle	surveyor's folding rule	18
1575	Humfrey Côle	surveyor's folding rule	19
1575	Humfrey Côle	surveyor's folding rule	20
1575	H. Côle	ring dial	21
1575	H. Côle	gunner's folding rule	22
1575	Humfridus Côle Londinensis	astrolabe	23
1575	Humfrey Côle	compendium	24
1575	Humfrey Côle	compendium	25
1579	Humfrey Côle	compendium	26
1579	Humfrey Côle	compendium	27
1579	H. Côle	horizontal sundial	28
1582	Humfræ Colle	horizontal sundial	29
1582	Humfrey Côle	nautical hemisphere	30
1586	H. Cole	theodolite	31
1588	H. Côle	plane-table alidade	32
1590	Humfrey Cole	compendium	33

★ Undated, probable date made.

Since a large part of Cole's working life was spent at the Mint, and it provided his regular source of income, it is worth giving some account of how the Mint was organized. Before 1544, the Mint in the Tower of London operated a contract system, with only the Warden, the Assay Master, and three other officials paid by the Crown. The moneyers, who fashioned the blanks and struck the coins, were organized into their own company under a Provost, who paid them on a piece-work system. In 1544, however, all this changed with the decision of Henry VIII to debase the coinage. The Tower Mint was divided into two, and secondary mints were established in other parts of London and in provincial cities. Each had an under-treasurer whose job was administrative, including accounting for the salaries of the staff, paid by the state. One of the early achievements of King Edward VI was to recall the debased coinage, and replace it with silver and gold coins to the old sterling standard. The debasement officially ended in 1551, and the task of replacement was completed by 1562. Henry VIII's staffing arrangements at the Mints were not rapidly changed, however, and remained effectively until 1571, when contracts were reinstated. It should be noted that the Tower of London, as well as housing the Mint, also included a foundry for the production of cannons and armour. For the Mint, see Challis (1975, 1989, 1992).

Cole's letter to Burghley of 1578 deals first with work undertaken at his patron's request, but ends with a plea to be granted a new position at the Mint, namely that held by Eloy Mestrell, which he claims was promised to him. Mestrell was a Frenchman who was brought to the Mint to introduce mechanized coining techniques, and worked in Tower II, the second Mint, in a self-contained unit, from about 1560. Considerable sums are recorded as being spent on machinery, and he certainly produced coins, but his success was not thought to justify replacement of the traditional techniques. Mestrell himself became involved in counterfeiting, and was eventually executed for this offence in 1578, at which point Cole attempted, unsuccessfully, to take over his position.

It was through his employment at the Mint that Cole came into contact with the men who were concerned with the introduction into England of mining for ores, and the manufacture of metals, to replace the importation of metals and metal goods from the Continent. This became a matter of increasing political urgency, given the precarious relations of England with the rest of Europe following Henry VIII's break with Rome, and later the war with Spain. Cole was among those occasionally employed by the Company of Mineral and Battery Works in prospecting for ores. This small company, together with the much larger Company of Mines Royal, both received their Royal Charters in 1568; see Chapter 3, Gunnery Tables. The task of both companies was to discover sources of marketable minerals, and to secure and maintain monopolies over their production. Burghley was a prime mover in this activity, and Cole's letter to him of December 1578 provides an example of the type of prospecting enquiries that were being undertaken, and of Cole's involvement in them (Gunther 1927, p. 314).

The earliest dated instrument by Cole to survive was made in 1568, but it is not known exactly when he started augmenting his income from the Mint by producing instruments. It is, however, clear that he was both highly skilled and well-established by the mid-1570s, for he was chosen to supply instruments for the 1576 Frobisher exploration voyage to find a north-west passage to the Far East. Michael Lok, a leading merchant adventurer and London agent of the Muscovy Company, joined Martin Frobisher in a project to attempt a north-west route, since the north-east route via Archangel had not been found. Both men secured powerful support for the venture, and the investment level was such that a new ship was built, and charts, navigation books, and instruments were ordered. Lok's account books, which have survived, reveal that 'Humfrye Cole and others' were paid the sum of £49 9s 2d for instruments, books, and maps, which are listed by Gunther (1927, pp. 315–16) and by Waters (1958, Appendices 10B, 10C).

The first Frobisher voyage, in 1576, returned with ore found on a small island at the mouth of what is now Frobisher Bay. This was analysed and stated to be rich in gold and silver. The exploration objective of the venture was forgotten and a second voyage, in 1577, was devoted to identifying and bringing back more samples of ore from the same area. These aroused such high hopes that the final expedition in 1578 consisted of 15 ships and 351 sailors, miners, and soldiers. Vast quantities of ore were brought back, but it was finally established as being worthless. As well as supplying instruments for the first voyage, Cole was one of those who certified the quality of the ore brought back from the second voyage in an assay which was used as the basis of plans for the third voyage. He was also involved in attempts to find, in the Newcastle area, a mineral required for the smelt to be presented at the assay.

What is known of Cole as an instrument maker, apart from the considerable quantity of fine instruments by him that have survived, is fragmentary. We know he worked for the Queen's printer, Richard Jugge (d. 1579), for whom he engraved the map of the Holy Land that appears in the 1572 edition of the Bishops' Bible. Jugge had premises near St Paul's, as did Cole. His name is the only one mentioned in connection with the instruments for the Frobisher expeditions. William Bourne, in his *Inventions or Devises* of 1578, describes a ship's log using cogs and dials that he says was invented by Humfrey Cole; the invention was not his, but he may well have made the instrument (Taylor 1930, p. 159). We learn of Cole's address from an advertisement in a book for surveyors, published by Edward Worsop (1582): 'Scales, compasses, and sundry sorts of Geometricall instruments in metall, are to be had in the house of Humfrey Cole,

neere vnto the North dore of Paules'. The Subsidy Roll for London for the year 1582 records Cole as living in the parish of St Faith, north of St Paul's. In his copy of John Blagrave's *The Mathematical Jewel* (1585), Gabriel Harvey noted in 1590 on the title page the names of his instrument makers, Kynvyn, and 'old Humfrie Cole, mie mathematical mechanicians'. Cole must certainly have been near 60 years of age at that time to warrant the epithet old, but, given the usual age span of the period, he was probably not much more. The estimated date given for his birth proposes that he was 66 when he died in 1591; he may have been older. We know the year of his death from administration of his estate being granted to his wife, Elizabeth, on 6 July 1591 (PROB 6/4 f.176).

Cole's Calligraphy

Humfrey is completely confident in his handling of the burin, or graving tool, and he tends to cut deeply, with precision. He does use guidelines, but they are not always apparent. The letters in a line of script are compact, but very legible, maintaining a consistently high quality throughout. The flourish is used, generally to fill a space. He favours the *st* ligature; the *ct* ligature is also found. Capitals are slightly shorter than the ascenders. Long letters, *b, l, h*, etc., have a diagonal serif to the left of the ascender. The *f* has a double serif at the top, and the bar extends to the right only. The *N* has a serif on the bottom right vertical. The *C* has a double serif at the top, while the *S* has a heavy double serif at each end. Cole maintains a uniform size for his numerals. His 6 and 9 keep a curve to the ascender and descender, unlike Whitwell who gives a long and straight cut. A good example of Cole's lettering is shown in the name 'Edward Williams' on **15**.

Cole's Ingenuity

Not only is Cole a remarkably precise craftsman, one who made the finest large astrolabe of the age (**23**) with faultless stereographic projections, he was also ingenious in small devices. In this regard he is superior to other makers. One may cite some examples.

The magnetic compass with the compendium, **8**, can be released by pressing the split-ring on the underside; there is a location stud for alignment.

The compass box on the compendium, **9**, has on the underside a circular extension with a shallow waist. This waist is gripped by a pair of sprung claws resembling those of a crab (Fig. 2.6). This mechanism is reached by

lifting up the hinged panel that crosses the next leaf in the compendium. It is easy to press the claws apart and so release the miniature theodolite.

On **16**, the alidade of the small astrolabe pivots on the central pin, and is held in place by another of Cole's ingenious devices. The pin has a circular groove, which is gripped by the pincer-shaped ends of two bars that are pegged to the base of the alidade. When closed, these bars are held by studs, and the alidade appears to be solid.

The large astrolabe, **23**, has plates in the mater, and to secure each in position, at the bottom and both sides, are tiny circular catches. These are inset into the limb and are pinned so they may be pulled out to press down on the plate (Fig. 2.7). A very small extension, with a hole, is to the side of each catch, and the plates are notched so they may be easily inserted over the extensions, which are used to extract the catches from their sockets.

In the case of the hemisphere, **30**, a large disc is attached by the pin holding an alidade

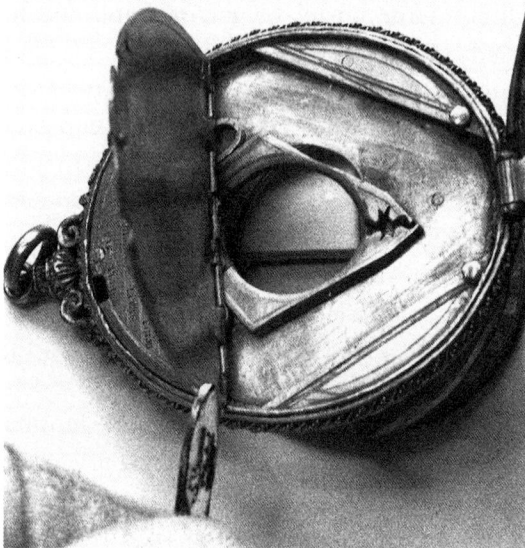

FIG. 2.6 Sprung claws to grip shaft of the Cole miniature theodolite in compendium **9**, 1569.

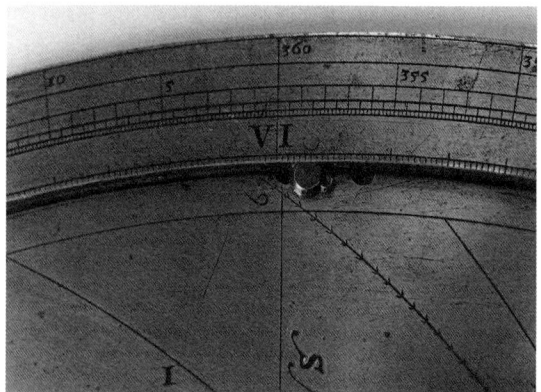

FIG. 2.7 Catch to hold a plate in the Cole astrolabe **23**, 1575.

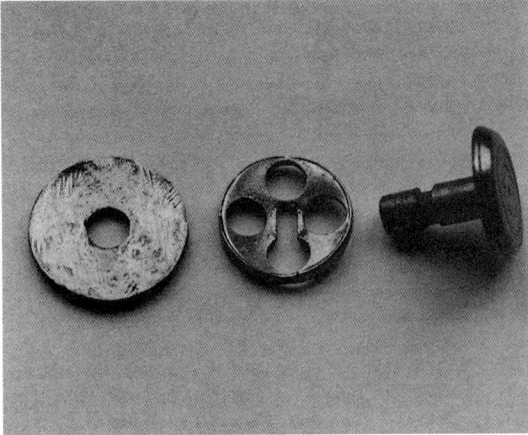

FIG. 2.8 Securing bolt for the alidade of the Cole
nautical hemisphere **30**, 1582.

on the front side by an ingenious small circular button with a central hole surrounded by four others. The central hole has a radial slot in the form of a key-hole (Fig. 2.8). This pushes onto the pin, and the slot engages in a cut made in the shaft of the pin. The button is attached over a small domed brass disc, or washer. This ensures that the alidade is free to rotate, but is held securely in the position to which it has been set.

JAMES KYNVYN
c. 1550–1615

James Kynvyn is somewhat of a mystery. We can place the other Elizabethan makers in context to some extent, though details of their lives are hard to find. Gemini came from Flanders already a skilled engraver, and settled in London. Cole worked at the Mint in the Tower of London as an engraver, and moved into instrument making. Ryther, Whitwell, and Allen form a succession of master/apprentice, and were all free of the Grocers' Company.

Kynvyn is known for a group of instruments made in the 1590s, and not, as all the others are, for maps or other engraved work. The dates on the instruments, both signed and attributed, span only ten years. His Will, in the name of Jacobus Kynvin, is dated September 1615 (PROB 11/126 ff. 362–362v). This document describes him as citizen and Merchant Tailor of London. The Court Minutes of the Company, dated 26 October 1573, record: 'First the Mr [Master] and Wardens aforesaide at the contemplacon of my Lord Mayor his Honour and the request of Sr Thomas Offley Knight have made free James Kynvyn per redempcon gratis'. Offley was a former Lord Mayor. This Minute reveals that Kynvyn was being honoured by the Company, on the recommendation of senior City figures, an indication of family or personal status. In January 1574, Kynvyn's marriage to Elizabeth Coke at the church of St Michael Bassishaw is recorded in the church registers. The two events may be connected.

The Will also reveals that Kynvyn was born in south Wales, and had brothers, John and Edward Job, and other family. The executor was his nephew, another James Kynvyn; no children are mentioned. His wife, Elizabeth, had predeceased him, and he wished a memorial to be erected to them both in St Bride's. Money was left to the poor of the parish of St Bride's, and also of his home parish of Llantilio, Monmouthshire. Reference is made to 'my dwellinge house in Fleet Street London'.

These facts together convey a different impression from that of the other instrument makers. These are skilled artisans, trained in their trade from an early age, and practising it all their lives. Kynvyn's family must have been of some status in Llantilio, for the Will of John William Kynvyn, dated 1617 (PROB 11/130 ff. 322–3), presumably James's brother, refers to him as 'gentleman'. James was well connected in the City of London and had money to leave, both in London and Wales.

How he came to practise instrument making must be conjectural, but his first instruments, judging by the date of his marriage, were produced in his forties. His name occurs in the Subsidy List for St Bride's parish, Fleet Street, in 1582, so he must have been in business then in the area. His burial is recorded in the registers of St Bride's on 19 December 1615. A possible explanation is

that engraving was an interest and skill that he decided in middle life could prove lucrative. Cole, whom he must have known, was growing old, as was Ryther, and Charles Whitwell was just ending his apprenticeship. It is tempting to speculate that Kynvyn seized his chance and filled a need, numbering Robert Dudley and Gabriel Harvey among his clients. The latter, in a hand-written note on his copy of Blagrave's *The Mathematical Jewel* (1585) refers to Kynvyn: 'His [Blagrave's] familiar staff, newly published this 1590. The Instrument itself made and sold by M. Kynuin, of London, neere Powles. A fine workman and mie kinde frend: first commended unto me bie M. Digges, & M. Blagraue himself.' And again: 'Mr Kynvyn selleth ye Instrument in

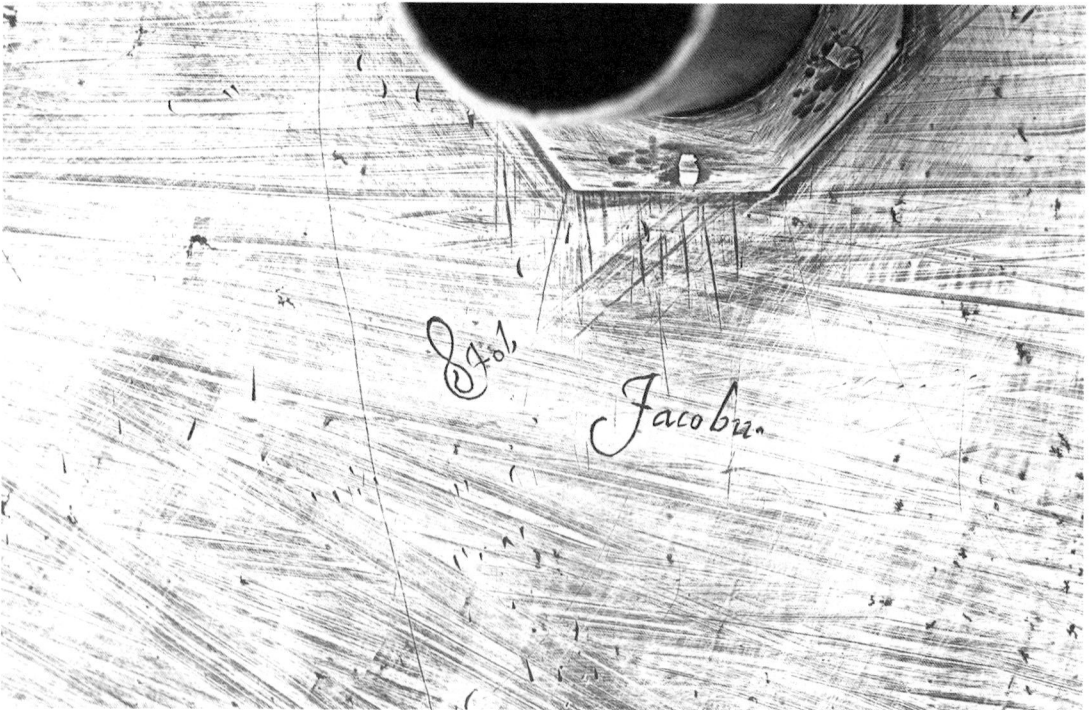

FIG. 2.9 Trials with the burin and two names on the Kynvyn theodolite 70, 1597.

brasse'. Harvey also refers to: 'old Humfrie Cole' in the same note. Harvey's copy of Blagrave is in the British Library, press mark C60.0.7. For comments on the whole of Harvey's annotation, see Bryden (1992, p. 305).

Kynvyn's Calligraphy

Of the 12 instruments of Kynvyn described in Part Two, four are signed, the earliest dated 1593, a compendium (**66**) made for Robert Devereux, 2nd Earl of Essex. Seven of his instruments were taken to Florence by Sir Robert Dudley in 1606. Fairly competent as an engraver, but not quite the quality of Cole, Ryther, or Whitwell, Kynvyn is careless. When numbering a scale, he quite frequently places a number in the wrong place, and has to abandon it, insert a caret mark, rub out the wrong portion, or over-engrave. Examples of this are seen on the large sector, **68**, which is signed, and dated 1595. His letters and numbers are better on a large instrument when there is plenty of space, as on **75**. It is evident that he is not comfortable with small writing, probably a pointer to lack of experience, and confirming he was not a map engraver.

Kynvyn uses guide lines, and these are visible. The spacing between letters and between numbers is variable. The ascenders on minuscules (*b*, *d*) are higher than majuscules; the minuscule *t* has its ascender bent sharply to the right above the bar; the numeral *1* has diagonal serifs top and bottom in rotational symmetry (I is also used). Double serifs are at the tops of *3*, *5*, *7*, and at the end of the bar on *4*. The ascender of *6* can be nearly upright or nearly horizontal. One has the impression that Kynvyn's control of the burin gradually increased as the years go by.

When preparing to engrave his theodolite, **70**, he made, on the rough-finished underside, many trial engravings with the burin to test the hardness of the brass and hence the strength he would need to make the cuts. Such practice is common. Amongst the scratches, there are two names cut in full. One is his own: *Jacobu*, while the other is his brother's: *Job* (Fig. 2.9). This clearly links James Kynvyn the instrument maker with the family of Llantilio through the Will.

AUGUSTINE RYTHER
c. 1550–93

Ryther became the first important English map engraver, and on occasion claimed his nationality by using the signature: 'Augustinus Ryther Anglus' to distinguish his work from that of the Flemish engravers. He was, primarily, a copperplate engraver, but he did make brass instruments, only two of which have survived, and these are of the highest quality.

The date and place of Ryther's birth are not known. It has been suggested, without any evidence, that he was born in Leeds. The only foundation for this is that he worked for, and therefore may have been a friend of, the surveyor Christopher Saxton, a Yorkshireman, for whom he engraved five maps. The first important information we have on Ryther comes from the title page of a book that he had published on his own account in 1590, a translation of an Italian treatise on the defeat of the Spanish Armada, by Petruccio Ubaldini. He now offered for sale together with the translation the ten detailed and one general charts of the progress of the Armada that he had engraved (Hind 1952, pp. 138–49, Plates

79–87). On the title page he gives his address: 'the shoppe of A. Ryther beinge a little from Leaden hall next to the Signe of the Tower', and completes the page with the arms of the Grocers' Company. So we know where he worked and that he was free of the Grocers', the first in a line of instrument makers to belong to the Company. Ryther had a number of apprentices, notably Charles Whitwell, who also became a Freeman of the Grocers' Company, and a leading instrument maker (J. Brown 1979, pp. 24, 58–60).

No record of a marriage for Ryther has been found, but we know that he died in 1593, since the burial of 'Augustine Ryther' is recorded in the registers of St Andrew Undershaft in the City of London on 30 August of that year. Administration of his estate is also recorded in 1593 in the name of August. Ryder of St Andrew Undershaft (Commissary Court records, vol. 14, f. 273v).

It has been accepted by some authors (Hind 1952, p. 138) that two letters referring to 'Mr Ryther' being held in the Fleet Prison, presumably for debt, in 1594 and 1595, concern Augustine Ryther. Since, however, his name is given in full in the burial register and not in the letters, and Ryther clearly had employment to the end of his life, this can be discounted. His last commission was probably the very large map of Cambridge on nine copper sheets, published in February 1592/93.

Ryther's Calligraphy

Ryther was an engraver of maps and charts (Fig. 2.10), of which some two dozen are known. This obliged him to cut for the titles large majuscules, commonly swashed, and tiny minuscules for the names of towns, of which there could be in excess of 500 on a map of a medium-sized county, such as Gloucestershire.

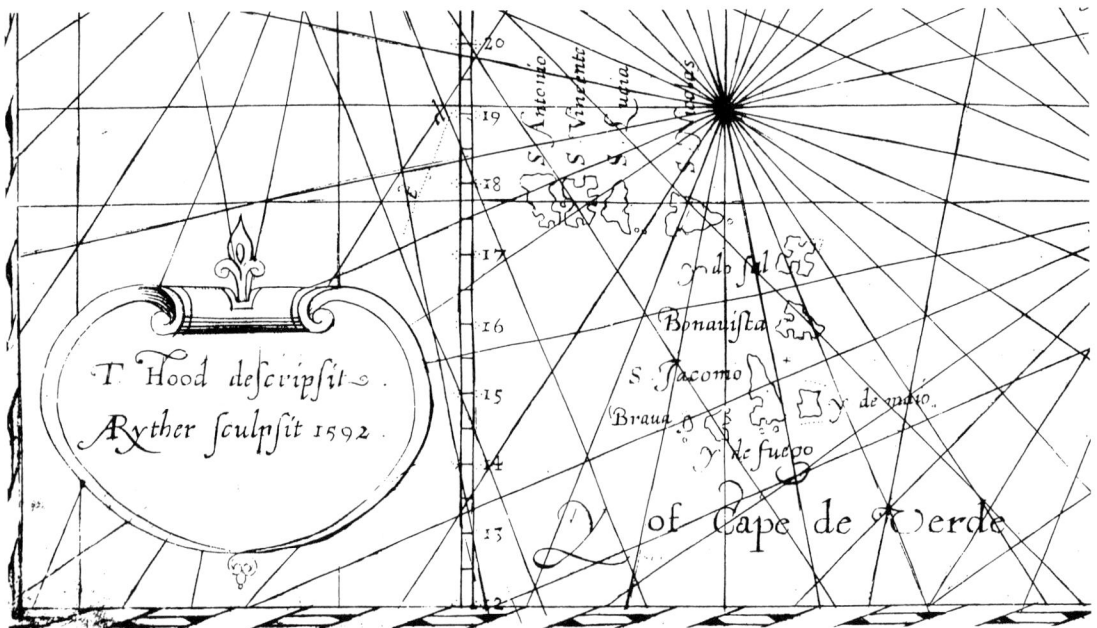

FIG. 2.10 Cartouche on Ryther's chart of the north-east Atlantic, 1592.

Very small lettering is used for Ryther's pack of 60 playing cards, **36**. The necessary control he achieved is also found on the compendium, **34**, and theodolite, **35**, both dated 1590. Noticeable is the extra width of the verticals in majuscules, such as T, H, E. His *t* is a long upright cut with a serif at the bottom and a short bar. The *l* has a flick to the left halfway up the ascender. This is of particular interest, as this seems to be the first use on an instrument, and is to be found on many instruments into the early eighteenth century.

This *l* with a flick could have been copied from the writing-book, *A Booke containing Divers Sortes of Hands*, by the Huguenot John de Beauchesne and John Baildon, printed at London in 1570. In fact, the samples of lettering, then current in England, in both the roman and italic script resemble the lettering used by Ryther, Whitwell, and others. A chapter on Beauchesne, written by Berthold Wolpe, is included in Osley (1980, pp. 227–32). According to Wolpe: 'Among the writing-masters of the sixteenth century John de Beauchesne occupies a special position in that, although he was a Frenchman, he gave the Elizabethans of England their first native printed writing-book. Before him, the only writing manuals available in this country were those imported from the Continent.'

CHARLES WHITWELL
c. 1568–1611

Charles Whitwell was apprenticed to Augustine Ryther, described as 'Rider' in the Grocers' Company records, on 17 December 1582. Assuming that his apprenticeship began at the age of 14, this makes the year of his birth about

1568. He became free of the Grocers' Company on 10 November 1590 (J. Brown 1979, pp. 24, 60–2). Whitwell engraved some maps in the early years of his career, but he is primarily an instrument maker, probably the most skilled and versatile of the Elizabethan period.

At the end of the sixteenth century, manuals on surveying and mariners' guides were being produced in considerable numbers, and the instruments described in them were naturally in demand. Whitwell made the most of this market. He engraved the plates for William Barlow's *The Navigators Supply* (1597) and advertised on the title page that he supplied the instruments. He also made the sector (**46**) devised by Thomas Hood, and advertised in Hood's book of 1598, giving his address as: 'without Temple Barre against St Clement's Church'. There are records at St Clement's of the births of his children between 1596 and 1609. This gives us some idea of when he married, and confirms that he died in middle age, and presumably suddenly.

In the Minutes of the Grocers' Company for 11 August 1606, it was recorded as 'Agreyd that Charles Whitwell grocer shall have the 50*l* [£50] for ii years wch his brother Robert Whitwell deceased latelie had. And William Whitwell and George Budd salters are alowed his sureties.' From this we learn that Charles had at least two brothers working in the City, and enjoyed patronage from the Grocers' Company. Remarkably, after 400 years, two watches have survived, one with the name of Charles, the other with the name of Robert (**50**).

Whitwell must have died in the latter part of 1611, since administration of his estate was granted to his widow, Elizabeth, in February 1611/12 (PROB 6/8 f.5).

Whitwell's Calligraphy

Having had a first-class craft master in Ryther, Whitwell achieved an excellent control over his engraving hand (Fig. 2.11). This is shown particularly well on the small brass chronological and astrological disc, **45**, only 67 mm in diameter, which, on both sides, manages to pack in four tables and an azimuth dial, employing letters and numbers about 0.6 mm in height. It is not signed, but datable to 1596, and the attribution rests on the letter forms and the dexterity. At the time, there was no other engraver who could have produced this

disc. The same view may be taken of the very large lunar computer, **51**, with a diameter of 726 mm, made for Sir Robert Dudley. This has a range of letter sizes, and painstakingly worked mathematical lines and scales. Of his non-instrument engravings, the large bird's-eye plan of Jerusalem completes the evidence for his range of talent.

Around 1595, Whitwell changed from an upright *l* with a flick to the left, to one with a wide curve to the right at the top; the ascenders on *b*, *d*, and *k*, also bend to the right. It is usual for Whitwell to engrave his E with a long foot, and he does not exaggerate the

FIG. 2.11 Signature on altitude dial by Whitwell **41**, 1593. MHS Oxford.

width of the verticals of majuscules. Unlike Ryther, Whitwell's *3* and *5* are attenuated on the bottom curve, and his *6* and *9* are not closed, and the ascender or descender are long straight cuts. This is also clear on his sigil for Cancer, which is exactly like both numerals lying horizontally one above the other. For the characteristic forms of his Zodiac sigils, see the Appendix.

ELIAS ALLEN
c. 1588–1653

Allen was the third in the succession of master/apprentice instrument makers who were Freemen of the Grocers' Company, and the most prolific of the group (J. Brown 1979, pp. 24–5, 62–5). He was born, judging from the record of his apprenticeship, in about 1588, and he began his training with Charles Whitwell in the last year of Elizabeth's reign, 1602. On this basis, he is not strictly an Elizabethan maker, since his productive years were in the age of the Stuarts, not the Tudors. Some of his early instruments are, however, described in Part Two because of the important inter-comparisons that can be made with the work of Elizabethan makers. The inscription beneath an engraving of 1666 from a portrait of Allen by H. van der Borcht tells us that he was born near Tonbridge in Kent. He died in March 1653, and was buried on 1 April 1653 at St Clement Dane's.

The end of Allen's apprenticeship coincided with the death of his master, Whitwell, in 1611. It seems that at some point he took over Whitwell's premises, and was making instruments even before he became a Freeman of the Grocers' Company in 1612 (see the 1606 sundial, **81**). Throughout his career, Allen was in close touch with the intellectual life of London; an influential contact was with Edmund Gunter (1581–1626) of Gresham College. Gunter's sector was first described in manuscript by 1606, the printed version being published in 1623 as *Use of the Sector, Crosse-staffe, and other Instruments*. The frontispiece is an engraving by Allen of the sector, which also has this advertisement: 'These instruments are wrought in brasse by Elias Allen dwelling without Tempel barre over against St Clements Church'. Allen was involved in the setting up of the Clockmakers' Company, incorporated in 1631, becoming its Master in 1636. In Allen's career, we see the early blossoming of the scientific instrument trade, whose influence is described in Higton (1996).

Allen's Calligraphy

It is not surprising that Allen's engraving hand follows closely that of Whitwell, and the difference is in small details. Allen has trouble with curves, especially the zero. He does not use one or two confident curves, but a series of short cuts, more often than not giving the appearance of the head of an owl. The *8* is similarly unsuccessful. His *4* is stretched vertically, the *5* likewise.

CHAPTER THREE
MAP, CHART, AND BOOK ENGRAVERS

By the mid-sixteenth century, copperplate engravers were in demand for book illustrations in conjunction with the new trade of printing with movable type. The level of skill became more exacting when maps and gores for globes were produced. The map maker had to learn to use his engraving tool, the burin, with extra care when cutting boundaries and rivers, and to cut letters and numbers that were very small, because a country map could have several hundred place names. A growing interest in geography was stimulated by printings of Ptolemy's *Geography* from 1477 (Moreland and Bannister 1989, pp. 301–2), and the interest increased vastly with the news of the discovery of America in 1492. The astrolabe, well known for centuries, was provided with a geographical plate by Peter Apian (1495–1552) in his *Cosmographicus liber* (1524). This book was 'adopted' by Gemma Frisius (1508–55), a mathematician, astronomer, and physician of Louvain, and he published editions of Apian's book with considerable additions of his own, in particular the longitude and latitude coordinates of hundreds of sites all over the known—and increasingly known—world.

Gemma Frisius required as high an accuracy and as high a clarity as could be had, and he obtained suitable craftsmen in the university town of Louvain. He had a terrestrial globe made in 1531, and in 1536–37 he commissioned a pair of globes, the gores being cut by Gerard Mercator (1512–95). Also a member of Louvain University, Mercator gave up the study of philosophy for astronomy and mathematics because he could thereby earn more money to keep his family. Mercator's first map was of Palestine, published in 1537, then in the next year came a world map, followed in 1541 by his terrestrial globe. By this time he was an exceptionally skilled engraver, a calligrapher who published in 1540 a manual on calligraphy. He also made mathematical instruments, such as astrolabes, quadrants, and dials.

It is argued elsewhere that Thomas Gemini (c. 1515–62), who came from Flanders, is most likely to have been trained with Mercator. The quality of his calligraphy virtually matches Mercator's both in skill and in form, and the design of his astrolabes shows a remarkable affinity. Gemini migrated to London, probably just after 1540. His first published work is dated London, 1545. The Mercator calligraphy was adopted by other craftsmen, and became the standard in England and the Low Countries for over two centuries.

The makers of the Elizabethan instruments described in Part Two additionally produced maps or charts, book illustrations, and even playing cards. Listed here are some of the non-instrument products of these men. All the items have been seen, some in facsimile, and the signatures transcribed. Writers on the subject seldom give the engraver's name, or the full name, or even list the charts in a collection. Two of the makers, Gemini and Ryther, established themselves as engravers, and moved on to instruments. Later, after 1600, the demand for instruments was such that Whitwell cut only a few maps or charts, and Elias Allen none. A chronological list of engravers working in England from 1540 to 1689 is published by Hind (1952, pp. 314–20).

Essential books for this study are: Taylor (1934); Hind (1952); Waters (1958); Karrow (1993).

THOMAS GEMINI

1545 Thomas Gemini, *Compendiosa totius anatomie delineatio, æra exarata, per Thomam Geminum. Londini.* Colophon: LONDINI in officina Ioanni[s] Herfordie: Anno Domini 1545. MENSE OCTOBRI. Gemini copied the wood engravings from Andreas Vesalius, *De humani corporis fabrica,* published at Basel in 1543, but his 40 copperplate engravings are of a much higher quality than the earlier wood engravings. Each plate is described and collated with the blocks printed by Vesalius by Hind (1952, pp. 49–52). Sir Sidney Colvin (1905, p. 14) considered this the 'first important instance of a title-page so engraved which occurs anywhere' (Frontispiece). Gemini dedicated the book to Henry VIII. From 1547 till his death, he had a pension of £10 a year from the royal purse. Locations given in Karrow (1993, p. 250).

1553 *idem.* Colophon: Imprynted at London by Nycholas Hyll dwellynge in Saynte Johns streate, for Thomas Geminus. Translated from Latin by Nicolas Udall; his epistle dated 20 July 1552. Dedicated to the king, Edward VI. A newly acquired copy is described by Keynes (1959), but his date for the death of the king is one year too early. Other locations in Karrow (1993, p. 252).

1559 *idem.* Colophon: Imprinted at London within the blacke fryars: by Thomas Gemini. *Anno salutis. 1559. Mense Septemb.* Dedicated to Queen Elizabeth. The portrait on the title page (Fig. 3.1) is commonly said to be of Queen Elizabeth, but it is of Queen Mary (see Chapter 2). Hind (1952, p. 46, plate 18a); Karrow (1993, p. 252).

1548 Thomas Gemini, *Morysse and Damashin renewed and encreased Very profitable for Goldsmythes and Embroderars by Thomas Geminus. at London Anno. 1548.* Edward VI is named on the title page, which was engraved by Gemini. Dodgson (1917); Hind (1952, pp. 55–6, plates 25–7).

1555 Map of the British Isles. BRITANNIÆ INSVLÆ QVÆ NVNC ANGLIÆ ET SCOTIÆ REGNA CONTINET CVM HIBERNIA ADIACENTE NOVA DESCRIPTIO. In the cartouche the end of the last line has been rubbed out and this substitution made: *Londini Anno 1555. T. Gemini.* The original wording is: ROMAE, *Anglorum studio & diligentia* MDXLVI. All Gemini did was to put his name to an original engraving, the plates having been taken to London at the time of the accession of Queen Mary, July

FIG. 3.1 Title page to
Gemini, *Anatomy*, 1559.
BL Oxford.

1553. The original was published in 1546 at Rome by George Lily. Hind (1952, pp. 57–8); Karrow (1993, pp. 253, 349–50).

1555   Map of Spain. *Nova descriptio Hispaniae ... Excusum Londini per Thomam Geminum· 1555.* Dedication is in a cartouche at lower right to Philip II and Queen Mary (reigned July 1554–November 1558). Gemini's first map, it is a close copy of the map published in 1553 by Hieronymus Cock, which was copied from Vincenzo Paletino, 1551; see Karrow (1993), p. 252. Schilder (1987, p. 98) has shown that the first edition of this map

FIG. 3.2 Gemini's portrait of Queen Elizabeth, 1560–62. ECL Eton.

was dedicated to the Holy Roman Emperor Charles V. Charles (1500–58) abdicated, in favour of Philip, the sovereignty of The Netherlands in October 1555, and his Spanish kingdom in January 1556. Always ready to attract royal attention, such a change can be expected of Gemini. Hind (1952, pp. 56–7, plate 28); Karrow (1993, p. 252).

1560–2 Portrait of Queen Elizabeth (Fig. 3.2), dedicated to Elizabeth in the same manner as Gemini's astrolabe, **6**, which is dated 1559. Signed: Thomas ♊ ✦ 156[]. The final number is lost through a torn portion of the print. Gemini died in May 1562, so there are only three possible years, however the most likely year is 1560. The only impression known is in the Storer Collection in Eton College Library, bequeathed by A.M. Storer in 1799 (Colvin 1905, pp. 15–16, fig. 7). Also in Hind (1952, pp. 47–8, plate 18b).

HUMFREY COLE

1572 Map of the Holy Land (Fig. 3.3), published in the second edition of the Bishops' Bible, 1572, signed: GRAVEN BI HUMFRAY COLE GOLDSMITH .... It is copied

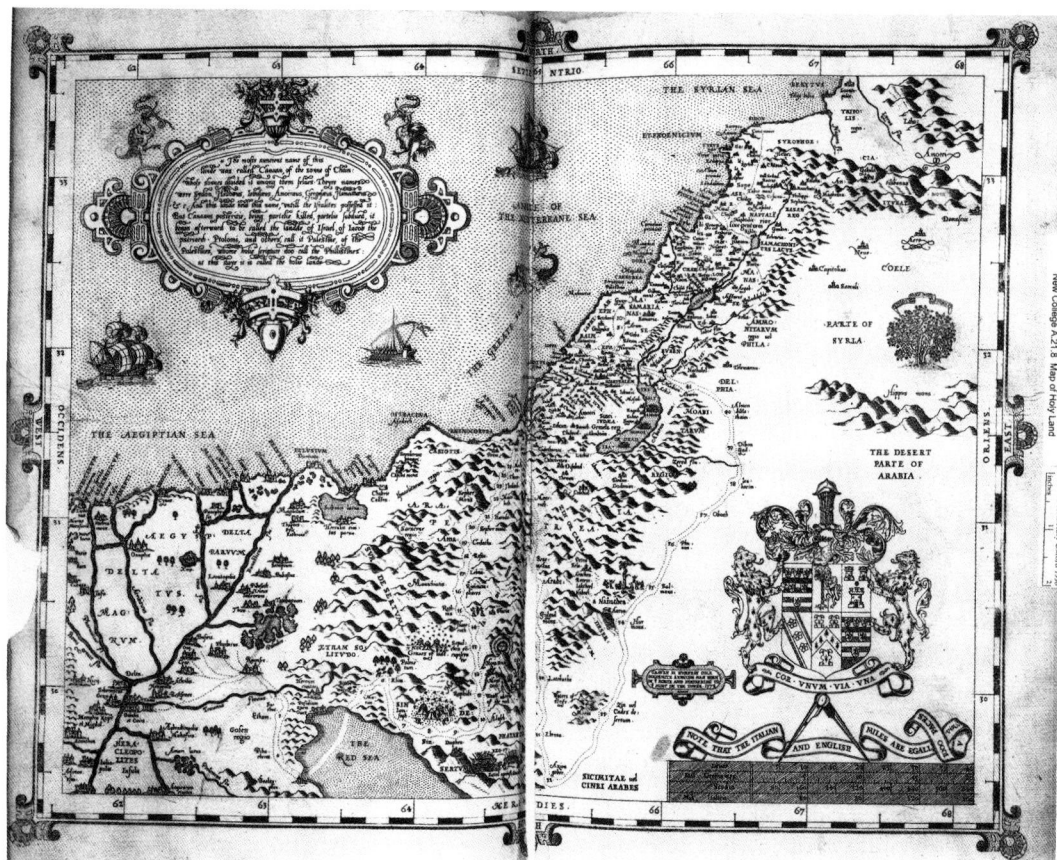

FIG. 3.3 Map of Canaan, or Holy Land, by Cole, 1572. NC Oxford.

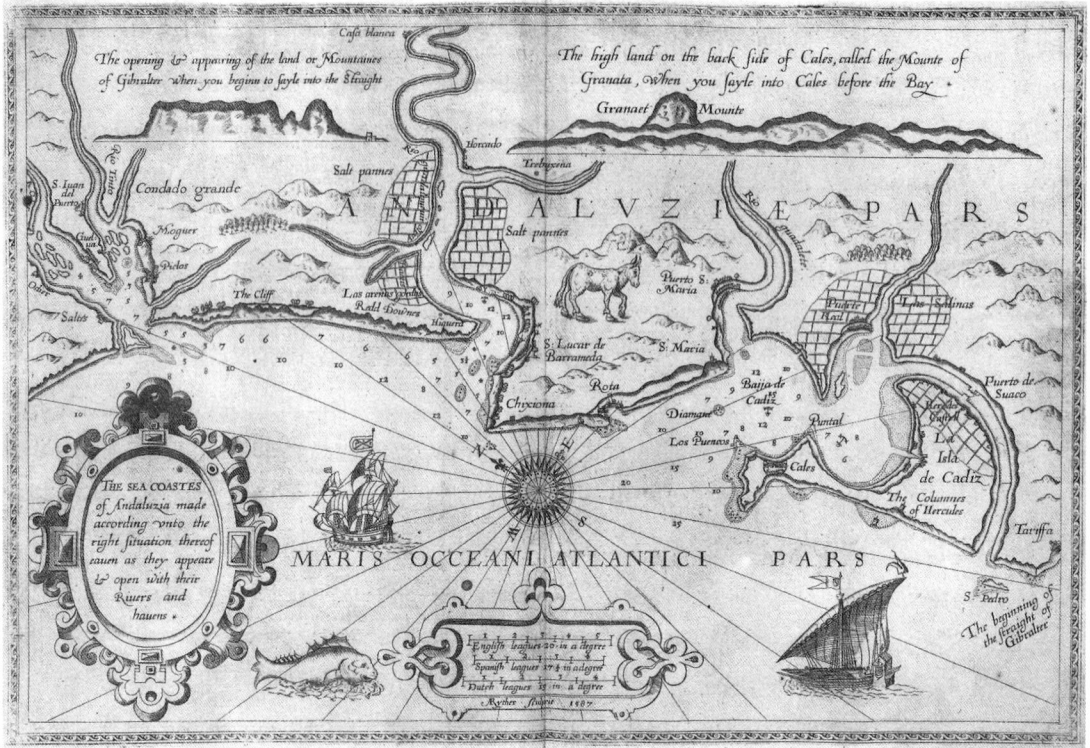

FIG. 3.4 Coast of Andalusia by Ryther, 1587. BL Oxford.

from the map of Tilmann Stolz engraved by Frans Hogenberg in Abraham Ortelius, *Theatrum orbis terrarum* (Antwerp, 1570). In the opinion of Skelton (1965, p. 56, plate 31), Cole's is the first map known to have been engraved on copper by an Englishman.

1576 A cosmographical diagram published by Thomas Digges in *A Perfit description of the Caelestiall Orbes*, that was added to the 1576 edition of his late father's *A Prognostication euerlastinge of righte good effecte*. The diagram is the first representation by an Englishman of the Copernican world system, showing for the first time an infinity of stars. The engraving is tentatively attributed to Humfrey Cole.

AUGUSTINE RYTHER

1576 Durham, in Christopher Saxton, *Atlas of the Counties of England and Wales*, 1579. Signed: *Christophorus Saxton descripsit. AVGVSTINVS RYTHER SCVLPSIT AnᵒDi 1576.* See Skelton (1970, pp. 14–16) for a list of Saxton's maps and engravers; Tyacke and Huddy (1980) for the project.

1576 Westmorland, with Cumberland, in Saxton; signed: *CHRISTOPHORVS SAXTON DESCRIPSIT. AVGVSTINVS RYTHER ANGLVS SCVLPSIT AnᵒDi 1576.*

1577 Gloucestershire, in Saxton; signed as for Westmorland; year 1577.

1577 Yorkshire, in Saxton; signed: *Christophorus Saxton descripsit. Augustinus Ryther Anglus Sculpsit An° D̄i 1577.*

1579 Anglia, in Saxton; signed as for Yorkshire; year 1579.

1587–88 Three 'sea coastes' in Lucas Jansz. Wagenaer, *The Mariners Mirrovr …*, translated with additions by A[nthony] Ashley, London [1588]. Contains new engravings for the English edition by Ryther and others. Biscay, signed: *ARyther sculpsit*; Galicia, signed: *ARyther sculpsit*; Andaluzia, signed: *ARyther sculpsit 1587* (Fig. 3.4). Skelton (1965, p. 47).

1588 Bird's-eye plan of Oxford by Ralph Agas (*c.* 1540–1621), drawn 1578, signed: AVGVSTINVS RYTHER ANGLVS DELINIAVIT 1588. In *Celeberrimae Oxoniensis Academiae … Descriptio*; see 'Old Plans of Oxford', *Oxford Historical Society*, **39**, 1899. Taylor (1934, p. 197); Hind (1952, pp. 139–41); Skelton (1970, p. 241).

1590 Eleven Armada charts after Robert Adams, Surveyor of Works to the Queen, with the engraved title page: EXPEDITIONIS HISPANORVM IN Angliam vera descriptio ANNO DO: MDLXXXVIII. Ten charts show the daily dispositions of the Spanish fleet, and the eleventh is a large map of the British Isles showing the complete course of the Armada. The eleven charts are described by Hind (1952, pp. 146–7). The Roxburghe Club published a full-size facsimile edition of the charts (H. Thompson 1919), with Ryther's translation of Ubaldini (see following entry). All are engraved: *Roberto Adamo authore 1588* (date on no. 1 only); and nos. 1, 6, 7, and the large map, are signed: *Augustinus Ryther Sculpsit.*

The ten daily charts at the National Maritime Museum, Greenwich are reproduced in colour in *Armada* (1988, pp. 243–8). See also Hind (1952, pp. 142–9, plates 80–5).

1590 Engraved title page of Petruccio Ubaldini, *A Discovrse concerninge the Spanishe fleete inuadinge Englande in the year 1588*, 1590. The arms of the City of London at the top, and those of the Grocers' Company at the bottom (Ryther was a Freeman of the Company and of the City). Ubaldini's Italian text is in manuscript only (British Library). Ryther advertised himself by engraving: 'These bookes with the tables belonginge to them are to be solde at the shoppe of A. Ryther beinge a little from Leaden hall next to the Sine of the Tower'. This small book and the set of plates were intended to be sold as a pair. H. Thompson (1919); Hind (1952, pp. 24–5, plates 79a, b, c); J. Brown (1979, plate 2).

1590 Playing cards designed by William Bowes; the suits are county maps of England and Wales. For the signature, see **36**. Hind (1952, pp. 182–6, plates 103–5) lists the wording on each card, but did not identify the engraver.

1590 Two polar projections of the constellations drawn by Thomas Hood; engraving signed: *Augustinus Ryther Anglus sculpsit 1590*. Explanation in a small book: *The Vse of the Celestial Globe in Plano, set foorth in two hemispheres*, printed by J. Windet for T. Cooke, London, 1590. On the title page is this notice: 'The Hemispheres are to be sold in Abchurchlane at the house of Th. Hood'. For Hood, see

Johnston (1991, pp. 330–41); for nearly full size reproductions of the projections in colour, see Stott (1991, pp. 54–7).

1592 Chart of the NE Atlantic to the Azores, for Thomas Hood, *The Marriners Guide. Set forth in forme of a dialogue* (1592). Separate, but published with Hood's edition of William Bourne, *A Regiment for the Sea.* Chart inscribed: *T Hood descripsit. ARyther sculpsit 1592* (Fig. 2.10). The *Guide* is a booklet to explain the use of the 'sea card', which doubtless was available at Hood's address. For a bibliographical description, see Taylor (1963, pp. 450–6).

1593 Bird's-eye plan of Cambridge by John Hamond (1558–1603/4). A very large plan (1.19 m × 877 mm) on nine copperplates, signed on sheet 6: *Augustinus Ryther & Petrus Muser sculpserunt.* On sheet 5: … *ex aula Clarensi die 22 mensis february 1592, Johāes Hamond.* (The date is in the Old Style calendar, thus 1593 in the historical year.) The only known copy is in the Bodleian Library, reproduced in facsimile in Clark and Gray, *Old Plans of Cambridge* (1921). For a note, see Clark, *Cantabrigia illustrata* (1905), Introduction, section 3, plate 28A. Hamond was a Fellow of Clare Hall (now College) 1582–99. Peter Muser probably engraved the trees, meadows, cows, etc., to spare Ryther, who died in August 1593. This is the only location for Muser known to Hind (1952, p. 150).

CHARLES WHITWELL

1594 Map of Surrey. Inscribed: *Jo: Nordenus deliniauit 1594 Carolus Whitwell sculpsit impensis Ro Nicolsoni gener.* Note that the *a* in *Carolus*

has a vertical stroke as *h* was started. Used to writing Charles, Whitwell sometimes writes *Charolus,* as on the maps of Jerusalem and of Asia, and on the instruments **41**, **42**, **43**. Copies are in the British Library and the Royal Geographical Society. Heawood (1932, sheet 7); Skelton (1970, pp. 21, 232).

1595 Map of Asia Minor. Inscribed: DESCRIPTION OF THE HITHER PARTE OF ASIA. Signed: *Charolus Whitwell sculpsit.* The date is taken from that of the publication of Minadoi, *History of the Warres between the Turkes and Persians* (1595).

1595 Bird's-eye plan of Jerusalem. Inscribed: *IERVSALEM with her suburbes, and the most principall places thereof, as it florished in CHRIST his tyme, most trewly described.* Signed: *Charolus Whitwell sculpsit.* Prepared for T. Tymme; nearly 300 features are numbered, which lead to explanations in a book. Copy in the British Library. This plan is based on that by Christiaen Adrichom, published in Georg Braun and Frans Hogenberg, *Civitates orbis terrarum,* Vol. 4, Cologne, 1588; see Skelton (1965, p. 49, plate 21).

1596 Map of Kent. Inscribed: *Engraven by Charles Whitwell.* The design was by Philip Symonson. Copy at the Royal Geographical Society. Heawood (1932, sheet 10); J. Brown (1979, p. x, and plate 4).

1597 The illustrations in William Barlow, *Navigators Svpply,* 1597.

1. Title page cut of azimuth compass and advertisement: 'If any man desire more ample instructions concerning the vse of these instruments, hee may repavre vnto

Ihon Goodwin dewillinge in Bucklersburye teacher of the grownds of these artes. The instruments are made by Charles Whitwell, over agaynste Essex howse, maker of all sorts of mathematicall instruments, and the graver of these portraytures'.

2. Sig. B4 cut of compass in its case, labelled: 'The Compasse of variation'.

3. Sig. C2 another cut of azimuth compass (larger), labelled: 'The traveylors iewell'.

4. Sig. D2, folded plate, of Nonius, labelled at top right: 'The face of the Pantometer sene from the inferior part of the horizontal'.

5. Before Sig. E2, folded plate, labelled: 'The Navigators Hemisphere'.

6. After Sig. E2, folded plate, labelled with names of the seven parts of above.

7. Before Sig. H, traverse board, ruler and quadrant, parts named on cut.

8. After Sig. K4 and before Sig. L, three sketches to show use of Mercator projection.

1598   The illustration of his sector in Thomas Hood, *The Making and vse of the Geometricall Instrument, called a Sector* (1598). On the title page, set in type, is the advertisement: 'The Instrument is made by Charles Whitwell, dwelling without Temple Barre against S. Clements Church'. Before Sig. B4: 'The description of the Sector and the partes thereof', clearly in Whitwell's hand.

1599   Chart of the NE Atlantic for Edward Wright, *Certaine Errors*, 1599 and 1610. The world chart drawn on the Mercator projection as improved by Edward Wright, was included with Richard Hakluyt, *Principal Navigations*, Vol. 2 (1599). Wright published another chart,

of the north-east Atlantic, with his *Certaine Errors* (1599), which is named in a cartouche at the top: 'The voyage of the right honorable the Earle of Cumberland to the Ilands Azores A.D. 1589'. Waters (1958, p. xxiv, plate 61) points out that a manuscript version of the printed chart is at Hatfield House; this is believed to have been drawn in about 1595, and is the earliest example of Wright's projection. The MS chart is fully described by Waters in Appendix 18A, pp. 550–1. The printed chart has been published by Waters (1958, p. xxiii, plate 58), and all the caption panels are transcribed in his Appendix 18B, p. 552. The chart in a second state was published in Wright's second edition of *Certaine Errors* (1610). There has been one change: the wording in the cartouche has been rubbed out and this substitution made: 'A particular sea Chart for the Ilands Azores'. On the copy at the Museum of the History of Science, Oxford, the remains of the old lettering are discernable (Fig. 3.5). The engraving includes many words and numbers, so there is ample evidence to show that this chart can, with confidence, be attributed to Charles Whitwell. The rather ornate compass rose in the middle is the same as supplied with the Whitwell compendium dated 1604 (**60**). See the compass rose type C in Chapter 4, Fig. 4.4.

c. 1600   The miniature map of England, attributed to Charles Whitwell, on an ivory diptych dial; see **56**.

Early seventeenth century. Map of France. Inscribed: *P. Stent exc. C.W. Scu.* The map is after the Dutch geographer, Petrus Plancius (1552–1622), and is dated conjecturally in the British Museum catalogue to 1660. Stent died

A particular Sea Chart for the Ilands Azores.

FIG. 3.5 Panel on chart of the Azores by Whitwell; 1610 state of the 1599 issue. MHS Oxford.

in 1665, and Hind (1952, p. 227) considered it may be a revision of a map engraved nearer to 1600. The hundreds of names on the map are not in the hand of Charles Whitwell, in spite of the initials *C.W.*

ROBERT BECKIT

1598 Five maps in Jan Huygen van Linschoten, *Discours of Voyages into yᵉ Easte & West Indies. Deuided into Foure Bookes*, trans-lated by William Phillip, printed by John Wolfe, London, 1598. Contains maps and charts by Beckit: Madagascar (signed: *Robertus Beckit apud Londinum sculpsit. 1598*), Indian Ocean (signed: *Grauen by Robert Beckit 1598*), Sumatra, South America, New Guinea (all signed as for Indian Ocean). Other maps by William Rogers and Renold Elstrack. For Beckit's brass sector to the Thomas Hood design, dated 1597, see **37**. Beckit is noticed in Hind (1952, pp. 221–2).

CHAPTER FOUR
THE INSTRUMENTS

INTRODUCTION

The survival of an artefact over centuries is bound to be fortuitous. Chance will play the largest part, though there are other influencing factors, including size, durability of material, what it was used for, and by whom. In the case of mathematical instruments, those made of metal, readily portable, and intended for personal use are proved to have the highest survival rate. Ownership by old-established institutions also offers a good chance of preservation. This is the reason for the group of instruments taken to Florence by Sir Robert Dudley having remained together and intact, since they passed into the care of the Grand Dukes, and became part of the Medici heritage. In the same way, a theodolite belonging to an Oxford college did not become expendable when it ceased to be in use, and survived.

Over half the instruments described in this book are for time telling, and were often personal possessions: compendia, sundials, nocturnals, and astrolabes. The rest were what have been called professional instruments, in that they were used by seamen, surveyors, or gunners (Table 4.1). Many more of these will have been made, and discarded when replaced by a later model, while cheaper versions in wood will have perished even more readily. It

TABLE 4.1 Types of instrument made

Compendia	24
Sundials	19
Sectors	9
Navigation instruments	8
Astrolabes	7
Theodolites	7
Quadrants	5
Surveyors' rules	4
Protractors	4
Nocturnals (separate)	3
Gunners' rules	3
Plane table alidades	2
Gunners' level	1
Drawing set	1
Armillary sphere	1
Globe	1
Calendar	1
Watch face	1

is impossible to draw any accurate conclusions about production rates from the instruments described.

The general commentaries on some of the instruments that follow in this chapter supplement the detailed individual descriptions given in Part Two. The accounts of the sources of the Easter tables and latitudes on the compendia shed light on the books available to instrument makers, as well as the requirements of customers. Surveyors' and gunners' rules have tables that require explanation, which it is difficult to find elsewhere. The development of the theodolite and the sector occurred during the Elizabethan period, and is, therefore, appropriately included here. The astrolabe, on the other hand, was drawing towards the end of its life as a serious astronomical instrument by the end of the sixteenth century, and its use and development have been fully described elsewhere. The best explanation for a general audience of the structure and working of the astrolabe is that published by the National Maritime Museum, Greenwich, *The Planispheric Astrolabe* (see NMM 1976). Also refer to Gunther (1932); Michel (1947); Saunders (1984).

THE COMPENDIUM

Compendia form the largest group of a single type among the recorded Elizabethan instruments, being a quarter of the total. These compendia are in brass, are small, and fit the pocket in a way similar to the gold watch and chain of modern times. Being a very personal possession accounts both for the apparently large number made, and the high survival rate.

The single feature of the Elizabethan compendium that immediately distinguishes it is the presence of an equinoctial sundial. The Continental market, especially in Germany, favoured the diptych dial, obviously produced in very large numbers judging from those that have survived. Such dials are commonly made in ivory, and require nothing like the mechanical skills of the London craftsmen. Only two English ivory diptych dials are known, **56**, **99**.

Of the compendia described in Part Two, one is rectangular in form (**8**), and two are

oval (**9**, **103**); **8** and **9** are very ornate, and packed with information. The remainder are circular, and become less ornate, and equipped with only the more essential information as the seventeenth century is reached.

A typical compendium is composed of all or some of the following parts:

(a) equinoctial dial

(b) Church calendar

(c) Easter perpetual calendar

(d) table of latitudes

(e) magnetic compass

(f) nocturnal

(g) tide computer

(h) establishment of the port.

All of these features are the subjects of the separate accounts that follow.

The Equinoctial Dial

For telling the time, the chapter ring of the dial has to be set with the noon-to-midnight line parallel to the meridian by using the magnetic compass. The plane of the ring has to be made parallel with the plane of the equator (equinoctial plane) by setting the attached quadrant to the latitude of use. This quadrant is fixed to the gnomon, which is parallel to the polar axis of the Earth when these procedures have been completed. The angular settings will not be accurate unless the instrument is on the level. To effect this, the quadrant has a small hole to take a plummet; this has never survived, presumably because it had a hook loop and not a closed ring to attach it in the hole.

The equinoctial dial was known in the fifteenth century. The disposition of the hours matches the mechanical clock, the earliest records for which date from the late thirteenth century. The regularity of its going produces hours all of the same length throughout the day. This made the ancient hour systems, such as planetary hours, redundant. An early Elizabethan description of an equinoctial dial is that by Leonard Digges in *A Prognostication*, printed by Thomas Gemini in 1555, section 20, how to make and use 'A perfecte instrument for the day, and the night'.

The chapter ring on an equinoctial dial is easy to mark out because every hour of the 24 in the day occupies 15°. The upper side, for the summer half of the year, is marked in 24 hours; the lower side, for the winter half of the year, is marked from 6 am to 6 pm. This leaves a blank semicircle, which is customarily filled by the signature of the maker, and the date. The ability to use this form of dial at any location meant that it was universal, and, for a traveller, greatly superior to a diptych dial that served one latitude, or perhaps three, by means of an adjustable gnomon.

The Church Calendar

The feast days of the Church are tabulated on the compendia made by Humfrey Cole, but not on those by Ryther (**34**), Whitwell (**58–62**), or Allen (**83–87**). They are, however, included on one compendium by Kynvyn, dated 1593 (**66**). It seems that Kynvyn used an edition of Leonard Digges, *A Prognostication*, either the first edition, which does not include the names of the four seasons, or a later one, and deliberately omitted the seasons. He also omitted nine feast days, quite likely through lack of space on a small and congested disc.

In Table 4.2 a comparison is made between the calendars in the first and second editions of

Digges, *A Prognostication*. The original edition was published by Thomas Gemini in 1555 and 1556, the following published by Thomas Marshe in 1564, 1567, 1574, 1576, 1578, 1583, 1585 (later publishers 1592, 1596, 1605; see Adams and Waters 1995, pp. 454–7). The Gemini printings contain 'A general Kalendar' (Fig. 4.1), which is different from the Marshe editions in a few details. The Marshe editions all print the same revised version of what is now named as 'The Generall Kalendar', pressed from the same block of type (Fig. 4.2). They all include *Spring*, 12 March; *Summer*, 13 June; *Heruest*, 15 September; *Wynter*, 13 December; and many days not in the early editions. The comparable calendars from Cole's earliest surviving compendia, dated 1568 (**8**) and 1569 (**9**) are added to Table 4.1. Cole's calendar on **8** was split over two leaves (as is that of Digges, printed over two pages), and the second is lost, so the contents of the missing leaf may be conjectured from the printed list from July to December, bearing in mind the omissions in **9**.

In Table 4.3 a comparison is made between the Church calendars engraved on seven Humfrey Cole instruments, six compendia, and one ring dial. It is noticeable that Cole soon abandons the four seasons, and makes a few changes, some produced through the amount of space available.

Commentary on Tables 4.2, 4.3

The feast days are based on Paul Harvey, *The Oxford Companion to English Literature*, 4th edition (Oxford, 1967), Appendix III, Table IV. The other columns give the wording on seven instruments (six compendia and one ring dial, **21**), but without the contraction and punctuation marks, which can be found under the catalogue entries.

TABLE 4.2 Feast days in Digges, *A Prognostication*, 1555 and 1564 editions, and Cole's earliest compendia

Date	1555 edn	1564 edn	8 1568	9 1569
Jan: 1 Circumc.	o	o	o	o
6 Ephiph.	o	o	o	o
11 ☉	10 Jan.	11 Jan.	11 Jan	11 Jan
13 Hilar.	o	o	o	o
25 Con. Pau.	o	o	o	o
Feb: 2 Purifi.	o	o	o	o
9 ☉	o	o	10 Feb	10 Feb
14 Valent.	o	o	o	o
24 Mat.	o	o	o	o
Mar: 11 ☉	o	o	o	o
12 Spring	●	o	o	o
25 Annun.	o	o	o	o
Apr: 4 Ambro.	●	●	o	●
11 ☉	o	o	o	o
23 Georg.	o	o	o	o
25 Marc.	o	o	o	o
May: 1 Ph. Iac.	o	o	o	o
6 Io Eua [ante Port. Lat.]	●	●	o	●
12 ☉	o	o	o	o
26 August.	●	●	o	●
Jun: 11 Barna.	o	o	o	o
12 ☉	o	o	o	O
13 Summer	●	o	o	●
24 Ioan bap.	o	o	o	o
29 Pe. Pa.	o	o	o	o
Jul: 6 Dog begin	o	o		o
14 ☉	o	o		o
22 Ma. Mag.	o	o		o
25 Iac. Apo.	o	o		o
Aug: 1 Pet. Vin.	o	o		o
14 ☉	o	o		o
17 Dog end	o	o		o
24 Bartho.	o	o		o
29 decol. Io.	o	o		o
Sep: 8 Nat. ma.	o	o		o
14 ☉	o	o		o
15 Heruest	●	o		●
21 Mathe.	o	o		●
29 Micha.	o	o		o
Oct: 14 ☉	o	o		o
18 Luc.	o	o		o

TABLE 4.2 Feast days in Digges, *A Prognostication*, 1555 and 1564 editions, and Cole's earliest compendia (continued)

Date	1555 edn	1564 edn	8 1568	9 1569
28 Si. Iud.	o	o		o
Nov: 1 Om. sa.	o	o		o
2 Om. ani.	o	o		o
13 ☉	o	o		o
30 Andr.	o	o		o
Dec: 6 Nicol.	o	o		o
8 Conce. ma.	o	o		o
12 ☉	o	o		o
13 Wynter	●	o		●
21 Tho. ap.	o	o		o
25 Nat. do.	o	o		o
26 Steph.	o	o		●
27 Io. eua.	o	o		●
28 Innocen.	o	o		●
29 Tho. [à Becket]	o	o		●

o Present; ● not present; **8** lacks the engraved plate for July to December.

The instruments in the first three columns contain seasonal reminders: Spring on 12 March; Summer on 13 June; the helical rising (6 July) and setting (17 August) of *Sirius*, known as the Dog Star. For an explanation of the latter, see the entry for **9**.

The engraved list on **24** is virtually the same as on **25**: both have the date of 16 February for St Valentine. The 1579 compendium, **27**, names the saint on 14 February, but on **26**, the date is 16 February. Alone of this group of seven instruments, both the compendia dated 1575 have the day of St James as 23 July, and not 25 July.

The three instruments dated 1575 (**21**, **24**, **25**) have the date 25 February for St Matthias, whereas the others all have the date 24 February. Harvey (1967, p. 959), notes that 25 February applies in leap years. The two compendia dated 1575 have their perpetual calendar beginning at 1572 (a leap year), but the ring dial, also 1575, has 25 February even though its calendar begins at 1575. Perhaps Cole was using some list during 1575 that was not entirely suitable for that year.

Easter Perpetual Calendar

Some compendia include a table for finding the date of Easter (e.g. **66**), from which follow all the other movable feasts, such as Septuagesima Sunday, Ash Wednesday, Whit Sunday, and so on (Fig. 4.3). Easter should be celebrated on the Sunday following the first full Moon on or after 21 March. It may fall on any of 35 days between 22 March and 24 April, both included. For calculating the date of the paschal moon the Prime or Golden Number is used. This is a Lunar Cycle of 19 years, numbered sequentially 1–19.

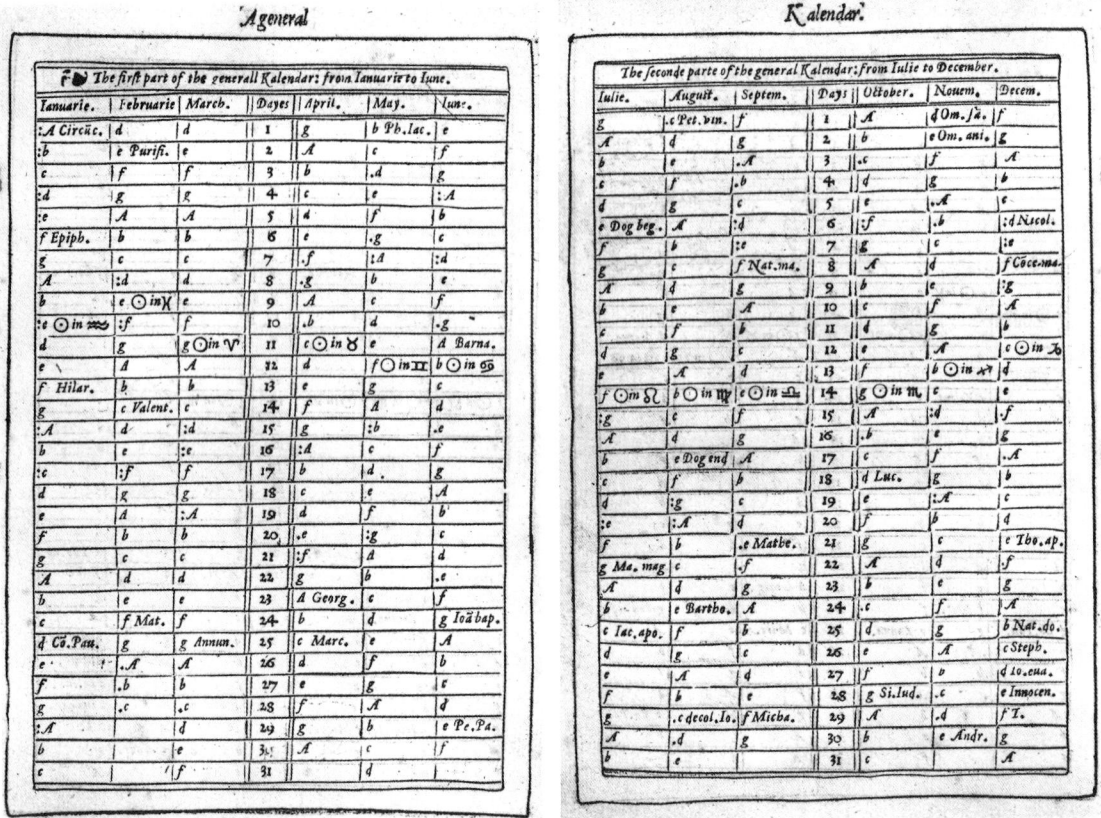

FIG. 4.1 Church calendar in L. Digges,
A Prognostication, 1555. BL Oxford.

It is also necessary to know the place of the Moon in the Solar Cycle. The lunar year of 12 lunations is 11 days shorter than the solar year, so the Moon is 11 days older at the beginning of each year. This figure of the age of the Moon is called the Epact. Leonard Digges, in his *A Prognostication* (1555 and later editions), has a table connecting Prime and the Epact. He explains that the Epact changes every year on the first day of March, whereas the Prime and the Dominical Letter change on the first of January.

The Dominical Letter gives the day in the week, and is the third element in finding the date of Easter, because one has to know the sequence following the paschal full Moon, each day of the week indicated by the letters A–G. The Letter for the first Sunday in January is then the same for all the Sundays in the year. In an Easter table, the Prime points to the day of the paschal full Moon, then the Dominical Letter (also called the Sunday Letter) following gives Easter Day.

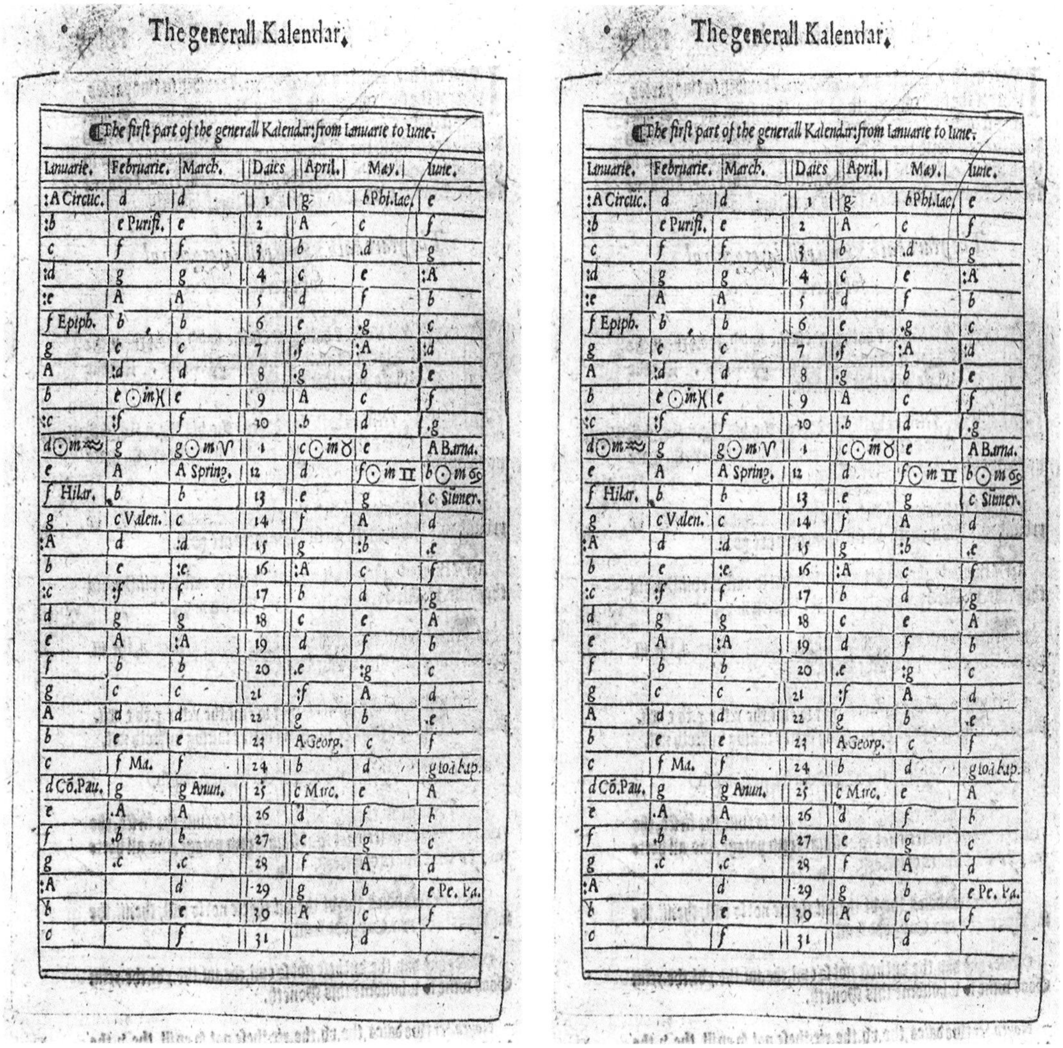

The generall Kalendar.   The generall Kalendar.

❧ The first part of the generall Kalendar; from Ianuarie to Iune.

Ianuarie.	Februarie.	March.	Daies	April.	May.	Iune.
:A Circuc.	d	d	1	g	bPhi.Iac.	e
:b	e Purifi.	e	2	A	c	f
c	f	f.	3	b	.d	g
:d	g	g	4	c	e	:A
:e	A	A	5	d	f	b
f Epiph.	b	b	6	e	.g	c
g	c	c	7	f	:A	:d
A	:d	d	8	.g	b	e
b	e ⊙ in ♓	e	9	A	c	f
:c	:f	f	10	.b	d	.g
d ⊙ m ♒	g	g ⊙ m ♈	11	c ⊙ m ♉	e	A Barna.
e	A	A Spring.	12	d	f ⊙ m ♊	b ⊙ m ⊙
f Hilar.	b	b	13	.e	g	c Sunner.
g	c Valen.	c	14	f	A	d
:A	d	:d	15	g	:b	.e
b	e	:e.	16	:A	c.	f
:c	:f	f.	17	b	d	.g
d	g	g	18	c	e	A
e	A	:A	19	d	f	b
f	b	b	20	.e	:g	c
g	c	c	21	:f	A	d
A	d	d	22	g	b	.e
b	e	e	23	A Georp.	c	f
c	f Ma.	f	24	b	d	g to A kip.
d Cô.Pau.	g	g Anun.	25	c Marc.	e	A
e	.A	A	26	d	f	b
f	.b	b.	27	e	:g	c
g	.c	.c	28	f	:A	d
:A		d	29	g	b	e Pe. Ka.
♉		e	30	A	c	f
c		f	31		d	

FIG. 4.2 Church calendar in L. Digges,
A Prognostication, 1564. BL Oxford.

The Epact is an important adjunct to the tide computer, because it enables the seaman to know the date of the full Moon and hence the age of the Moon in its lunation of $29\frac{1}{2}$ (near enough 30) days. For a given day, one adds to the Epact the number of the months passed, counting March as 1, April as 2, etc., and the day in the month chosen (subtract 30 from a total over 30). The somewhat simpler compendia from 1600, exemplified by Charles Whitwell's (e.g. **58**), are provided with both an Epact/Prime table and a tide computer.

TABLE 4.3 The Church calendar as found on instruments by Humfrey Cole

Date		Feast	8 1568	9 1569	21 1575	24 1575	25 1575	26 1579	27 1579
J	1	Circumcision	Circu	Circu	Circu	Circu	Circu	Circu	Circu
	6	Epiphany	Epiph	Epiph	Epiph	Epiph	Epiph	Epiph	Epiph
	13	St Hilary	Hilar	Hil	Hila	Hill	Hill	Hill	Hill
	25	Conversion of St Paul	Co Pau	Co P	Co Pa	Co Pau	Co Pau	Con Pau	Co Pau
F	2	Purification of BVM	Purifi	Purifi	Purifi Mari	Puri Ma	Puri Ma	Puri Ma	Puri Ma
	14	St Valentine	Valent	Valen	Vale	16 Vale	16 Vale	16 Vale	Vali
	24	St Matthias Apostle	Mat	Math	25 Mat	25 Mat Ap	25 Mat Ap	Mat Apo	Ma Apo
M	12	Spring / St Gregory	Sprig	Spring	Gregori	Grego	Grego	Grigo	Grigo
	25	Annunciation of BVM	Annu	Anunci	Anun Mar	Annu Ma	Annu Ma	Annu Mary	Anu Ma
A	4	St Ambrose	Ambro	–	–	Ambro	Ambro	Ambro	Ambro
	23	St George	Georg	George	George	George	George	George	George
	25	St Mark	Marc	Marcke	Marcus	Marcus	Marcus	Marcus	Marcu
M	1	St Philip & St James	Phi Iac	Phillip Iacob	Philli Iaco	Phi Iac	Phi Iac	Phi Iac	Phi Iac
	6	St John ante Portam Latinam	Io Eua	–	Ioan Euan	Io Eua	Io Eua	Io Eua	Io Eua
	26	St Augustine Archbishop	Augus	–	–	Augu	Augus	Augus	Aug
J	11	St Barnabas	Barna	Barn	Barn	Barn	Barn	Barn	Bar
	13	Summer	Sumer	–	–	–	–	–	–
	24	St John Baptist	Ioa bap	Io ba	Io Bap	Io Ba	Io Ba	Io Ba	Io Ba
	29	St Peter & St Paul	Pe Pa	Pe	Peter Paul	Pe Pa	Pe Pa	Pe Pa	Pe Pa
J	2	Visitation of BVM		–	–	Visi Ma	Visi Ma	Visi Ma	Visi Ma
	6	*Sirius* rises		Dog Beg	Doge begin	–	–	–	–
	22	St Mary Magdalen		Ma Ma	Mag	Ma Mag	Ma Mag	Ma Mag	Ma Mag

TABLE 4.3 The Church calendar as found on instruments by Humfrey Cole (continued)

Date	Feast	8 1568	9 1569	21 1575	24 1575	25 1575	26 1579	27 1579
25	St James		Ia Apo	Ia Apo	23 Ia Ap	23 Ia Ap	Ia Ap	Iac Apo
A 1	St Peter ad Vincula		Pe Vi	Pe Vin	Pe Vi	Pe Vi	Pe V in	Pe V in
10	St Laurence		–	–	Lau	Lau	–	–
17	*Sirius* sets		Do en	Dog en				
24	St Bartholomew		Ba	Bart	Bar	Bar	Barth	Bart
29	St John Baptist decapitated		D.I	Dec Io	De Io	De Io	Dec Ione	Dec Io
S 8	Nativity of BVM		Na Ma	Nati Mari	Nat Ma	Nat Ma	Nat Ma	Na Ma
21	St Matthew		–	Mat	Mat	Mat	Mat	Mat
29	St Michael		Mi	Mich	Mic	Mic	Mich	Mi
O 13	Translation of King Edward		–	Edward Con	Edwa Co	Edwa Co	–	Edw
18	St Luke Evangelist		Luke	Luc	Lu	Lu	Lucke Eua	Luc
28	St Simon & St Jude		Simon Iude	Si Iud	Si Iude	Sim Iude	Simon Iude	Simo Iud
N 1	All Saints' Day		Om Sa	Omni San	Om Sa	Om Sa	Al saintes	Om sa
2	All Souls' Day		Om An	Om An	Om Ani	Om Ani	–	Om an
30	St Andrew Apostle		Andr	Andr	Andr	Andr	Andro Apos	Andr
D 6	St Nicolas		Ni	Nico	Ni	Ni	Ni	–
8	Conception of BVM		Co Ma	Con Ma	Co Ma	Co Ma	Co Ma	Co Ma
21	St Thomas Apostle		Tho A	Tho	Tho	Tho	Tho ap	Tho
25	Christmas (*Natale domini*)		Na	Nat Do	Na	Na	Na	Nat

Profitable rules.

The prime.	The sodaies letter.	The first Lent sondaie.	Easter daie.	Rogation	whitson tide.	Betwixt whitson ad midso.	
16	D	Februa.	March	April.	Maie.	wek.	dais
	d	8	22	26	10	6	3
	e	9	23	27	11	6	2
13	f	10	24	28	12	6	1
2	g	11	25	29	13	6	0
	A	12	26	30	14	5	6
10	b	13	27	May.1.	15	5	5
	c	14	28	2	16	5	4
18	d	15	29	3	17	5	3
7	e	16	30	4	18	5	2
	f	17	31	5	19	5	1
15	g	18	Aprile.1,	6	20	5	0
4	A	19	2.	7	21	4	6
	b	20	3	8	22	4	5
12	c	21	4	9	23	4	4
1	d	22	5	10	24	4	3
	e	23	6	11	25	4	2
9	f	24	7	12	26	4	1
	g	25	8	13	27	4	0
17	A	26	9	14	28	3	6
6	b	27	10	15	29	3	5
	c	28	11	16	30	3	4
14	d	Marche.1.	12	17	31	3	3
3	e	2	13	18	Iune.1	3	2
	f	3	14.	19	2	3	1
11	g	4	15	20	3	3	0
	A	5	16	21	4	2	6
19	b	6	17	22	5	2	5
8	c	7	18	23	6	2	4
	d	8	19	24	7	2	3
	e,	9	20	25	8	2	2
	f	10	21	26	9	2	1
	g	11	22	27	10	2	0
	A	12	23	28	11	1	6
	b	13	24	29	12	1	5
	c	14.	25	30	13	1	4

FIG. 4.3 Easter calendar in L. Digges, *A Prognostication*, 1555. BL Oxford.

This procedure is explained, with a table, by William Bourne in his *Almanacke ... 1571*, edited by Taylor (1963, pp. 60–2).

Latitudes

The reason why lists of towns with their latitudes occur on compendia is simple: the latitude was required to set the equinoctial dial, in order to tell the time when on a journey. Two queries immediately arise as a result of this explanation. One is why such exotic loca-

tions as Damascus, Nineveh, or Hangzhou should have been included. The other is, why give the latitudes in minutes as well as degrees, such a level of accuracy being needless. The answer to both questions lies in the origin of the latitude tables used by the instrument makers.

It is not difficult to measure the latitude; this may be readily done by taking the height of the Sun's meridian passage at one of the equinoxes. It was a measure of longitude that gave the geographers and astronomers considerable trouble. For centuries, the works of Ptolemy were the sole guide for the Islamic regions, and a revision was made in Muslim Spain at Toledo in the eleventh century. The Toledan Tables were translated by Gerard of Cremona (*c.* 1114–87), and these were superseded by the Alphonsine Tables of 1255–60 under the Christian king, Alfonso el Sabio. These tables entered the medieval universities of Europe, so that the coordinates of many towns and regions became common knowledge. See Wright (1923), Lemay (1978), Harley and Woodward (1987). That latitudes were used on dials from a very early period is proved by the survival of two classical dials with latitudes and an adjustable suspension. One is Roman, ?AD 250, the other Byzantine from the late fifth century AD (Field and Wright 1985). They are not, however, equinoctial dials, and can have had no influence on the Renaissance craftsmen.

With the first crossing of the Atlantic, and the opening up of trade to the Far East, ocean navigation had become of great economic importance, as had land surveying. Mathematical astronomy was the key to greater accuracy in working out a position at sea, and making charts and maps of an unfamiliar coastline or tract of unexplored territory. As maps ceased to be largely works of imagination, and were drawn with ever greater precision, the accuracy of latitudes had to be regularly refined, and the maps revised accordingly. This occurred constantly with the maps of the great cartographer, Mercator. Also in need of constant revision were the latitude and longitude lists in books of reference used by seamen and surveyors, and travellers.

Of these works the most famous was Gemma Frisius, *Cosmographicus liber Petri Apiani*, 1529 and many later editions. Gemma added an *Appendix* to Apian's work, which lists towns all over the world. Gemma's version went through at least a dozen editions in the century, and was the key source for latitudes and longitudes. In Gemma's tables, accuracy was of importance, and changes were made from edition to edition (Table 4.4); hence the level of accuracy represented by the use of degrees and minutes. The tables were also very comprehensive geographically, including many places in the Middle and Far East. So instrument makers, wishing to include lists of latitudes, would copy from the easiest available source book, and select locations either by whim or at the suggestion of the customer.

The European locations, and those further afield, chosen by Humfrey Cole, James Kynvyn, and others, certainly came from Gemma, as a comparison of the names and latitudes in Gemma's tables with those on the compendia makes clear. One of the later editions was used, possibly that of 1564. In the space of some 100 pages, several hundred locations are given, usually with their Latin and common names, and covering the known world. On two or three pages in the mid-century editions is printed a list of up to 170

selected locations with their coordinates, and it is from this listing that the London craftsmen probably took their information. However, sometimes the main text has to be searched for places such as Malta and Babylon, and there are a few discrepancies between the main text and the table. With so many editions of Gemma's version of Apian in existence, and the opportunities for error, it is not worthwhile to try to identify a particular edition as being used by the makers.

What is not so readily discovered is where the latitudes of British towns used on the compendia were found. Very few locations in Britain occur in Gemma's tables. York is placed in Scotland, with a latitude of 57° 0′, barely south of Edinburgh at 57° 13′. Comparing the 1564 and 1584 editions, one finds Oxford at 52° 41′ or 51° 51′, and London 52° 30′ or 51° 40′. Only two other towns are named. It seems that for this part of the world Gemma was using the Ptolemaic map, a proposal confirmed by the incredible latitude for Dublin, 59° 30′.

The figures Gemma gives do not correspond to those of Cole. A listing heavily based on Gemma is published in an English work, William Cuningham, *Cosmographical Glasse* (1559), but Cuningham also includes British latitudes, and it might be expected that these would have been used by instrument makers. However, this is not so, since Cole's values for half his English towns differ from those of Cuningham. Born in Norwich, Cuningham was educated at Cambridge University and then at Heidelberg, where no doubt he became familiar with Gemma's work. He made astronomical observations, and surveyed and mapped his native city, as well as practising as a physician. His book was a popular and important one, but he made some serious

TABLE 4.4 Latitude comparisons: modern, Cuningham, Gemma 1553, 1564, 1584

Modern name	Modern latitude	Cuningham 1559	Gemma 1553	Gemma 1564	Gemma 1584
Alexandria	31. 0	31. 0	31. 0	31. 0	31. 0
Amsterdam	52. 23		52. 39	52. 39	52. 40
Antakya, *Antioch*	36. 14	35. 30	35. 30	35. 30	35. 30
Antwerp	51. 13	51. 28	51. 28	51. 28	51. 20
Athens	37. 58	37. 15	37. 15	37. 15	37. 15
Augsburg	48. 22	48. 15	48. 15	48. 15	48. 8
Babylon	32. 40	35. 00	35. 0	35. 0	35. 0
Barcelona, *Barchino*	41. 21	41. 35	41. 35	41. 35	41. 30
Basel, *Brasilia*	47. 35	47. 45	47. 41	47. 41	47. 40
Bordeaux, *Burdigala*	44. 50	44. 30	45. 30	45. 30	46. 0
Braga	41. 35	43. 40	43. 40	43. 40	43. 40
Brisighella, *Briscils*	44. 14	44. 30			
Bruges	51. 13	51. 20	51. 30	51. 30	51. 22
Brussels	50. 51	51. 28	51. 4	51. 4	51. 4
Burgos	42. 21	42. 48	42. 48	42. 48	43. 40
Cartagena	37. 38	38. 00	38. 0	38. 0	38. 0
Caesarea	32. 30		32. 30	32. 30	32. 30
Cologne	50. 56	51. 00	51. 0	51	50. 58
Compostella	42. 52	42. 15	44. 13	44. 13	44. 20
Constantinople	41. 0	43. 05	43. 5	43. 5	43. 05
Corinth	37. 56	36. 55	36. 55	36. 55	36. 55
Cracow	50. 04	50. 15	50. 12	50. 12	50. 12
Cuba, *Isabella*	22	23. 30	23. 30	15. 30	23. 30
Damascus	33. 30	33. 0	33. 0	33. 0	33. 0
Danzig	54. 22	54. 55	54. 54	54. 54	54. 54
Edinburgh	55. 57	59. 20	59. 20	57. 13	57. 13
Emden	53. 22		53. 28	53. 28	53. 42
Florence	43. 47	42. 45	43. 4	43. 4	42. 56
Frankfurt	50. 07	50. 10	50. 12	50. 12	50. 12
Ghent	51. 02	51. 15	51. 24	51. 24	51. 16
Gibraltar	36. 07	36. 15	36. 15	36. 15	36. 15
Granada	37. 10	34. 20	37. 50	37. 50	37. 50
Hangzhou, *Quinsey*	31. 15	37. 40	25. 15	25. 15	37. 40

TABLE 4.4 Latitude comparisons: modern, Cuningham, Gemma 1553, 1564, 1584 (continued)

Modern name	Modern latitude	Cuningham 1559	Gemma 1553	Gemma 1564	Gemma 1584
Jerusalem	31. 47	31. 22	31. 40	31. 40	31. 40
Lisbon	38. 42	36. 40	39. 38	39. 38	39. 38
London	51. 30	51. 30	52. 30	52. 30	51. 40
Louvain	50. 22	50. 59	50. 59	50. 59	50. 50
Lubeck	53. 52	54. 50	54. 48	54. 48	54. 50
Lyons	45. 46	45. 15	45. 10	45. 10	45. 10
Mainz, *Ments*	50. 0	50. 0	50. 8	50. 8	50. 8
Malta, *Melita*	35. 50		34. 40	31. 20	34. 40
Marseilles	43. 18	42. 05	43. 6	43. 6	43. 6
Middelburg	51. 30	51. 48	51. 50	51. 48	51. 40
Montpellier	43. 37	42. 50	43. 5	43. 5	43. 5
Nantes	47. 12	47. 15	48. 12	48. 12	47. 15
Naples	40. 50	39. 55	40. 0	40. 0	40. 50
Nineveh, *Niniui*	36. 25	36. 0	36. 40	36. 40	36. 40
Nuremberg	49. 26	49. 30	49. 24	49. 24	49. 22
Orleans	47. 54	47. 30	47. 13	47. 13	47. 13
Oxford	51. 45	51. 50	52.41	52.41	51. 51
Paris	48. 50	48. 40	47. 55	47. 55	48. 0
Padua, *Patavia*	45. 24	44. 45	44. 46	44. 46	45. 24
Perugia, *Perusia*	43. 6		42. 56	42. 56	42. 20
Prague	50. 05	50. 6	50. 6	50. 6	50. 6
Rome	41. 54	42. 0	41. 40	41. 40	42. 2
Rouen	49. 27	49. 30	49	49	49. 10
Toledo	39. 50	37. 0	39. 55	39. 55	39. 56
Toulouse	43. 37	42. 0	43. 30	43. 30	43. 30
Tournai	50. 35	50. 10	51. 40	51. 40	51. 32
Tours	47. 22	47. 30	47. 28	47. 28	47. 20
Venice	45. 27	44. 45	44. 50	44. 50	44. 50
Vienna	48. 12		48. 22	48. 22	48. 22

Of the several cities named Antioch, Cuningham selects one, and gives Gemma's latitude for St Peter's city on the river Oronte, but provides a gloss for St Luke's city near Mount Taurus, a degree further north. Gemma lists the Antiochs of both saints, and they are correctly named. Some variations of latitude values between editions arise through errors in type-setting. Caesarea is a common name; it is clear that Caesarea Stratonis, Judea, is intended.

errors in his latitudes of British towns. Indeed, as Table 4.4 shows, Cole was often nearer the correct latitude than Cuningham.

Errors, of course, are inevitable with the sort of lengthy, complex lists produced by Gemma and others. These could occur in the manuscript, in the printing process, and certainly on instruments in copying by the engraver from a long list. Some will be obvious in the following tables. Apian/Gemma give the coordinates, firstly longitude from the Azores, secondly the latitude. Only the latitude is listed here, in degrees and minutes of arc.

Table 4.4 compares three editions of Gemma over 30 years with Cuningham's list.

Tables 4.5 and 4.6 deal with the latitudes on eight Cole compendia, divided into British and Continental locations. Table 4.5 compares Cole's British latitudes with those of Cuningham, while Table 4.6 relates the European latitudes to those of the 1564 edition of Gemma, and gives Gemma's names for the towns.

Table 4.7 compares the latitudes on Kynvyn's two compendia (**65**, **66**) with two Gemma editions, of 1564 and 1584.

Commentary on Tables 4.4–4.6

Modern values are taken from *Philip's Atlas of the World*, 2nd edition (London, 1992). Gemma's town names and latitudes are from Gemma Frisius, *Cosmographia* (1564).

In general, Cole follows the spelling of Gemma Frisius for the names of the towns. He has also used the values for the latitudes. Cole did not take his latitude values for Britain from Gemma's table, where there are very few British locations listed, and the latitudes do not match those of Cole.

Cole is remarkably consistent with the values of latitude from 1568 to 1579, the exception being Bristol, which is 52° in only four out of seven occurrences. If one omits London, where one would expect to find an exact value in minutes of arc, then on the four compendia dated 1575 or 1579, one finds that of the 15 towns, the latitude is an exact number of degrees in seven cases, to 50 minutes in 5 cases, and for 10, 15, 30 minutes in one case each. It is tempting to see here a rounding of the values to save an amount of trouble. When one looks at the last known compendium by Cole, dated 1590, of 15 values of latitude, 6 are to exact number of degrees, and 9 have an apparent accuracy to one minute of arc, for example: 14, 18, 26, 44, 47. It seems likely that for this compendium Cole was taking values from Saxton's survey of England published in 1579.

The identification of the English towns is straightforward except for *Harforde*. Cole, on eight instruments, is consistent with the latitude of 52° 50′, and he has the same value for Leicester and Northampton. Ryther, on his compendium of 1588, has *Hereford* 52° 2′. It is probably fortuitous that this is within 2 minutes of the modern value. Books on British place names say that Hereford is from the Saxon, Harford, a crossing suitable for an army. Flavell Edmunds (1872, p. 225) claims that the true etymology is from the British Caer-ffawydd, Harford to the Saxons, hence Hereford. John Leland (?1506–52), in his *Itinerary*, writes: Hereforde, Heneforth, Heriforde, Herford, Hereford (Smith 1906).

Commentary on Table 4.7

In Table 4.7, the name as engraved is printed in italics when there is sufficient difference

TABLE 4.5 List of British towns with their latitudes on eight instruments by Humfrey Cole, 1568–90

Modern name	Modern latitude	Cuningham 1559	8 1568	9 1569	21 1575	24 1575	25 1575	26 1579	27 1579	33 1590
Basingstoke	51. 15									
Bath	51. 22	53. 40								51. 26
Berwick	55. 47	56. 50	56. 50	56. 50	56. 0	56. 50	56. 50	56. 50	56. 50	56
Bristol	51. 26	56. 50	52	52. 0	51. 20	51. 0	51. 0	52. 0	52. 0	
Cambridge	52. 13	52. 0		52. 0						52. 14
Canterbury	51. 17	51. 10								51. 18
Chester	53. 12	52. 10	53. 10	53. 10	53. 10	53. 10	53. 10	53. 10	53. 10	
Colchester	51. 54	51. 40								52
Coventry	52. 25	52. 0								52. 28
Dover	51. 07	51	51	51	51. 0	51. 0	51. 0	51. 0	51. 0	
Edinburgh	55. 57	59. 20	57	57		57. 0	57. 0	57. 0	57. 0	57
Exeter	50. 43	52. 15	51	51	51. 0	51. 0	51. 0	51. 0	51. 0	50. 45
Hereford	52. 04	52. 50	52. 50	52. 50	52. 50	52. 50	52. 50	52. 50	52. 50	
Leicester	52. 39	52. 50	52. 50	52. 50	52. 50	52. 50	52. 50	52. 50	52. 50	
Lincoln	53. 14	55. 10	53. 15	53. 15	53. 15	53. 15	53. 15	53. 15	53. 15	
London	51. 30	51. 30	51. 34	51. 34	51. 34	51. 34	51. 34	51. 30	51. 34	51. 33
Newcastle	54. 13	58. 55	55	55. 0	55. 0	55. 0	55. 0	55. 0	55. 0	55
Northampton	52. 14	52. 15	52. 50	52. 50	52. 50	52. 50	52. 50	52. 50	52. 50	52. 15
Norwich	52. 38	52. 10	52. 30	52. 30	52. 30	52. 30	52. 30	52. 30	52. 30	52. 44
Nottingham	52. 57	55. 30	53	53. 0	53	53. 0	53. 0	53. 0	53. 0	53
Oxford	51. 45	51. 50	51. 50	51. 50	51. 50	51. 50	51. 50	51. 50	51. 50	51. 47
Salisbury	51. 04	51. 50		51. 34	51. 15					
Winchester	51. 04	50. 15			51. 0					
York	53. 58	54. 20	54	54. 0	54. 0	54. 0	54. 0	54. 0	54. 0	54

TABLE 4.6 List of Continental towns with their latitudes on eight instruments by Humfrey Cole, 1568–90

Modern name	Modern latitude	Gemma name	Gemma 1564	8 1568	9 1569	21 1575	24 1575	25 1575	27 1579	26 1579	33 1590
Amsterdam	52. 23	Amsterodamum	52. 39	52. 40							
Antwerp	51. 13	Antwerpia	51. 28	51. 28	51. 28	51. 28	51. 28	51. 28	51. 28	51. 28	51. 28
Augsburg	48. 22	Augusta Vindelic.	48. 15	47. 32	48. 0						
Barcelona	41. 21	Barchino	41. 35				41. 35	41. 35	41. 35	41. 35	
Basel	47. 35	Basilæa	47. 41		48. 0						
Bruges	51. 13	Brugæ Flandr	51. 30				51. 30	51. 30	51. 30	51. 30	
Brussels	50. 51	Bruxella Brab.	51. 4	51			51. 0	51. 0	51. 0	51. 0	
Cologne	50. 56	Colonia Agrippin.	51	51	51. 0						
Compostella	42. 52	Compostella	44. 13		45. 0		44. 20	44. 20	44. 20	44. 20	
Istanbul	41. 0	Constantinopolis	43. 5				43. 0	43. 0			
Cracow	50. 04	Cracouia	50. 12				51. 0	51. 0	51. 0	51. 0	
Danzig	54. 22	Dantiscum Prussiæ	54. 54	54			55. 0	55. 0	55. 0	55. 0	
Emden	53. 22	Emde	53. 28				53. 42	53. 42	53. 42	53. 42	
Florence	43. 47	Florentia	43. 4				43. 4	43. 4			
Frankfurt	50. 07	Francfordia	50. 12	50. 10			50. 12	50. 12			
Ghent	51. 02	Gandanum	51. 24	51. 24			51. 24	51. 24			
Lisbon	38. 42	Lisbona	39. 38	39. 38	38. 0		39. 38	39. 38	39. 38	39. 38	
Louvain	50. 52	Louanium	50. 59			50. 58	50. 58	50. 58	50. 58	50. 58	
Lubeck	53. 52	Lubeca	53. 48	54. 48			54. 48	54. 48			
Lyons	45. 46	Lugdunum Gallia	45. 10		45. 0						
Marseilles	43. 18	Massilia	43. 6	43. 6			43. 6	43. 6	43. 6	43. 6	43. 6
Middelburg	51. 30	Middelburgum	51. 48	51. 48							

TABLE 4.6 List of Continental towns with their latitudes on eight instruments by Humfrey Cole, 1568–90 (continued)

Modern name	Modern latitude	Gemma name	Gemma 1564	8 1568	9 1569	21 1575	24 1575	25 1575	27 1579	26 1579	33 1590
Montpellier	43. 37	Mons Pessulanus	43. 5	42. 5							
Nantes	47. 12	Nannetis	48. 12				48. 12	48. 12			
Naples	40. 50	Neapolis	40. 0		39. 0		41. 0	41. 0			
Paris	48. 50	Parisÿ	47. 55	47. 55	47. 55		47. 55	47. 55	47. 55	47. 55	
Prague	50. 05	Praga	50. 6						50. 4	50. 4	
Rome	41. 54	Roma	41. 40		42		41. 50	41. 50			
Rouen	49. 27	Rothomagus	49	49	48. 0		49. 0	49. 0	49. 0	49. 0	
Toledo	39. 50	Toletum	39. 55	39. 56	39. 0						
Toulouse	43. 37	Tolosa	43. 30	43. 30			43. 30	43. 30	43. 30	43. 30	
Tournai	50. 35	Tornacum	51. 40				51. 40	51. 40			
Venice	45. 27	Venetiæ	44. 50		44. 50		44. 50	44. 50	44. 50	44. 50	

TABLE 4.7 List of towns with their latitudes on instruments by James Kynvyn, *c.* 1590, 1593

Modern name	Modern latitude	Gemma 1564	Gemma 1584	65 *c.* 1590	66 1593
Alexandria	31. 0	31. 0	31. 0	31	31
Antakya, *Antioch*	36. 14	35. 30	37. 20	37. 20	37. 20
Antwerp	51. 13	51. 28	51. 20		51. 28
Athens	37. 58	37. 15	37. 15	37	37. 15
Babylon	32. 40	35. 0	35. 0	35	35
Basel, *Brasilia*	47. 35	47. 41	47. 40	47. 41	47. 41
Bordeaux, *Burdensi*	44. 50	45. 30	46. 0	45. 30	45. 30
Braga	41. 35	43. 40	43. 40		43
Brisighella, *Briscils*	44. 14			44. 5	44. 5
Burgos	42. 21	42. 48	43. 40	42. 48	42. 48
Cartagena [Spain]	37. 38	38. 0	38. 0		38
Caesarea [Judea]	32. 30	32. 30	32. 30	31. 40	31. 40
Compostella	42. 52	44. 13	44. 20		42. 15
Constantinople	41. 0	43. 5	43. 05	43	43
Corinth	37. 56	36. 55	36. 55	35. 55	35. 55
Cuba	22. 0	15. 30	23. 30		23½
Damascus	33. 30	33. 0	33. 0	33	33
Edinburgh	55. 57	57. 13	57. 13	58	57
Florence	43. 47	43. 4	42. 56	43. 40	45. 40*
Gibraltar, *Heercules pillers*	36. 07	36. 15	36. 15		36. 15
Granada	37. 10	37. 50	37. 50	37	37
Hangzhou, *Quinsey*	30. 18	25. 15	37. 40		37. 40
Jerusalem	31. 47	31. 40	31. 40	36. 40	36. 40*
Lisbon	38. 42	39. 38	39. 38	39. 38	39. 38
London	51. 30	52. 30	51. 40	51. 30	51. 33
Lyons	45. 46	45. 10	45. 10	45. 10	45. 10
Malta	35. 50	31. 20	34. 40	34	34
Mainz, *Ments*	50. 0	50. 8	50. 8	50. 8	50. 8
Naples	40. 50	40. 0	40. 50	40. 36	40. 36
Nineveh, *Niniui*	36. 25	36. 40	36. 40	41. 40	41. 40*
Nuremberg	49. 26	49. 24	49. 22	46. 24	40. 24*
Orleans	47. 54	47. 13	47. 13	47. 13	47
Paris	48. 50	47. 55	48. 0	48. 30	48. 30
Padua, *Patauia*	45. 24	44. 46	45. 24	44. 28	44. 28
Perugia, *Perusia*	43. 6	42. 56	42. 20	42. 30	42. 30
Rome	41. 54	41. 40	42. 2	41. 40	44. 40*
Tours	47. 22	47. 28	47. 20		47. 30
Venice	45. 27	44. 50	44. 50	45. 18	45. 18
Vienna	48. 12	48. 22	48. 22	48. 22	48. 20

from the modern spelling. It is clear that the towns with their latitudes are taken from the Gemma Frisius editions of *Cosmographia*.

Bordeaux: *Burdegala* in Gemma's table, from the Roman *Burdigala*.

Cartagena: Roman *Carthago nova*. Gemma has this, and the common name *Cartagena*.

Mainz: *Moguntia* in Gemma's table; *Moguntiacum* is the latinized Celtic name.

Quinsey: Gemma has *Quinsay.vrbs* in his table, with latitude 25° 15′ (1553, 1564), or 37° 40′ (1584). Gemma's text reads: 'Quinsay Ciuitas totius mundi maxima, quæ nostro idiomate dicitur Cæli ciuitas: in medio lacus est habens in circuitu 1 200 pontes: 226.15 [Long.] 37.40 [Lat]'. The city is Hangzhou (Mouth of the Han). Marco Polo described the city as the largest in China, with many bridges, and a very large population.

A close examination of all the instruments made by James Kynvyn shows that he did not take sufficient care when engraving numbers. Some discrepancies, marked with an asterisk (★), are as follows. Florence is correctly 43° 40′ on **65**, but 45° 40′ on **66**. One should assume this is simply an erroneous 5 replacing a 3. Compared with Gemma, both Jerusalem and Nineveh are 5° too high, and Rome 3° too high. The value for Nuremberg of 40° on **66**, and 46° on **65**, is a case of engraving the *o* and then forgetting to finish with the curl. In fact, he should have engraved 49°, so both entries are in error and it is easy to see how the mistakes arose.

Edinburgh and London are the only towns from the British Isles on these two compendia; the rest cover a wide selection of the Continent and Near East. There is a marked preference for cathedral cities and Biblical places.

The Compass Rose

The origin of the magnetic compass, with a magnetic needle above a circular, marked-out base known as a compass rose, remains conjectural. The Norsemen divided the horizon into the four cardinal points, representing North (the Pole Star), South (midday), East (sunrise), and West (sunset). Classical Greek pilots regarded the wind directions as an important navigational aid, which accounts for the names of the winds in Greek, Latin, or Italian being used, as well as the language of the place of manufacture, on Renaissance instruments. The Tower of the Winds at Athens has sides facing eight winds. Subdivision then occurred, and the oldest surviving chart of the thirteenth century shows the division into 32 directions, or rhumbs of the winds. Now established, this configuration continued through the sixteenth century and beyond.

This division of the compass rose was taken over when the magnetic compass was to be used on land, even though division into 360° could have served better. Some of the instruments described below have the compass rose divided into degrees as well as into wind directions. The magnetic compass was required on land as well as at sea to orient most portable sundials, especially the equinoctial dial found on most Elizabethan compendia. It was not until the early seventeenth century that self-orienting dials, notably the universal equinoctial ring dial, became general. A compass was also needed by surveyors in the field, and was fitted to some Renaissance astrolabes for horizontal use. This led to the development of the simple theodolite (see below), on which, and on all later forms of the instrument, a magnetic compass is always found. See Taylor (1951a, 1951b); Waters (1958, pp. 20–30).

The compass needle points to magnetic North, not to the geographic North Pole. The difference is known as the declination of the magnetic needle. Its value depends on the location and the year. At London (and much of northern Europe), the needle can move through 35° in 240 years. In 1560, the declination at London was about 11° East; in 1660 it was zero; in 1700 it was 6° West (Malin and Bullard 1981).

Some Elizabethan compass bases simply have the cardinal points scribed in the brass. However, most of the compendia and pocket dials are fitted with printed paper, hand-coloured roses, normally only some 50 mm in diameter. In Table 4.8 are listed 23 instruments: 17 compendia, 2 diptych dials, 2 compass dials, and 2 theodolites. There is a progressive change in pattern, and a simplification, and it becomes clear that the same design of rose and the same printing can be found on instruments signed by different craftsmen. These compass roses, when mounted in an instrument serve, therefore, to point to the place of origin—London—rather than to a particular craftsman, and the change in pattern is helpful in dating.

Classification of Compass Roses

A-1 32 parallel-band pointers; diameter 50
Used by Humfrey Cole, 1570s, and James Kynvyn 1593, **66**

A-2 engraving similar to type A-1; diameter 56
Used by Charles Whitwell c. 1600, 1602, 1606, and Elias Allen from c. 1610

B-1 16 triangular pointers, partitioned, small version; diameter 41

Used on the two diptych dials, **99** and **56**, the latter having a coloured one below a plain one
A large hand-drawn version, with 8 partitioned triangles, is on Augustine Ryther's theodolite, **35**, dated 1590
A small hand-drawn version on the compendium, **65**, is used by James Kynvyn c. 1590

B-2 16 triangular pointers, partitioned, large version; diameter 69
Used by Charles Whitwell 1600, **58**, and Elias Allen, c. 1610

B-3 16 triangular pointers, partitioned, smallest version; diameter 38
Used on **98** by T.W., 1602

C 16 pointers: 8 triangles with curved sides, 4 triangles with serrated sides, 4 baluster-shaped, an ornate version; diameter 60
Used by Charles Whitwell, 1604, 1610, and by Elias Allen, 1617

FIG. 4.4 Compass rose type C on Whitwell's chart of the Azores, 1599. MHS Oxford.

TABLE 4.8 Compass roses on paper

No.	Maker	Year	Type	Diameter	Scale	Instrument
24	Cole	1575	A-1	< 50	trimmed	compendium
25	Cole	1575	A-1	50	0–90° × 4	compendium
26	Cole	1579	A-1	< 50	nil	compendium
34	Ryther	1588	hand	50	nil	compendium
35	Ryther	1590	B-1 hand	< 82	120	theodolite
65	Kynvyn	c. 590	B-1 hand	< 58	nil	compendium
66	Kynvyn	1593	A-1	55	0–360° R	compendium
56	Whitwell	c. 1600	B-1 upper	41	nil	diptych
56	"	"	B-1 below	41	nil	diptych
57	Whitwell	c. 1600	A-2	56	0–90° × 4	compass dial
99	RG	c. 1600	B-1	41	nil	diptych
58	Whitwell	1600	B-2	69	0–360°	compendium
98	TW	1602	B-3	38	nil	compendium
59	Whitwell	1602	A-2	56	0–90° × 4	compendium
60	Whitwell	1604	C	60	0–90° × 4	compendium
61	Whitwell	1606	A-2	56	0–90° × 4	compendium
62	Whitwell	1610	C	60	0–90° × 4	compendium
82	Allen	c. 1610	A-2	56	0–90° × 4	compass dial
83	Allen	"	B-2	69	0–360°	compendium
84	Allen	"	A-2	57	0–90° × 4	compendium
85	Allen	"	A-2	57	0–90° × 4	compendium
86	Allen	"	A-2	47	trimmed	compendium
87	Allen	1617	C	60	0–90° × 4	compendium
88	Allen	c. 1620	azimuth	c. 91	120	theodolite
	Hood	1592	A-2	62	nil	chart
	Wright	1610	C	60	nil	chart
	Hopton	1611	azimuth	91	120	engraving

Azimuth = magnetic azimuth dial; R = engraved on a brass ring above glass cover; 120 = not degrees, but a surveyor's usage; hand = hand drawn; entries showing > and < symbols are estimates because some cards are difficult to measure.

All these types of compass rose are illustrated in Plate 14.

Some of these designs are found on printed charts. Thomas Hood's chart of the NE Atlantic, dated 1592, and engraved by Augustine Ryther, has an example of type A-2. The Armada charts, engraved by Ryther, have many variations on the B-1 type (*Armada* 1988). Edward Wright's chart of the Azores (Fig. 3.6), with his *Certaine Errors* (1599/1610), the engraving attributed to Whitwell, has the identical printing of type C (Fig. 4.4). The theodolite of *c.* 1620 by Allen, **88**, has a magnetic azimuth compass card, which is the same as that published in 1611 by Arthur Hopton in his *Speculum Topographicum*.

The Nocturnal

The nocturnal was devised at the end of the fifteenth century in order to tell the time by the stars during the hours of darkness. The larger, single nocturnals were made in brass, or frequently in wood, for the use of seamen. The Cole nocturnal (**12**), with a tide computer on the reverse, is a fine brass navigator's instrument. The nocturnal was mostly used, however, for travellers at night, and indeed for domestic purposes, and was therefore incorporated into a number of sixteenth-century compendia.

The basic instrument consists of a disc fitted with a handle, a smaller disc (volvelle) that fits over the first and is equipped with one or two pointers, and lastly another disc with an extended index arm that moves over the second disc. All these parts are held together by a nut and bolt pierced by a hole. The outer disc is marked with the months and days in the year, and often, especially in sixteenth-century nocturnals, with a scale showing the 12 Signs of the Zodiac. The smaller disc has its edge divided into 24 hours, twice 1 to 12. The pointer over this disc is set to the date when the observation is being made. The index arm is used to line up the stars in either the Little Bear or the Great Bear. The index arm is aligned with the stars by holding the instrument vertically by its handle, and then looking through the hole in the bolt at the Pole Star. The time can then be read from the scale on the second disc. To make it easier to read the time from the disc in darkness, teeth were usually cut in each hour position, so that a count could be made with the fingers, starting from the position of the pointer, which was at midnight.

The nocturnal was designed to use the fact that the stars appear to rotate about the celestial North Pole once in 24 hours, so providing 'a clock in the sky'. The star nearest to the

Pole suitable to be taken as the centre of the 'clock' is *Polaris* (α UMi), called the Pole Star. The angular distance of *Polaris* from the Pole was about 3° in Elizabethan times (see **96**), but is three-quarters of a degree in AD 2000. The 'hour hand' of this 'clock' is the line from the Pole Star to the bright star *Kochab* (β UMi) in the constellation of the Little Bear (*Ursa Minor*). An alternative 'hour hand' is provided by the constellation of the Great Bear (*Ursa Major*), which has the popular names of the Plough and the Big Dipper. In the latter case, the two bright stars at the front of the Plough (*Dubhe* and *Merak*, α UMa and β UMa) are employed. These stars line up with the Pole Star, and are known as the pointers, because they always point to the Pole, or as the guards, because they watch over the Pole.

An allowance has to be made for the difference between time as measured by the Sun (the solar day) and as measured by the stars (sidereal day). A sidereal year is one day longer than a solar year. The same stars rise 3 minutes and 56 seconds earlier on successive nights, so appearing to lose an hour every 15 days. The necessary allowance was made by setting the pointer to the day in the year on which the instrument was being used.

One of the earliest manuscripts that deal with the nocturnal is by John of Gmunden (*c.* 1380–1442), Professor of Astronomy at the University of Vienna. It may be no coincidence that the earliest known nocturnal, preserved in the British Museum, is signed and dated by Hans Dorn of Vienna, 1491. In the National Maritime Museum, Greenwich, is another early nocturnal, dated 1516, that was made in Florence by the Volpaia family. It is in brass, like many early nocturnals, and can give the time in Italian hours. A popular work among students of astronomy in the sixteenth century was Peter Apian, *Cosmographicus liber* (1524), more especially the several editions of Gemma Frisius, entitled *Cosmographia*. These contain an illustration of a nocturnal.

A most influential manual for seamen was the comprehensive work by the Spaniard, Martin Cortés, *Breve compendio de la esfera y de la Arte de Navegar* (Seville, 1551), which discussed astronomy and instruments, including the nocturnal. This was the first text chiefly concerned with navigation to publish an illustration of the nocturnal, and was translated into English from 1561. After this an increasing number of instructional texts for seamen included the use of the nocturnal. It is often depicted on the trade cards of mathematical instrument makers during the eighteenth century, a late example being the card of Dudley Adams of about 1800. This, however, does not mean that they continued to be sold.

The pocket compendium, because of its small size and the need to fit into a case for the pocket, required some modifications in the design of the nocturnal. Initially, the extended pointer was hinged for convenient packing. The next step was to reduce the length of the pointer, requiring more care and skill in lining it up. Humfrey Cole therefore devised an alternative for his compendium dated 1590 (**33**), the use of a slit that enabled both *Polaris* and *Kochab* to be viewed together, instead of the usual central hole and index arm. This design was adopted by Charles Whitwell on his compendium of 1600 (**58**), and was used by him on a sequence of compendia made in 1602 (**59**), 1604 (**60**), 1606 (**61**), and 1610 (**62**).

Elias Allen, most likely in conjunction with Edmund Gunter, introduced further varia-

tions. After the compendium with a slit (**84**), the first of his modifications was the selection of 12 stars, one for each month. From the centre, 12 radial lines are cut to intersect each month, and the star identified by name. When one of the named stars is on the meridian, the volvelle is turned to match, when the day of the month gives the time. This enables the nocturnal to be used in a manner similar to the slit, but offers greater choice of stars to suit the time of year. The declination of a star at the meridian transit can give the latitude. It is possible that this scheme is derived from the 'notable fixed starres for nauigation, with tabels of their shining' published as the fourteenth Rule by William Bourne in his *An Almanacke and Prognostication for three yeares* [1567]; see Taylor (1963, pp. 99–101).

A modification made later, so far as one can judge, for the Allen instruments are seldom dated, was the pictorial type. This incorporated pictorial representations of six constellations, with a simplified star pattern on each. The time is found by matching the orientation of an easily observed constellation on the instrument with that visible in the sky, and then reading off the time from the date, in the manner of the modern Philips' Planisphere. This form of nocturnal was illustrated by Allen in Edmund Gunter, *The Description and vse of the Sector* (1623, Sig. A4, pp. 60–1), and recommended for the use of seamen. The usage is described: 'For looke vp to the pole, and see what starres are neare the meridian, then place the rundle to the like situation, so the day of the moneth will shew the houre of the night'. The pictorial design is used on two extant Allen compendia (**83** and **86**). It continued in use throughout the seventeenth century and into the eighteenth century. Note that in some modern descriptions of the nocturnal it is named as a planispheric nocturnal.

It is clear that with the nocturnals on compendia, the chief requirement was ease of use, allied with compactness, and that accuracy was not of great importance; an approximate time was all that was expected.

The Tide Computer

Since the Mediterranean Sea is almost cut off from the Atlantic, the rise and fall of the tide is small, and of little consequence. Around the British Isles the range of the tides may be 40 feet (12 metres). The tidal effect of the Sun is less than half that of the Moon, but as the Moon progresses around the Earth in $29\frac{1}{2}$ days, the combined effect of their gravitational pull is marked. At new moon the two bodies are in conjunction, and the pull is at a maximum. There is a considerable effect, though not as great, at the full moon, when the two bodies are in opposition. These tides are known as spring tides. When the Moon is at its first or last quarter, the rise and fall of the tides are least, and are known as neap tides.

A tide computer has scales for the 32 wind directions, another of 24 hours, a volvelle with the $29\frac{1}{2}$ days of the Moon's age, and a further volvelle with an index arm that is put to the age of the Moon. This top disc has an eccentric circular aperture that passes over a bright/dark pattern to show the configuration of the phase of the Moon. There is commonly the addition of the aspects: opposition, trine, quartile, and sextile.

Humfrey Cole fitted his compendium of 1569 (**9**) with such a device, which he called an instrument to know the ebb and floods.

See Waters (1958, pp. 16–18), Taylor (1963, pp. 1–2), Taylor (1971, pp. 22, 197).

The Establishment of the Port

The time of high tide is critical for entering or leaving port. The highest tides occur when the Moon is new or full, and the time of such a tide depends on the geographical location of the port. Therefore seamen needed to know the compass bearing of the Moon at the time of high water, known as 'full sea'. Each port has its own bearing of the Moon at high water, so the establishment of the port is either listed as a table in columns, or in a circle with the names of the ports associated with their compass bearing. Humfrey Cole's compendium of 1569 (**9**) has a table with the columns headed by the names of the principal ports and havens of Europe and what Moon makes a full sea. These names are arranged under headings of the wind directions. On James Kynvyn's compendium of 1597 (**66**) the equivalent table is in the round, the ports arranged in the order of the compass bearings. See Waters (1958, pp. 31–3), Taylor (1963, pp. 1–2), Taylor (1971, pp. 22, 197).

THE FOLDING RULE

The plane table alidade and the surveyors' folding rule may be used with a plane table for small scale plotting of ground. The folding rules have other functions, such as estimating heights or declivities, and one of the following can be used as an altitude sundial (**17**). The rules are also engraved with what is regarded as useful information, which includes timber and board measure, and mensuration definitions.

Humfrey Cole made all the folding rules (**10**, **11**, **17–20**, **22**), including gunners', and the alidade (**32**) described in Part Two. Thomas Digges added to the 1591 edition of his father's *A Geometrical Practise named Pantometria* (1571), an appendix at the end of the First Book called *Longimetra*:

> A note touching a Platting Instrument for such as are ignorant of Arithmeticall Calculations. Some take the Semicircle of my Topographicall Instrument [theodolite], setting the Perpendiculare thereof vpon a straighte long Ruler deuided into a thousande or more equall parts, and insteade of the Horizontall Circle vse only a plaine Table or boarde: whereon a large Sheet of Parchement or Paper may be fastened… . This being an Instrument onelye for the ignorante and vnlearned, that haue no knowledge of Noumbers, and not to be practized but in fayre weather, and where yee may haue time sufficient euen in the Fielde to pricke and set downe the Charte.

The divisions of an inch require an explanation. Plane tables are not particularly large, and can measure 15 × 12 inches, which can take a full sheet of printing paper measuring 17 × 13½ inches. A surveyor would wish his plan to fill the sheet of paper, so he would, accordingly, need to choose a convenient scale. For example:

110 yards	in 15 inches	can use a scale of	1:7
300 yards	" "	" "	1:20
400 perches	" "	" "	1:27
440 yards	" "	" "	1:30
600 yards	" "	" "	1:40
960 feet	" "	" "	1:64

This is an explanation for the series of divisions of the inch engraved on the alidade and folding rules. The choice of the various inch divisions depends on the space available, the alidade having 7 different scales, the rules nor-mally 12, and one rule 24. The 1000-part scale, that is 24 inches divided into 1000, is mentioned by Digges, and this is engraved on the inside edge of the rules for use when the two arms are set in line. The sights on the arms, a backsight at the hinge, and the protractor at the hinge are all used in simple surveying.

It is normal to find on the folding rules the following definitions of linear measure:

> 3 Barly cornes, to an ynche: 12 ynches, a foote: 3 foote a yarde: 5 yardes and ½ a pearche: 40 pearches in lengthe and 4 in breadthe an acre. So an acre contayneth 160 pearches the halfe acre 80 pearches a roode 40 pearches.

Timber Measure and *Borde measure* are scales intended for the use of surveyors, gunners, and carpenters. The line of timber measure is used with the inch scale, so that, knowing what the square measurement of the balk is, the length can be found to give a cubic foot. For example, if the piece is 17 inches square, this figure on the timber scale is opposite 6, so 6 inches is (very nearly) the length to make a cubic foot (1 cubic foot equals 1728 cubic inches). The line of board measure gives the length to make a square foot from measuring the width of a board. If the width is 18 inches, this figure is opposite 8 inches.

Sometimes a table for the area of land is engraved, which is a copy of the table in Leonard Digges, *A boke named Tectonicon briefe-lye shewynge the exacte measurynge … all maner lande, squared tymber, stone …* (1556). This gives the length and breadth, in perches, which multiplied together come to 1 acre in area. (A perch measures 16.5 feet, or 5 metres.) Thus the extent of a plot 8 perches in breadth and 20 in length comes to 160 square perches, or 1 acre.

THE SECTOR

Land surveyors, fortification draughtsmen, and navigators are often obliged to work in inhospitable surroundings, and consequently any aid to quick calculation would be welcome. Such a device is the sector. It is a jointed rule with two radial arms engraved with graduated lines. With the addition of a pair of dividers, and using the principle of similar triangles, scaling ratios, linear or square, could be found. An early version was published in 1598 by Thomas Hood (*c.* 1557–1620) with the title, *The Making and use of the Geometricall Instrument, called a Sector* (Fig. 4.5). John Napier (1550–1617) devised logarithms, which he published in 1614, and within the next decade the slide-rule had been invented. The sector became the standard item in any set of drawing instruments until the middle of the nineteenth century, while the slide rule was a convenient accompanying device, particularly for engineers.

The formative period for the sector (French *compas de proportion*) and the proportional compass (French *compas de reduction*) was 1570 to 1600. The latter is a different instrument serving a similar purpose, which causes some confusion. Contributions to the development of both instruments were made by Michiel Coignet (1549–1623), Fabritio Mordente, and Galileo Galilei. During its initial phase, the popularity of the sector was due to Galileo, who published a description of his *compasso geometrico e militare* in 1606 (Drake 1978). In spite of Hood's book, and the existence of a sector, the earliest known, signed by James Kynvyn and dated 1595 (**68**), Galileo has commonly been too readily credited with the invention of the instrument.

If it is true that Galileo constructed his first sector in 1598, this may well have been because he had received news from London, just as he heard in 1609 of the new telescope just invented in the Low Countries in 1608. There were precursors of the Galilean sector, one of which was the *reigle platte* of Michiel Coignet, another the compass of Mordente, and, of course, the sector described by Hood. This period has been ably studied by Meskens (1997, 1998), whose work should be searched for references; and by Schneider (1970), who pertinently comments that Galileo stands at the end, not the beginning, of the line of development of the sector.

Thomas Hood was an MA, elected a Fellow of Trinity College, Cambridge, in 1581. Following a Privy Council recommendation that citizens should be instructed in military matters, Hood was appointed the Mathematical Lecturer to the City of London in 1582. Later, he taught privately, dealing with navigation, drawing and explaining charts, and designing instruments. As well as the sector, he devised a form of cross-staff. The best account of Hood's mathematical practice is that by Johnston (1991, pp. 330–41); see also *DNB*.

Of the Elizabethan sectors described in this book, that by Kynvyn, dated 1595 (**68**), lacking the arc, may well have been an early trial commissioned by Hood. It is reasonable to conjecture that the three sectors that resemble Hood's closely (**37**, **46**, **69**), two dated 1597, the year before the date on Hood's title page, may be the result of Hood asking leading craftsmen to make sectors to his design.

The next development of the sector in England, particularly for mariners, was the achievement of Edmund Gunter (1581–1626),

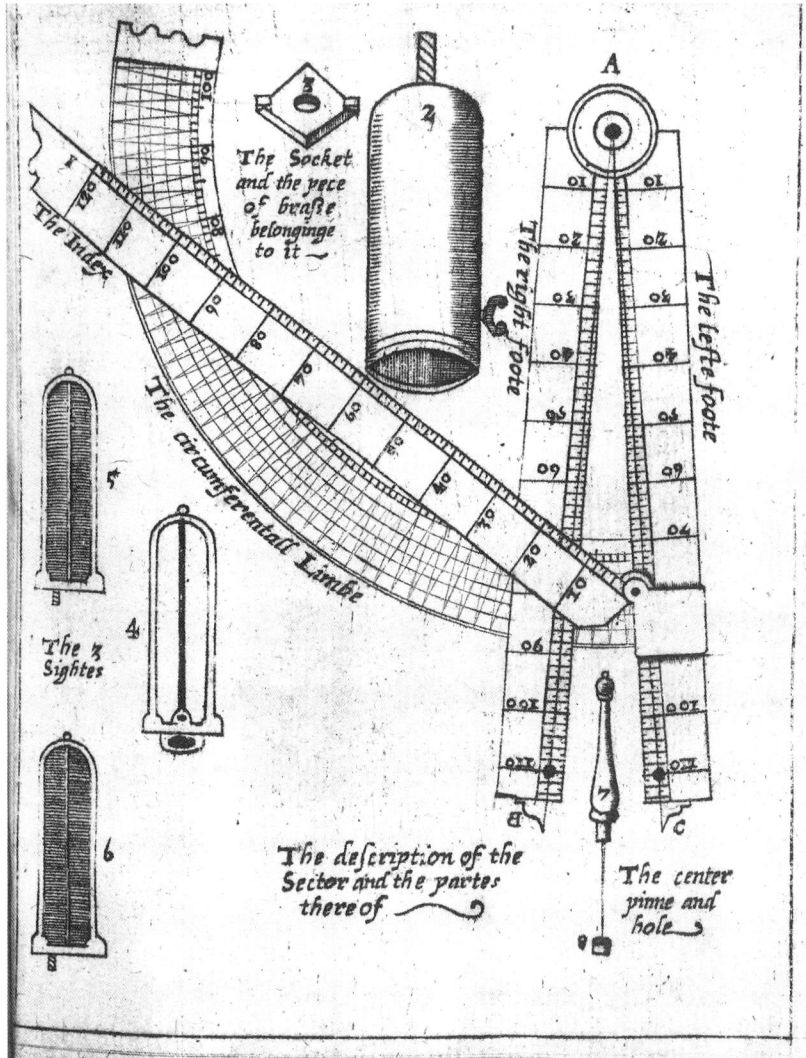

FIG. 4.5 Whitwell's engraving of the parts of Hood's sector, 1598. BL Oxford.

Professor of Astronomy at Gresham College from 1619. His book, *The Description and vse of the Sector*, was published in London in 1623. The engraved frontispiece shows the sector, and bears the notice: 'These instruments are wrought in brass by Elias Allen ... and in wood by Iohn Thompson ... and Nathaniell Gos.' However, as early as 1606, Gunter had produced a manuscript account in Latin of his sector and how to use it (Waters 1958, p. 358). Taylor (1954, p. 339) comments: 'As was customary, these hand-written notes on the use of his newly-designed instrument were circulated among the author's pupils and friends. Only when the demand made hand-copying irksome were they sent to the printer (in an English translation).' Therefore, from before 1610, anyone interested in mathematics and scientific instruments could have known of Gunter's innovations.

Two early sectors (**63**, **64**) of the Gunter type are those signed *CW*, for Charles Whitwell, and undated. They were made shortly before Whitwell's death in September 1611, presumably as a result of his being aware of Gunter's manuscript description of his sector. There are three sectors of the Gunter type, all inscribed *John Goodwin scu:*, but undated. One of these came to the Museum of the History of Science, Oxford, in 1930 (**100**); another to the National Maritime Museum, Greenwich, in 1937 (**101**); a third came on the market in the 1980s and is in private ownership (**102**). The Maritime Museum instrument is mentioned by Taylor (1954, p. 194), where there is a biographical entry under: 'Go(o)dwyn, John (fl. 1597–1600)'. John Goodwin was a teacher of arithmetic and geometry in the City of London. William Barlow, in *The Navigator's Supply* (London, 1597), Sig. K2, wrote of him: 'yet I know but one, whose name is Iohn Goodwyn, dwelling in Bucklers-burie: A man unskilful in the Latin tongue, yet having proper knowledge in Arithmetike, and Land-measuring, in the use of the Globe, and sundry other Instruments'. Arthur Hopton (1588–1614) refers to Goodwin in *Speculum topographicum or the Topographical Glasse* (London, 1611), p. 126, at the heading of the chapter on the circumferentor, 'whose use was first practized by *I.G.* and now published'. Goodwin is also mentioned in Aaron Rathborne, *The Surveyor* (1616, p. 125), where he refers to 'M. IOHN GOODWYNS inuention [plane table ruler], that excellent and honest Artist, whose liuing Name (though himselfe be dead) I cannot remember without good respect'.

These references to Goodwin tell us that he was one of London's respected mathematical practitioners, was a man who knew about instruments, and was even one who 'invented' them. He is very likely to have known of Gunter's design for a sector, even though he did not read Latin, and was dead some years before the publication of Gunter's book on the subject in 1623. Gunter will have referred to his design in his lectures, which had to be in English as well as Latin, and a manuscript English version may well have been produced.

THE THEODOLITE

The appropriate names for theodolites are explained by Bennett (1987, pp. 40–1). Considering the principles of triangulation, William Cuningham, in his *Cosmographical Glasse* (1559, p. 136), described an instrument with a circular plate 'made muche like the backe parte of an Astrolabe'. The edge was divided in degrees, with four quarters of 90°, with an alidade pivoted at the centre, and a magnetic compass. Also provided was a diagram of the 32 wind directions. As Bennett remarks: 'With Cuningham's instrument, the rationalization of the astrolabe into a surveying instrument was complete.'

Such instruments were described by Thomas Digges in his edition of *A Geometrical Practise named Pantometria* (1571) written by his father, Leonard. His *Theodelitus* included a double shadow square. From the nineteenth century on, the word theodolite denotes an altazimuth instrument, whereas in earlier usage it meant an azimuth instrument. To avoid confusion when referring to these devices in modern historical texts, Bennett proposed calling the azimuth instrument a simple theodolite. Modern secondary literature has also used the word 'circumferentor' for the simple theodolite, even though the word had been used from the seventeenth century for a different instrument.

Humfrey Cole is certainly the maker of the first extant theodolite. Three of these surveying instruments by him have survived, the one with the earliest date, 1569, being a miniature simple theodolite measuring 2 inches in diameter, with central compass and alidade, incorporated in a compendium (9). Closely similar in the information provided, and in the layout, is the simple theodolite dated 1574 (15), $6\frac{3}{4}$ inches in diameter, to which a modern altitude semicircle has been added. The complete, full-size, altazimuth theodolite by Cole is dated 1586 (31), 8 inches in diameter. There are two other English Elizabethan theodolites, one by Augustine Ryther, dated 1590 (35) and similar to Cole's instrument, and the other by James Kynvyn, dated 1597 (70, 71), both at the Museo di Storia della Scienza, Florence.

It is instructive to read the entry under theodolite in the *Oxford English Dictionary*, which adds a gloss on the origins of the word, a rare event for the compilers of the dictionary. It was thought necessary because the origin is unknown. After describing the construction (as above), we are told that the original *theodelitus* of Digges was for horizontal angles only, and that many quotations down to the nineteenth century use this sense. Digges also described a compound instrument having a vertical semicircle for taking altitudes, but this he called his *topographicall instrument*, restricting *theodelitus* to the horizontal circle. Therefore, according to the dictionary:

The name, alike in the Latinized form *theodelitus* and the vernacular *theodelite* (subseq. -*dolite*), originated in England, and is not known in French and German until the 19th C. Its first user, and probable inventor, L. or T. Digges, has left no account of its composition, as to which various futile conjectures, incompatible with its early history and use, have been offered; such is the notion that it arose in some way out of *alhidada* or its corruption *athelida* occurring in Bourne's *Treasure for Travailers* 1578, which an examination of the works of Digges and Bourne, where both words occur in their proper senses, shows to be absurd. *Theodelite* has the look of a formation from Greek; can it have been (like many modern names of inventions) an unscholarly formation from θεάομαι 'I view' or θεῶ 'behold' and δῆλ-ος 'visible, clear, manifest' with a meaningless termination?

THE GUNNERY TABLE

The principal duty of any government is to defend the realm. For Tudor England, a navy was of crucial importance, and both ships and coastal defences needed guns and trained gunners. During the early part of the sixteenth century ordnance was imported from the excellent foundries of Flanders and Germany. Henry VIII founded his special Armoury at Greenwich in 1515, with some 20 craftsmen from Germany and Flanders; the Master Workman was paid £26 and the hammermen £20 in the year, plus allowances for livery, food, and lodgings (Williams and De Reuck 1995, pp. 26–30). With the Spanish occupying Flanders, and the loss of the English enclave at Calais in 1558, the naval potential of Spain to do damage to England increased. It was necessary to arm and to reduce the reliance on imports. Sources in Britain had to be found for copper ore and calamine (zinc carbonate) to make brass and bronze, and German miners and metal workers were brought to the country; the subject is thoroughly treated by Hamilton (1967). As a consequence two companies were formed in the 1560s to obtain the raw materials and to manufacture guns. These were the Company of Mines Royal and the Company of Mineral and Battery Works. The early years of these companies, and the manner in which they set about their affairs, is described in two books by M.B. Donald (1955 on the Mines Royal, and 1961 on the Mineral and Battery).

Gunners required a knowledge of how their guns would perform, and tables were produced that linked to each named gun its weight, calibre, diameter of the shot, the weight of the shot, and the weight of powder necessary to fire the gun. Sometimes another set of figures was provided, the range of the shot at point blank, which means with the barrel level. This distance was given in paces (5 feet) or in scores of paces (100 feet).

There are just three folding rules with gunners' tables described in Part Two, and these are all by Humfrey Cole, c. 1569–75 (**10**, **11**, **22**). For each the table is transcribed; **10** and **22** are similar, but **11** has several variations. Since Cole worked at the Royal Mint, which was part of the complex at the Tower that also included the Armouries, it is not surprising that he made rules for the use of gunners. The problem is to find the source for the table that was used by Cole.

An important and thorough work on ordnance is published by the Armouries in the Tower of London under the direction of Blackmore (1976). An appendix lists the sizes and range of guns, and it is this section that is of importance for any consideration of gunnery

tables. As Blackmore remarks, there is a 'wide divergence of opinion amongst contemporary writers as to the sizes of the different types of cannon'. Blackmore publishes five tables from the sixteenth century, and only one corresponds to Cole's. Hall (1952, p. 166) has one English table from this period, and this, too, varies from the others. One may conjecture that variations may occur through the differences in the powder used, and in the practice of the various gunners. One might have expected a fairly close match between the table of Cole and that published by William Bourne in his *The Arte of Shooting in great Ordnauance* (1578), but it is quite different. Bourne's book was based on *Quesiti* by Niccolò Tartaglia, published in Venice in 1546.

The table on Cole's rule described under **22** is virtually identical to that published by William Harrison (1534–93), transcribed in Blackmore (1976, p. 393). After the death of Holinshed in 1580, his *Chronicles* (1577) was revised by Harrison, and republished in 1587. Harrison's information is not all in tabular form, and the names of the guns are spelled slightly differently (Falcon, Sacre for Cole's Fawconet, Sacar). Only four entries in Cole's table are not exactly the same as in Harrison's (Cole first): the width of the demi culverin is $4\frac{1}{4}$ / $4\frac{1}{2}$, the diameter of shot for the Bazilis is $8\frac{1}{2}$ / $8\frac{1}{4}$, the calibre of the Bazilis is $8\frac{3}{4}$ / $8\frac{1}{4}$ (an error here because the diameter of shot cannot be the same as the calibre), and the demi canon point blank range is 28 / 38. The last is an obvious typographical error probably not in Cole's source. One gun in all three of Cole's tables, the 'E Canon', seems out of sequence. This was known as 'Queen Elizabeth's Pocket Pistol', kept at Dover. It was made at Utrecht in 1544, and presented to Henry VIII by the

Emperor's general, Maximilian van Egmond, Count Buren (Blackmore 1976, pp. 7–8).

Cole had use of the figures in the Holinshed/Harrison table by the late 1560s, and certainly by 1575, the date on one of the three rules. There must have been a common source, and perhaps one need look no further than the Tower of London.

CHAPTER FIVE

SIR ROBERT DUDLEY: AN EXCEPTIONAL PATRON

ROBERT DUDLEY
(1573–1649)

Agroup of 19 of the finest English six-teenth-century scientific instruments is to be found in the Museo di Storia della Scienza in Florence. They were taken there in 1605 by Sir Robert Dudley (Fig. 5.1), who settled in Italy, and was largely responsible for creating a seaport and a navy at Livorno (Leghorn) for the Grand Dukes of Tuscany. In the last 20 years of his long life, he wrote one of the great manuals of seafaring, *Dell'Arcano del Mare* (Secrets of the Sea). His father was the Earl of Leicester, favourite of Queen Elizabeth I.

Though described in his father's Will as 'my base son', Robert Dudley, bearing his father's name, believed himself to be legitimate, and left England for good when he failed to establish the legality of his parents' marriage. His mother was Lady Douglas Sheffield, a widow, who was a member of the Howard family. The Earl of Leicester's private life was distorted by his long and close association with the Queen, which made the disclosure of any marriage difficult. Each of the Earl's three 'marriages' (the first was to Amy Robsart) took place in secrecy. At first young Robert remained with his mother, but in 1578, when he was five, he was taken into his father's

household. Both his parents remarried, the Earl marrying in 1562 the former Lettice Knollys, widow of the first Earl of Essex. The boy was at this period recognized as his father's heir, and was privately educated. In 1587, he went up to Oxford University, where he became a member of Christ Church, his name recorded as *Comitis Filius* (son of an Earl) on 7 May 1588.

It was in 1588, when Dudley was 15, that his father died. In the Earl's Will, the boy was referred to throughout as 'my base son'. Lettice, his widow, was appointed sole executrix of a detailed and thorough document. She was fully provided for, but the Earl's brother, Ambrose, Earl of Warwick, and his son, Robert, were his chief heirs. Robert was to inherit Kenilworth Castle, and much other property, after his uncle's death, and provided he survived to the age of 21. Ambrose died in 1589, and Robert Dudley duly inherited his lands, but not his title, because of the illegitimacy. A residuary legatee was Lettice's son by her first marriage, 'my Son in Law the Earl of Essex', who was Robert Devereux, the second Earl and the ill-fated young favourite of the Queen, executed in 1601. There is no evidence that Lettice made any attempt to deprive Dudley of his inheritance, but the slur of illegitimacy embodied in the Will barred him

FIG. 5.1 Portrait of
Robert Dudley by
Nicholas Hilliard, 1591.
SK Stockholm.

from the titles he believed were his, and was certainly believed by him to be due to her influence. This injustice, as he saw it, was to affect the whole course of his life.

From an early age Dudley had an over-mastering interest in seafaring. He wrote in

the account of his voyage across the Atlantic in 1594, produced at the request of Richard Hakluyt: 'Having ever since I could conceive of anything bene delighted with the discoveries of Navigation, I fostered in myselfe that disposition till I was of more yeres and better ability to undertake such a matter' (Warner 1899, p. 67). In this study, which absorbed him throughout his life, he was both a practical seaman and a scholar. He lived at a time when navigation was in process of becoming a serious, mathematically based science under the imperative to compete for the riches of the New World. With the coming of ocean navigation, the problem of finding the longitude on board ship was a pressing one, as it remained for nearly two centuries. The method of calculating the longitude was one of the main themes of the *Arcano*; the other was the art of great circle sailing.

In the preface to his Italian sailing manual, 'Direttorio Marittimo' (never published; MS in possession of Leader 1895, p. 32, n. 1), Dudley describes how his passion for seafaring developed. Typically, for his pride in his ancestry and his own achievements was unbounded, he recalls that he is related to three Grand Admirals of England: his two grandfathers, the Duke of Northumberland and Lord Howard of Effingham, and his uncle, Lord Charles Howard. He dates his study of navigation from the age of 17. This interest will have been stimulated at Oxford University, where his tutor was Sir Thomas Chaloner (1561–1615), philosopher and experimentalist, who later became tutor to Prince Henry, son of James I. Chaloner introduced alum mining to Yorkshire, having discovered the similarity of terrain between his estate there and alum mining areas in Italy. More pertinently from Dudley's point of view, he was knowledgeable about the design and capacity of ships. Dudley goes on to describe how he had ships of war built and manned, seeking out the best pilots for them. For his voyage to the West Indies, he chose 'the famous mariner, Abram Kendall' to

be master, and from him 'learned enough navigation for an Admiral' (Leader 1895, p. 32). Kendall, who died in 1596, was one of the leading practitioners of mathematical navigation, sailing with Frobisher, Drake, and Hawkins.

Dudley's voyage to Trinidad and Guiana took place between November 1594 and May 1595. He sailed from Southampton with four ships, but soon became separated from his vice-admiral, and was obliged to sail alone to the Canaries, where he captured two Spanish ships, and was able to re-allocate space to his crew more comfortably. He then sailed to Trinidad, and explored the coastline of Guiana, of which he published a chart in the *Arcano*. His return voyage was via the Florida coast and the Azores. This expedition was successful in patriotic terms, the fleet having taken or sunk nine Spanish ships, and notably so because of the experience Dudley gained of seamanship and command. He had planned an expedition to China, but it was impossible for him to lead this in person because he was called to join the English fleet against Spain. He led the vanguard at the battle of Cadiz in 1596, when the Spanish Indies fleet was destroyed, and the city captured. For some reason we do not know, Dudley's name was not amongst the 61 knights created at Cadiz immediately after the victory on 27 June. However, after the fleet's return to Plymouth, he received his knighthood on Sunday, 8 August, 'in the open street when the Lords General came from the sermon' (Lee 1964, p. 85). Dudley's name appears at the very end of the list of knights in the Journal of the Queen's ship the *Mary Rose* that is transcribed and published as Appendix I in Usherwood (1983, pp. 124–58).

In 1596, Dudley had married Alice Leigh, the daughter of a Warwickshire landowner, and over the next seven years they had four daughters. During the years after Cadiz, Dudley divided his time between his Kenilworth estates and the court. He took no further active part in seafaring, but that he was still a student of navigation is proved by his acquisition during this time of the instruments which he later took with him to Italy. In the Florence museum are English scientific instruments made by all five of the leading London craftsmen: one by Humfrey Cole, two by Gemini, one by Ryther, eight by James Kynvyn and seven by Charles Whitwell. Among them are some of the finest and most complex artefacts of the age, such as Gemini's double quadrant, Ryther's theodolite, and Whitwell's nautical hemisphere. Most remarkable of all are the two great discs, which were bespoke by Dudley from Whitwell and Kynvyn. Whitwell's lunar computer (**51**) is the most complex instrument made during the sixteenth century, and Dudley's close involvement in its design is confirmed by the inscription, added later in Florence, that states it to be Dudley's invention. The astronomical disc (**75**), with two detachable hemispheres and a fret, was made by Kynvyn in London, but altered and added to after it was taken to Florence. Interestingly, one London-made instrument, an azimuth dial attributed to Whitwell (**53**), was taken to Florence by Dudley, later inscribed with his claim to invention, and subsequently returned to England to find its way into the British Museum.

Also in the Florence museum's collection are instruments associated with Dudley that were not made in London. Two of these, a wind rose, and a horary quadrant, also bear the inscription claiming invention, in both cases with dates; and these dates, 1596 and 1597, are in the period at present being discussed, nearly a decade before Dudley went to Italy. Either these instruments were made for him on the Continent while he was still living in England, or the invention pre-dated the actual making. Also in the Dudley group of instruments at Florence are a mariner's astrolabe by the Portuguese maker, Goes (Fig. 5.5), bought after Dudley had settled in Florence; and an astrolabe, now at the Adler Planetarium in Chicago (Figs 5.6–5.8), clearly of Italian workmanship, and closely copied from an astrolabe by Whitwell in the original group. The Adler astrolabe may well have been made in Florence to Dudley's order, as a gift for the Grand Duke. The claims to invention have to be seen in the light of Dudley's urge to self-advertisement that is apparent throughout the *Arcano*, and his other writings, and may well have been caused by the illegitimacy slur. In fact, of the four instruments described as his invention, two, the lunar computer (**51**) and the wind rose (Fig. 5.4), may well have been actually designed by him, as may the azimuth dial **53**. The same cannot reasonably be claimed for the non-English horary quadrant (Fig. 5.2).

In 1601, Dudley unwisely supported the young Earl of Essex in his defiance of the Queen, taking part in the rebellious march on London. He was briefly imprisoned and fell from favour at court. At this time, he also began the attempt to establish his legitimacy in law. However, his father's widow, Lettice Knollys, brought a charge of defamation against him in the Star Chamber, and following this, all the papers that Dudley had painstakingly assembled to establish his parents' marriage were suppressed. By 1605, no hope of success remained. This disaster, as he saw it, led him to throw up completely his life in England. He sought permission from King James I to travel abroad for three years, and, as reported in a contemporary document in the Medicean archives, may have transferred a considerable part of his wealth to France. In the summer of 1605, he left his wife and family and travelled to France. With him went his cousin, Elizabeth Southwell, one of the ladies in waiting to Queen Anne, causing a considerable scandal. The couple reached Lyons, where, having both become Catholics, a papal dispensation for their marriage was obtained. Once they were married, they travelled to Florence, where they must have arrived during 1606.

Dudley rapidly made himself useful at the Medici court, and the Grand Duke Ferdinand I, while making enquiries concerning him, clearly had no wish to lose this energetic foreigner who possessed the very skills he needed. Pisa, the only port in Tuscany, was heavily silted up, and inadequate for trade and defence. The transformation of the fishing village of Livorno into a fortified port had already begun, and it was to this work that Dudley realized he was qualified to contribute. By 1608, when Ferdinand died, to be succeeded by Cosimo II, Dudley was living at Livorno, and the first ship of his design had been launched. He was entrusted with the task of improving the port's fortifications, draining the marshes between it and Pisa, and creating a navy to rid the Mediterranean of Turkish pirates. The Medici were well aware of the seafaring skills of the English. Matthew Baker (*c.* 1530–1613), one of Queen Elizabeth's master-shipwrights, was

approached in 1607 by Lotti, the Florentine emissary in London, in an attempt to persuade him to work in Florence. Baker refused, pleading old age, but Dudley, pupil of Kendal, was able to offer similar skills. He also persuaded English merchants to trade with, and even to settle in the new *Porto Mediceo*, and encouraged the Grand Duke to make Livorno open to international business; eventually it became a free port.

At this time, it is possible that Dudley had some hopes of returning to England. He was engaged in selling his Kenilworth estates to Prince Henry, eldest son of James I, working with his former tutor and friend, Sir Thomas Chaloner, who was now in charge of the prince's education and household. He also sent designs for warships to the prince and to his father. Any real hope of reconciliation ended, however, with the premature death of Prince Henry in 1612. Nor had Dudley's reinstatement ever been very likely, for he was insisting not only on recognition of his legitimacy and restoration of his titles, but also on acceptance of Elizabeth Southwell as his wife. Dudley never received all the money due to him for the sale of Kenilworth. However, King Charles I, in Letters Patent of May 1644 (Leader 1895, appendix VI), endeavoured to make amends to the Dudleys. He accepted Robert Dudley's claim to be legitimate, his right to the titles and estates of his father, and the fact that he had never been fully paid for Kenilworth. In restitution, since Dudley himself had by this time received on the Continent the title of Duke of Northumberland, his former wife, Alice, was created Duchess Dudley in her own right.

By 1614, Dudley had clearly accepted that Florence was to be his home, and had bought a house in the city, on a fine wedge-shaped site between the Via della Spada and the Via della Vigna Nuova, where it can still be seen. Here Robert and Elizabeth brought up their seven sons and five daughters, Dudley working hard to ensure advantageous marriages and careers for them. In this he was aided by having the title of Duke of Northumberland conferred upon him by the Holy Roman Emperor in 1620, replacing the English titles that he believed to be his of right, Earl of Warwick and Earl of Leicester, these having been given to other claimants. Dudley's position at the Florentine court was unassailable, based on his achievements at Livorno and his skills as a courtier; he was chamberlain to three successive Grand Duchesses. That his milieu was exclusively the court is borne out by the fact that, despite his scientific tastes and skills, there is, surprisingly, no evidence that he had any dealings with the great Italian natural philosopher of the time, Galileo.

After the death of his wife, Elizabeth, in 1631, Dudley took a less active part in the life of the court, and devoted himself to writing. He had already produced manuscripts on naval matters, of which the most important, already referred to, was 'Direttorio Marittimo', an instruction manual for the Tuscan navy. But it was into his encyclopaedic *Dell'Arcano del Mare* that most of what he had hitherto written, and indeed the knowledge acquired throughout his long life, was subsumed. The *Arcano*, first published in 1646, appeared in three volumes, divided into six parts, or books. It was dedicated to the Grand Duke Ferdinand II, and was approved as not containing 'anything repugnant to the Catholic Religion or to good customs' by the papal inquisitor.

Book I of the *Arcano* deals with various ways to calculate the longitude, as 'invented by the author'. There is no question that Dudley, like many of his contemporary navigators, was deeply concerned with the problem of finding the longitude at sea, out of sight of land. He produced no striking innovations, but he detailed and refined the practical methods of calculation with the use of instruments. Book II contained charts and sailing directions, with sections on tides, currents, and winds. Book III embodied much of the material of the 'Direttorio', dealing with naval discipline, and also giving detailed proposals for the creation of a navy with five classes of ship according to tactical function. Shipbuilding is the subject of Book IV, which also includes the building of fortifications, containing Dudley's experience and practice when working at Livorno. Book V deals with great circle sailing, building on the work of Pedro Nunez, and again full of practical instruction for the sailor. The final Book is an atlas of 127 maps, the first atlas to use Mercator's projection.

Published in Italian, and not until the middle of the seventeenth century, the *Arcano* cannot be seen as a pioneer textbook for the ocean navigator. Much of what Dudley describes so painstakingly had been practised for many years by seamen, and had appeared in earlier manuals. Nevertheless, it is an extraordinary achievement by a man whose practical seafaring skills were proven, and whose mathematical knowledge and inventiveness were remarkable. The numerous and exquisite charts and maps, the engravings, by Antonio Francisco Lucini, of every kind of nautical and astronomical instrument, and the sheer scale and thoroughness of the volumes deserve far greater interest and study than they have received. Dudley died at the age of 75 in 1649, too early to know that the *Arcano* would be republished in 1661.

Sir Robert Dudley was a patron of nearly all the instrument makers whose work is the subject of this book. He combined the two qualities that made the Elizabethan period so remarkable in terms of exploration and seamanship, namely, the practical urge to 'sail beyond the sunset', and the realization that to do this with any success required the study of mathematics and astronomy. Such study, in turn, relied upon accurate instruments, both for astronomical observation and computation, and for practical navigation and surveying.

Sources include: Robert Dudley, *Dell' Arcano del Mare* (1646–47); John Temple Leader, *Life of Sir Robert Dudley* (1895); G.F. Warner, *The Voyage of Robert Dudley* (1899); C. Ciano, *I primi Medici e il mare* (1980); S. and E. Usherwood, *The Counter-Armada 1596* (1983).

DUDLEY'S NON-ENGLISH INSTRUMENTS

A group of instruments associated with Dudley, but not made by the London craftsmen, has been referred to. This group consists of a horary quadrant and a wind rose, not of English workmanship, which he claimed to have invented; the Gois sea astrolabe; and the Italian-made astrolabe now in the Adler Planetarium, Chicago. Since all these are of intrinsic interest, and are contemporary with the English Elizabethan instruments that are the subject of this book, they are described in the same manner below.

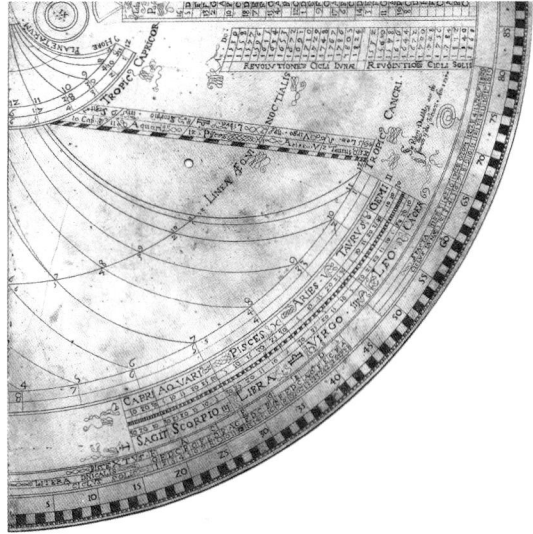

FIG. 5.2 Robert Dudley's horary quadrant, *c.* 1600, not in an English hand. IMSS Florence.

Horary Quadrant *c.* 1600

Unsigned and undated

Inscribed: *Sir Robert Duddeley was the Inuenter of this Instrument* ✶ *Ano 1597* ✶

Brass, radius 360

This instrument is not of English manufacture. All the engravings indicate non-English work (Fig. 5.2). The inscription, imitating an English hand (Fig. 5.3), is nearly the same as that on the lunar computer **51**, except for the addition of the year. The year 1597 occurs again within a tiny circle near the small quadrant for planetary hours.

Side 1

Here are engraved: a degree scale, 0–90°, a horary quadrant for equal hours, a small horary quadrant for planetary hours, a solar declination scale divided into the signs of the Zodiac, a solar altitude scale, a table for the dates of Easter, and tables for the Lunar Cycle and for

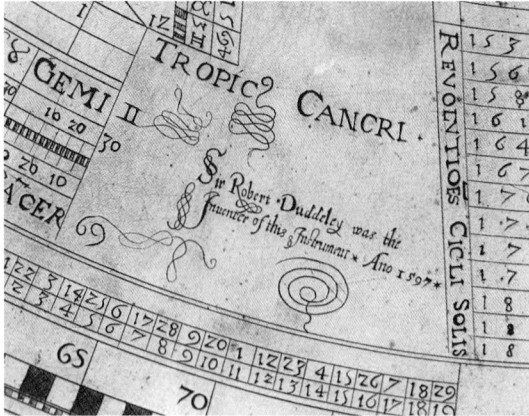

FIG. 5.3 Inscription on the horary quadrant in imitation of an English hand. IMSS Florence.

the Solar Cycle, both marked out in years. In two arcs above the degree scale are tables for Dominical Letters and the Solar Cycle 1–28, and for the Epact and Lunar Cycle 1–19.

The Easter table seems to be a copy from the Gemini quadrant in the possession of Sir Robert Dudley, **3**, and from the Gemini quadrant dated 1551 in the British Museum, **2**. The latter has the same tables, in an identical form, for the Lunar and Solar Cycles in years, and for the Dominical Letters and Epacts as the present instrument. Both have a small quadrant for planetary hours.

Side 2

The arc has a degree scale, 0–90°, a shadow square to 60 at the apex, and within this is a square of the winds with the directions numbered 1–32 at the North (*Septemtrio*). There follow a Zodiac circle and a calendar circle, and finally a wind rose with the points numbered to 32 as before. At the centre is a small circle with five trial engravings, which would have been hidden by volvelles, now missing. These will have been a solar and a lunar

volvelle to complete the tide predictor.

Side 1 shows that the diagram for equal hours is designed for the latitude of $51\frac{1}{2}°$ (correct for London). Side 2 shows that the First Point of Aries is at 10 March. One can only speculate as to whether the Gemini quadrant at the British Museum, or one similar now lost, was the inspiration for the design of the present quadrant; the similarities are, however, remarkable. The claim to invention occurs on four instruments in all, with different dates.

The horary quadrant would not have been usable at Florence, the latitude of which is about 43°, and where the Gregorian calendar was in use from 1582. It is possible that it was made for Dudley on the Continent of Europe before he left England in 1606. It is probable that the inscription was added later in Florence.

Location
Museo di Storia della Scienza, Florence (3365).

Wind Rose *after* 1606

Unsigned and undated; Italian;

Inscribed: SUR ROBERT DVDDELEY WAS THE INVENTER OF THIS INSTRVMENT 1596

Copper; diameter 338; diameter of central circle 132; thickness 1.6–1.9

The middle of the copperplate has four arcs cut out of the metal, leaving four straps to connect the central circle to the rest of the plate (Fig. 5.4). This central part has the eight main directions scribed, and each has a hole at the outer edge. These holes, together with a centre hole, are reinforced by copper sleeves. One may assume that a magnetic compass was fixed to this portion of the instrument. Outside this central part the plate is engraved with the 32 points of the wind rose. Each wind is emphasized by an elongated triangle,

FIG. 5.4 Robert Dudley's wind rose, after 1606, not in an English hand. IMSS Florence.

the right-hand half being shaded. The winds are named by initials, except for the four cardinal points, which are contractions, e.g., NO, WE, SO, EA. Symbolic marks are at each of these points (respectively): fleur-de-lis, bee, barbed arrow head, bee. The winds are also numbered from 1 (NE) to 32 (NO). Each wind is divided into 11 'degrees', so the total in the complete circle is 352, and not 360 normal degrees. Each point of the compass occupies in reality $11\frac{1}{4}$ degrees. The divisions at the rim of the instrument are alternately shaded.

The underside of the plate is roughly finished by filing. At the South point is fixed, with copper rivets, an attachment that holds a ring. At the North point is fixed, by brass screws, a bracket that curls over to the upper side for holding a wooden handle (length 114, diameter at foot 35). The handle has a baluster form, and is in wood naturally dark or stained.

It is obviously a later attachment.

The engraved letters and numbers on the main part of the instrument are reasonably competent, but not the work of an English craftsman. All known English instruments of the period are not made in copper. Again, the inscription is an anomaly. This is in a circle immediately outside the straps holding the central disc, seemingly an afterthought. The upper-case letters, and numbers, are punched, and the height of the letters is only 1.5 mm. They are arranged in a very incompetent way, with variation in level and in rotation, with much double hitting of the punch. W is formed by striking V twice. The 1 in the date has a bifurcated foot, as found on some Continental instruments, and in the present case it is upside down. The spelling of the inscription suggests a non-English production. The same punched numbers are to be found on a supplementary piece to **75**. It is probable that Sir Robert Dudley had this wind rose made for him in Florence. The inscription might be considered as a boast, but it may be the actualization of an invention he did make in 1596. It is this instrument that is shown in Dudley, *Dell'Arcano* (1647), Vol. 3, Book V, p. 18, fig. 47. The cardinal points have the two bees, arrow head, and fleur-de-lis. It has a compass and alidade, and stands on a central post. The purpose is given in proposition VIII: To find the amplitude of the sun through a horizontal instrument. A similar instrument is shown in fig. 34 earlier in the same volume. This has a hand holding a side handle, and a hand holding a bracket under the instrument.

Location

Museo di Storia della Scienza, Florence (3372).

Mariner's Astrolabe 1608

Signed and dated: GOIS / ✳ 1608 ✳

Brass; diameter 195; thickness 15; weight
2 kg 260 gm

This is the conventional form of Portuguese sea astrolabe, with a solid semicircle below (Fig. 5.5). The suspension ring has a pair of curved supports to a bracket and a double shackle of cast brass. The alidade, held by a large wing-nut, has long tapered ends. The sight vanes (51 × 39) have single pin-hole sights, countersunk on the outer side. The scale on both upper quadrants of the instrument is in half degrees, marked in punched numerals every 5°. The name and date are also punched.

The date is surrounded by four, punched, six-pointed stars, the mark of Francisco de Goes, who received his licence as an instrument maker on 13 July 1587 (Stimson 1988, pp. 74–5).

A mariner's astrolabe is depicted in Dudley, *Dell'Arcano* (1647), Vol. 3, Book V, p. 21, fig. 55, but its form is not the same as the present instrument.

Location

Museo di Storia della Scienza, Florence (1119).

Astrolabe *c.* 1609

Unsigned and undated; probably Italian

Brass; diameter 475

I have not been able to see this instrument. The description is based on photographs, kindly supplied by Dr Bruce Stephenson of the Adler Planetarium, and on the description published in Webster (1998), pp. 85–9.

This is the second extant example of the unusual astrolabe published by John Blagrave (*c.* 1558–1611) in his book *The Mathematical Jewel* (London, 1585). For the earlier example by Charles Whitwell, dated 1595, see **43**.

Throne

At first sight this is the most impressive part of the instrument. The curved panel is made from two sheets of brass soldered together, engraved with a crown on the front (see below), and quite plain on the back. It is dovetailed to the limb on the front, and has an extension band at the back which is soldered. A swivel bracket holds the suspension ring.

FIG. 5.5 Mariner's astrolabe by Gois, 1608. IMSS Florence.

Limb

The annulus forming the limb is attached by rivets to the front of a circular plate so forming the mater. The outer edge is divided into months, with '21' March at the top (but see below on the numbering). Each month is divided into units of one day, alternately shaded, and apparently numbered in tens (28 or 31). The next scale is of 24 hours, twice I–XII, with VI at the top. There follows the Zodiac, labelled with sigil and full name, with the First Point of Aries at the top. Finally, the degree scale is divided to 1° units, alternately shaded, numbered in tens from the top 90°–0–90°–0–90°. This scale serves for angles, Zodiac, and time.

Mater

The whole of the area within the limb annulus is engraved with the universal stereographic projection known as the *Saphea*, which was favoured by Gemma Frisius (Fig. 5.6). Lines of longitude and latitude are at 1° intervals. The equator is emphasized by arrows, and lines of latitude by ticks every 10°, as are the tropics. The lines of longitude are emphasized by ticks every 15°, that is at intervals of 1 hour. The hours are marked along the tropics twice 12 in arabic numerals. The ecliptic is shown by a series of parallel lines every degree, with the lines of various lengths to show the 5°-, 10°-, and 30°-intervals. The signs of the Zodiac are indicated by their sigils.

Rete

The rete is in two halves, the lower with foliate star pointers, the upper half a fret corresponding to the universal stereographic projection (Fig. 5.8). This follows the design of John Blagrave. The stars are listed in Webster

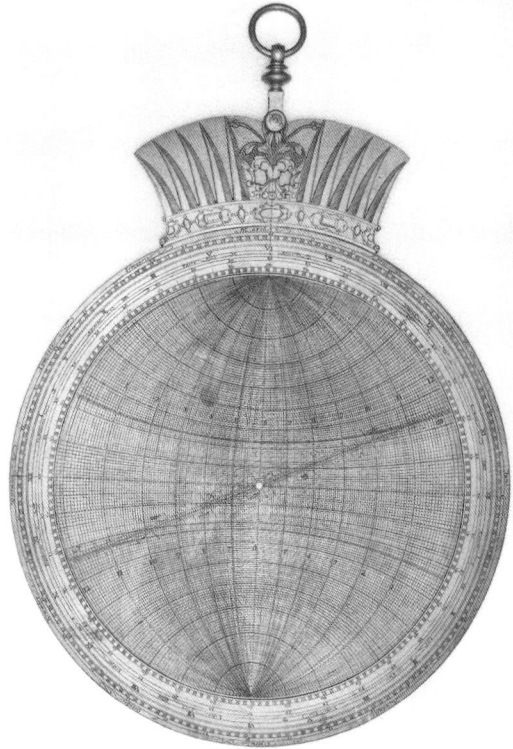

FIG. 5.6 Blagrave-type astrolabe, *c.* 1609, probably Italian; front. AP Chicago.

(1998), 88–9, who give 34 stars on the lower half, and 28 on the upper, but not all are properly transcribed or identified. It seems that Blagrave's published table has not been consulted. Blagrave numbered his stars across the complete face of the rete, and achieved a total of 76. The star names on the present astrolabe should be compared with those on the Whitwell 1595 astrolabe, **43**, to which they closely correspond.

Back

The outer edge has a degree scale in 1° units, labelled in tens from 90° at the top both ways to 0 and then to 90° at the bottom (Fig. 5.7). After a blank annulus are two scales that are

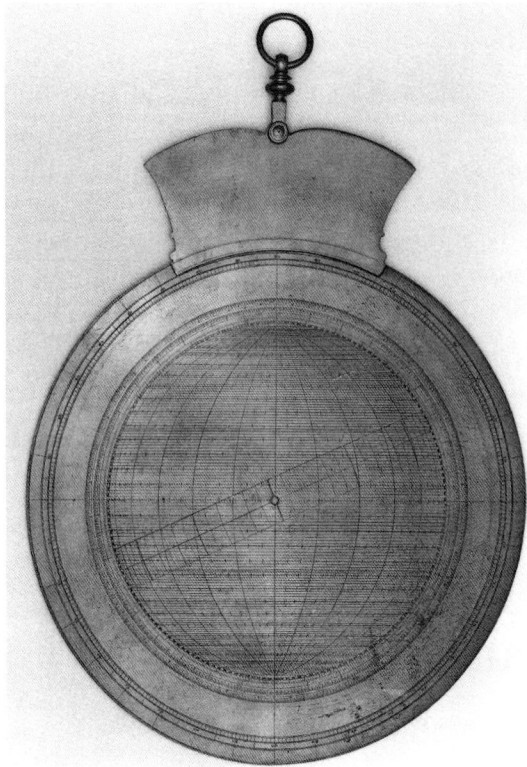

FIG. 5.7 Blagrave-type astrolabe; back. AP Chicago.

not numbered. The inner space is occupied by an orthographic universal projection of the Rojas type. An ecliptic band is provided; there is no labelling. The metal on this side shows considerable fissures, indicating a poor casting technique.

Commentary

The maker and place of manufacture have not been identified. At first sight it appears to be English of about 1600. However, one can say who it was not made by: Augustine Ryther, Charles Whitwell, James Kynvyn, Elias Allen. Nor is it by Benjamin Wright, who engraved a copperplate for the printing of Blagrave's

Uranical Astrolabe of 1596 (Gunther 1929, pp. 64–70). Although there are some similarities with the English engraving hand, there are anomalies that one does not expect to find. The divisions into days on the limb do not include markings for the beginnings and ends of the months. The 365 days have faint indication lines every 5 days, and the number of days in each month placed only roughly in the right place. On Whitwell's astrolabe the month divisions are absolutely clear. On the present instrument at the top, the 21 in March is placed above and between two divisions, whereas the scale counts to 20 March. This position is at the First Point of Aries, so it seems the Gregorian calendar is intended, which would explain the number 21 rather than 20 that one finds in the other months. In Italy, the Gregorian calendar was used from 1582, in England from 1752.

The engraved letters and numbers all show some personal quirks. Because the engraving on the rete's ecliptic (although not on the limb) is done in restricted space, problems have occurred. This has caused trouble with those letters with descenders. The g gave trouble, so a small capital G has been used in: SaGittarius, VirGo, AuGuste. The p in ŚrPio and CaPricornus are further examples. The engraver shows another of his quirks by the frequent placing of small double serifs at the ends of numbers, such as 6, 9, and at the ends of Zodiac sigils, such as Leo, Libra, and Cancer. None of these practices has been found on English instruments. The numbering of the days in the months on the calendar is unexpectedly careless. The First Point of Libra is always 2 to 3 days later than the March figure, so $23\frac{1}{2}$ September should be expected; it is at $17\frac{1}{2}$ September, which shows carelessness.

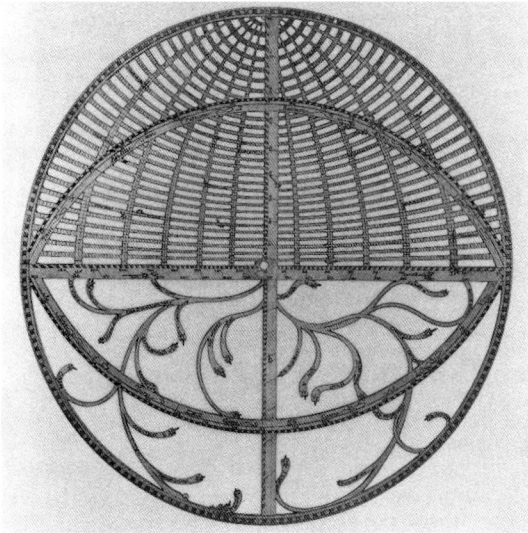

FIG. 5.8 Blagrave-type astrolabe; rete. AP Chicago.

If not in England, where could this instrument have been made? Apart from the throne, the appearance is so like that of the Whitwell astrolabe that a copy of it can be imagined. The star name Caput Medulse is found on both, whereas Caput Medusæ is general; the same goes for Iuba Ceti, Fidicula, and many others. If

a copy were made, and not in England, then Italy is the likely place. Sir Robert Dudley took his instruments with him in 1606, so the Whitwell astrolabe was in Florence from that time. Dudley may have wanted to make a presentation of an unusual design of astrolabe that had never been seen in the brass by anyone but himself. A suitable occasion came in 1609, when the Grand Duke Cosimo II succeeded Ferdinand I. The astrolabe is surmounted by the characteristic crown of the Grand Dukes of Tuscany, with a circlet of gold set in front with a large fleur-de-lis, the rest of the rim ornamented with blades of iris leaves. The same crown motif is found on the cartouche of the map of the Orinoco estuary in Dudley's *Dell'Arcano del Mare*, and in many places in Florence, some of which are illustrated in Harold Acton, *The Last Medici*, revised edition (London, 1980), plates 60, 89, 111.

Location

Part of the Anton W.M. Mensing Collection, catalogued in 1924. Purchased in 1929 by Max Adler, founder of the Adler Planetarium, Chicago, Illinois. Inventory no. M-33. IC 309.

PART TWO
THE CATALOGUE

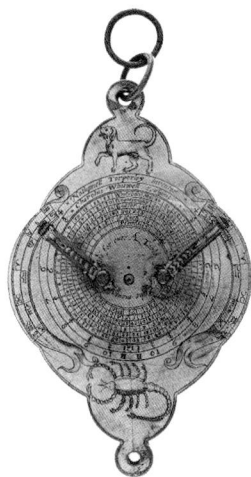

.

ADVICE TO THE READER

Numbering

Each item is given a number printed in bold throughout this book (e.g. **87**). It appears in the heading of each entry, below the accompanying illustrations, and in cross references. After the number is the name of the item, and the date.

Signature or inscription

If the item is signed, or bears other wording indicating the source of manufacture or the ownership, this is reproduced immediately below the heading. The inscriptions are reproduced as faithfully as possible, in roman, italic, or upper case, and the original punctuation is adhered to. If the item can be attributed, on stylistic or other grounds, to a maker, this name is substituted for the signature.

Material and measurements

The main materials of construction come next, and then the dimensions in millimetres, without being so specified. On a very few occasions the units of length are given to avoid confusion with other numbers.

Description

The body of the entry comprises a description of the instrument or instrument-related object.

As well as providing a physical description of an artefact, its function and historical development are given. The description is detailed, because it is in the detail that the craftsmen can reveal themselves. For example, the alternate shading of the units in a degree scale may point to one maker, and its absence to the avoidance of a tedious task as a business grew.

References

Within the body of the text references are given in a short form, with a key-word followed by a date, for example, Bond (1875). This is then found in the Bibliography under Bond, J.J. (1875), where the full citation is given. The key-word usually refers to the author of the work, but not in every case; with some museum catalogues the name of the museum or the subject of a special exhibition is used.

Provenance

The ownership of an object is taken back as far as possible, but for some it is not possible to go further than the auction record. Fortunately, for some 20 instruments it is known that they belonged to Sir Robert Dudley, and are now at Florence.

Literature

References near the end of the entry are specifically to the object concerned.

Location

At the end of each entry is the name of the present owner (except for a private collection). The institution's Register, Inventory, Accession, or Acquisition Number (the word depends on local usage) follows in parentheses. For astrolabes, the International Checklist Number is given thus: IC 304 (see Price 1955; Gibbs *et al.* 1973).

Dating

Particular attention has been given to dating. The convention used is as follows:

specific date when known	1597
close approximation	*c.* 1600
span, if deducible	1590–1600
terminal date	*before* 1592
	after 1570
estimate to the quarter century	4/4 16th C

THE INSTRUMENTS DESCRIBED

1 **Astrolabe** *c.* 1551

Unsigned and undated; attributable to Thomas Gemini

Brass; diameter 352; limb width 9, thickness 4; back plate thickness *c.* 1.4

The limb, throne, and back remain from the original astrolabe; rete, plates, alidade, and rule have been discarded, and it is likely that the maker's name was on one of these parts. It seems that the attractions of a good large disc of brass were such that an obsolete instrument was used on which to scribe, on the blank mater, the spiral calculator, which gives sines to 90°, and tangents to 45°. This is signed: *H: Sutton fecit 1655.*

The limb, an original part, is divided to 360°, beginning at the left, in single units alternately shaded, and numbered every 5°. A different labelling is in hours, twice 12 in arabic numerals. The throne has a pair of scroll supports to a band holding the shackle and suspension ring. The form of the throne is the same as that on Gemini's 1552 astrolabe made for Edward VI (**5**).

The upper central part of the back has a strapwork frame enclosing the Tudor royal arms in a cartouche inscribed with the cypher E R, surrounded by the Garter, and surmounted by a royal crown. The arms are flanked by two bunches of fruits typical of Gemini's decorative style (e.g., Fig. 3.2), and in the semicircular frames, the double rose and the portcullis Tudor badges. The cypher denotes King Edward VI, who ruled from January 1547 to July 1553 (see below). Lower central is a conventional shadow square, divided in units alternately shaded to 12 at the 45° position.

The outermost scales round the edge are of degrees, Zodiac, and calendar. The degrees and Zodiac share the same units, alternately shaded, labelled, respectively, every five degrees 0–90° in quarters, and every five degrees 0–30° in every sign. The separate calendar scale is marked out in a similar way, labelled every five days to 30, 31, or 28 as appropriate.

The unique feature on the back is the set of ten concentric bands comprising astrological information, which is described in Gunther (1927, pp. 302, 304). The outermost band has 28 sections for the Lunar Mansions. The remaining bands are in 12 segments. The dispositions of the planets and the significance for good or bad fortune are displayed. The form of this table and its contents duplicate that published by Gerard Mercator at Louvain in May 1551. An example of Mercator's disc is preserved at the Historisches Museum, Basel

1 Back.

instrument engraver, Gemini's work is of the highest quality, only exceeded by his fellow-countryman, Gerard Mercator. See Chapter 3.

George Gabb once owned the instrument, which he exhibited to the Royal Society on 11 May 1927; a photograph with a brief note were published in *The Illustrated London News* on 14 May 1927. Gabb assumed the astrolabe was made for Queen Elizabeth by Humfrey Cole. Consequently, Gunther, in his paper on Cole's instruments read to the Society of Antiquaries on 17 June 1926, inserted a Note dated 17 June 1927 in the published version (Gunther 1927). On 12 November 1936, Gabb read a brief paper to the Society of Antiquaries on what he called 'The Astrological Astrolabe of Queen Elizabeth' (Gabb 1937). He thought the initials E R by the royal arms referred to Elizabeth, and to explain the astrology he proposed John Dee (1527–1608) as the source. He pointed out that Dee had picked 14 January 1559 as a good day for the Queen's coronation. There are reasons for preferring an earlier date

(inv. no. 1876–20); see Broecke (2001). The table was engraved by Mercator on a copper-plate, and the printed sheet is attached to a board. Rotating about the centre are nine blades with type-set material. On the back of the disc is a type-set printing of the instructions for use, headed: *CANDIDO LECTORI*, and signed: *Louvany MDLI Mense Maio. Gerardus Mercator Rupelmondanus*. See Figs 2.3 and 2.4.

Dating

The present astrolabe was certainly made by Thomas Gemini, who came to England from Flanders by 1540. He dedicated his 1545, 1553, and 1559 editions of the *Compendiosa totius Anatomie delineatio* of Vesalius to successive monarchs of England. His quadrant dated 1551 is named for Edward VI (2), as is a small astrolabe of 1552 (5). His astrolabe for Queen Elizabeth is dated 1559 (6). As a map and

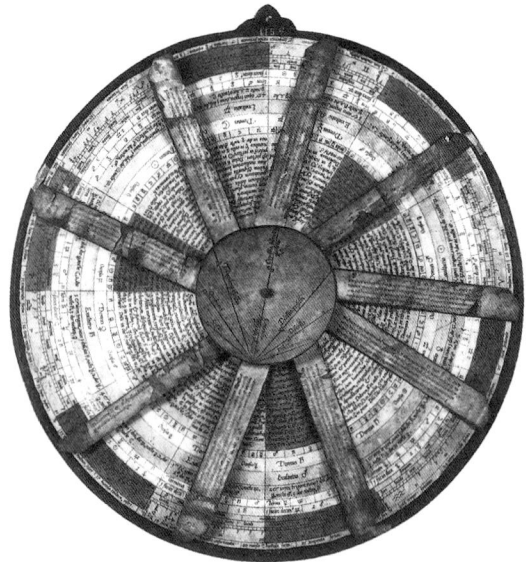

1 Mercator disc, 1551.

for the instrument, and therefore Edward VI as the recipient. One must remember that Gemini made an astrolabe dated 1559 for Elizabeth naming her in full (6). Edward was interested in astrology, as is to be expected at that period. Gemini had received a pension from Henry VIII for the *Anatomy*, which was continued under Edward. A large astrolabe with the latest astrological features as published by Mercator would be most attractive and educational. Thus a likely date for the astrolabe is 1551 when Mercator dated his disc. Gemini's small astrolabe of 1552 has the universal projection of Juan de Rojas, which was first published at Paris in 1550, reprinted in 1551, so the knowledge of it soon reached London. Rojas was a student of Gemma Frisius at Louvain in the 1540s, and in his book credits his contemporary at Louvain, Hugo Helt, a Frisian, with the invention and the description that Rojas published (Maddison 1966, p. 38). This Louvain nexus points to Mercator's new astrological disc as the model for the Gemini astrolabe because of the remarkable identity between the astrological tables. For a full discussion of the link, see G. Turner and Cleempoel (2000).

Provenance

George Gabb (1868–1948) said that he had 'discovered' the astrolabe in 1926. He had acquired it through a member of the Wollaston family, and it had, apparently, belonged to the physicist William Hyde Wollaston (1766–1828), President of the Royal Society in 1820. Wollaston was no doubt attracted by the spiral calculator, for he added to it. Gabb had the grime removed, and found traces of gilding. Gabb's collection was purchased in February 1937 by Sir James Caird for the National Maritime Museum.

Literature

Gunther (1927); Gabb (1937).

Location

National Maritime Museum, Greenwich (AST0567; old no. A.36/37–92C) IC 425.

2 Horary quadrant 1551

Signed: T.G. [Thomas Gemini]; dated: *Anno Domini. Polus 51. 34. 1551.*

Brass; sides 268, 271; radius 270; thickness: 0.6–1.3

This well-made instrument is associated with King Edward VI through the inscription: *Edwardus Rex*. He succeeded his father, Henry VIII, on 28 January 1547, and reigned until his death on 6 July 1553, at the age of only 16. Members of the court are identified by initials. J.C. (apex of shadow square) is Sir John Cheke (1514–57), Professor of Greek at Cambridge, 1540–51, appointed Royal Tutor in 1548 and Provost of King's College, Cambridge by the King's mandate. He was knighted the following year. W.B. (foot of Solar Cycle) is William Buckley (died 1570?), mathematician and Fellow of King's College, Cambridge, tutor to Court Pages from 1551. This is probably a companion piece to the horizontal dial by Gemini for the same latitude and the same year (4).

Side 1

There are two large features, both requiring sights, now missing, and a plumb-line, not present. One is a shadow square, with sides divided in units, alternately shaded, and numbered 0–60 at the apex; the sides are named: *Vmbra Versa* and *Vmbra Recta*. The other is a horary quadrant for equal hours. The hours are marked in two places, along the summer solstitial line, 4–12–8, and along the equinoctial line 6–12–6. Above and below this line is

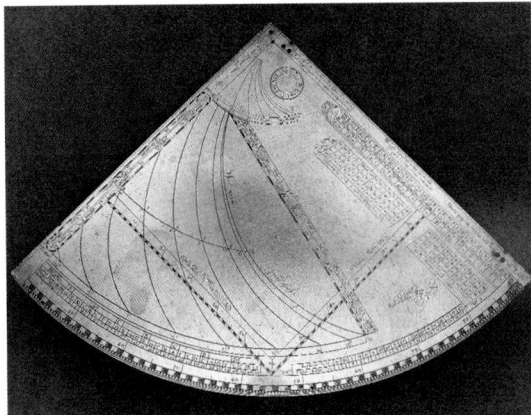

2 Side 1.

inscribed: *Horæ hyemales.* (winter hours), and *Horæ æstiuales.* (summer hours). There are two solar declination scales: to the left, the months in the year, named and numbered (March is 1), and to the right the Zodiac, with names, sigils, and numbers (Aries is 1). On both scales the days and degrees are in units of five, alternately shaded. To the right of this face is a perpetual Easter calendar: *Tab. Pas.*; a table of the Lunar Cycle of 19 years: *Reuolutiones Cycli Lunæ* (1539 to 1824); a table of the Solar Cycle of 28 years: *Reuolutiones Cycli Solis* (1532 to 1868). The curved edge of the quadrant is divided into units of a quarter of a degree, and additionally to single degrees, both sets alternately shaded. Numbering is in tens to 90°. In two sections above the degree scale are four calendrical scales. (1) *Litera Dominicalis*, including *Bisextilis*; (2) *Ciclus Solis*, 1 to 28; (3) *Epacta*, beginning *11, 22, 3* ...; (4) *Ciclus Lunæ*, 1 to 19. At the apex is a circle showing the sigils of the seven planets in the descending Ptolemaic order, Sun, Venus, Mercury, Moon, Saturn, Jupiter, Mars. In repeated sequence, these govern the planetary (unequal) hours of day and night. Adjacent to

the circle is a small horary quadrant, seemingly for unequal hours, but terminating too soon.

Along the right-hand edge are mottos in Latin.

1. *Fluxus aquæ celer est, celer est et Fulminis ictus, At magis hijs tacitum tempus utrisque celer.* (The flow of water is swift, swift is a bolt of thunder, but silent time is swifter than either.)

2. *Illud metiri quadrans tamen iste docebit. Et quota sit fias certior hora facit.* (Yet this quadrant will teach how to measure it [time] and makes you become more certain of what hour it is.)

3. *Omni negotio tempus est et op[]tu[]tas* [sic]. *Salom.* (For every activity there is an appropriate time. Solomon.)

Side 2

The whole of this face is occupied by a an array of parallel lines forming a sinical quadrant. The arc is divided into units of 1°, alternately shaded, labelled in tens from 0 to 90°. The edge is divided 0–90° (85°–90° imperfectly), and in reverse 90°–0, for ratios between

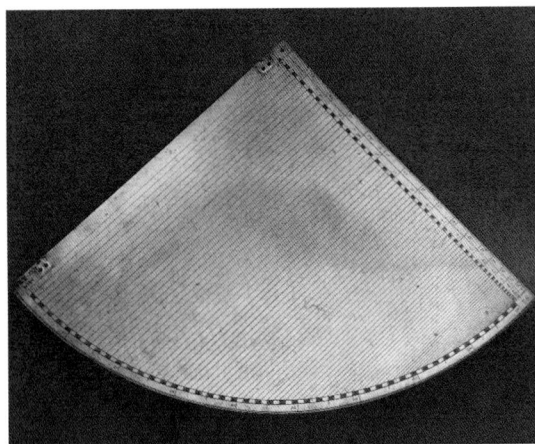

2 Side 2.

sines or cosines of given angles. At the apex is a small hole into which to peg a cord to span the face. With a bead running on the cord it would be possible to find the solar declination corresponding to a solar longitude. At the end of the edge are the maker's initials: T.G. This quadrant is to the design of Peter Apian, published in his *Instrvmentvm sinvvm* (1534; 2nd edition 1541), and is one of his variant sinical instruments. For the Islamic use of a such a quadrant see King (1975, esp. pp. 109–18). See **3** for a similar Gemini quadrant.

Provenance
Purchased by the British Museum in 1858.

Literature
Ward (1981, pp. 56–7), no. 149, plate 19.

Location
The British Museum, London (1858, 8–21.1).

3 Compendium of quadrants
c. 1550

Signed: *Thoma* ♊; undated

Brass, iron, and wood; overall radius 273; sides 270; thickness: plate A 0.7–1.4; plate B 0.6–1.3

The instrument comprises two double-sided quadrant plates separated by a walnut board (thickness 8), bound around the edges with a brass band (width 13; thickness 1.5–2). The plates are attached by iron bolts, with tapered, squared ends. This allows the external faces of the combined quadrants to be selected as desired. Along one edge is a pair of sights attached to the brass band, each with a peg sight and a pinhole. This pair of quadrants should be compared with the double-sided

one at the British Museum dated 1551, **2**. On this, the calendrical information is identical with that on Plate B, side 1, and the sinical quadrant corresponds to Plate A, side 2. The calligraphy is exactly the same; the only difference worth noting is in the horary quadrant, which is for equal hours. One cannot read anything into this. Gemini was capable of producing this compendium of quadrants between 1545 and just before his death in 1560, but closer to 1550 seems reasonable.

Plate A, side 1
Engraved with Peter Apian's invention for reading sines and cosines, described in his book *Instrvmentvm primi mobilis* (Nuremberg, 1534), with the arcs labelled SINVS✦ RECTVS✦ and SINVS✦ VERSVS✦. For a description of the purpose of the device, see Röttel (1995), 226, 331. Also drawn is a shadow square, labelled *Vmbra Recta*✦ and *Vmbra Versa*✦. Both elements are divided 0–60, the units alternately shaded, with half units indicated by dots in a band below. Along the quadrant arc is a degree scale 0–90°, units alternately shaded

3 Plate A, side 1.

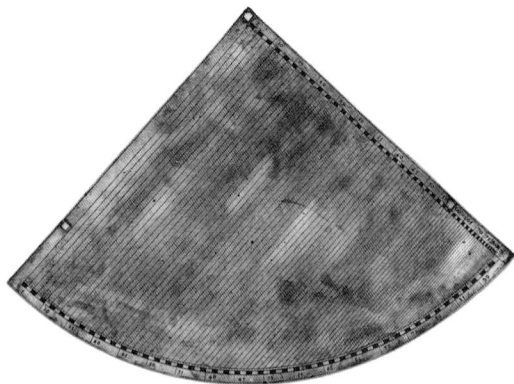

3 Plate A, side 2.

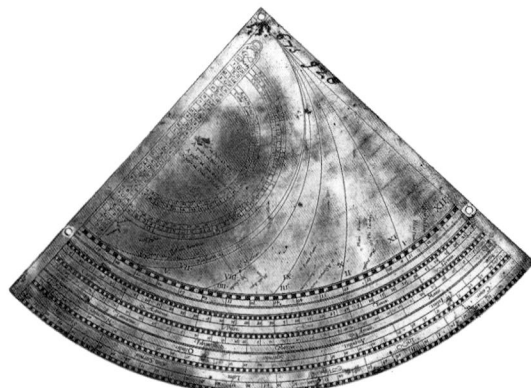

3 Plate B, side 1.

with diagonal lines or cross hatched. Signed at the 45° position with the maker's name and sigil of Gemini. At the apex is a hole for the peg of a plumb-line (not present).

Plate A, side 2

Engraved with parallel lines forming a sinical quadrant. The arc is divided in degrees, 0–90°, units alternately shaded, with halves indicated by dots in another band. The edge is divided 0–85, units alternately shaded. Both scales are numbered every 5. Unusually, these numbers are placed at the beginning of a section, not the end. A cord with bead would have been pegged into the small hole at the apex. For the use of such a quadrant, see **2**.

Plate B, side 1

This side is engraved with a horary quadrant for planetary, or unequal, hours. For use with the hours is a degree scale 90°–0, alternately shaded and numbered in 5° intervals. There is a calendar for Easter and other consequent movable feasts. Two other tables are drawn as arcs; the upper begins: *Bisextilis* | F ... / *Litera Domenicalis* | G ... / *Ciclus Solis* | 1 [to 28].

The lower begins: *Epacta* | 11 [to 29] / *Ciclus Lunæ* | 1 [to 19].

There follow two scales for the Zodiac and the calendar; both are folded once so that twice six months fit into the quadrant arc. Degrees and days are in single units, numbered every five (the days to 28 or 31). The First Point of Aries is at $10\frac{1}{2}$ March, and of Libra $13\frac{1}{2}$ September. Because of the folding, the names of the Zodiac signs and of the months are written recto or inverted, as are the names of the saints, whose days are read from a plumb-line (not present) at the apex to the calendar scale. The saints' days are:

(recto) *Nats dmi[s], Epiphia[s], Purificatio ma[s], S.Mathias, Annu[s] Marie, Marc[s] Euagel, Phi.Iacobi*

(inverted) *Andreas, Fest.oim[s] Sanctoru[s], Simo[s] Iude, Michaelis, Math Apo, Bartholome, Assumptio marie, Mari Magda, Pet[s] et Pauli, Nat[s] Ioh Bap.*

Note that All Saints' Day is not inverted, but is in the correct place (1 November). Gemini sometimes places a contraction mark in the form of *s* over a name. These are shown as a superscript *s*.

Plate B, side 2

A Profatius astrolabe quadrant occupies this side, which requires the full circle to be folded twice to fit into a quadrant (see Dekker 1995). One side is inscribed: HORISON RECTVS, and the other side: LINEA✦ MERIDIANA. This has a solar altitude and declination scale in degrees (shaded), calibrated from the equator South to $23\frac{1}{2}°$, and North to 60° (latitude), also 90°–30° (co-latitude). Almucantars are at 5° intervals. The arc has a folded degree scale (unit degrees alternately shaded), numbered in four quadrants: 0–90°, 90°–180°, 180°–270°, 270°–360°. Across the face, along the appropriate lines, are the labels: LINEA✦ ÆQVINOC-TIALIS✦; TROPICVS✦ CANCRI; TROPICVS✦ CAPRICORNI. The positions of seven stars are plotted:

Caput Algol, Hircus, Hu.dexter.Orion, Canis maior, Canis minor, Lucida hÿdræ, Aquila.

At the very bottom of the arc is an hour scale 0–6 returning to 12, where 1 hour occupies 15°.

Provenance

One of the instruments taken to Florence by Sir Robert Dudley in 1606.

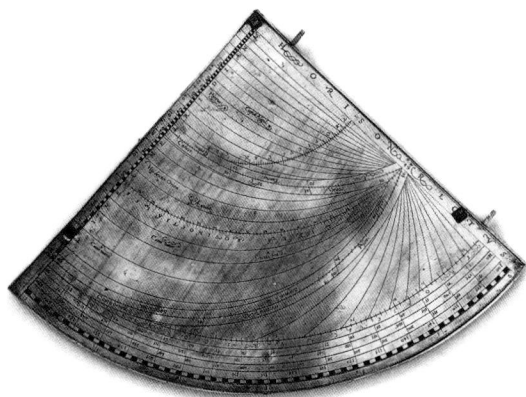

3 Plate B, side 2.

Literature

G. Turner (1991, pp. 122–3), all four sides illustrated.

Location

Museo di Storia della Scienza, Florence (2509).

4 Horizontal sundial 1551

Unsigned (?); dated: *Anno domini | 1551*; attributable to Thomas Gemini

Brass, diameter 308; vertical height of gnomon 134, thickness about 2

The surface details on this dial are particularly difficult to read because of corrosion caused by long exposure to the weather. Already identified, among a few other details, are the date, and the latitude for which one set of hour lines and the gnomon were constructed:

Polus | 51. 34

Around the edge are eight holes for fixing screws, and the dial sits on a circular base of slate, which is not an original fitting. At the outer edge are two concentric circles between which are the names of sixteen winds, in English, e.g., *North Wynd*. There follow the hours in roman and arabic numerals, a concentric double circle, a further concentric circle, and a pair of solar amplitude scales to the East and West sides. Space on the South side contains a motto over two lines, date, latitude, and initials: W | B.

The scale to the East comprises the Zodiac between the solstices, with alternately shaded divisions to 5°. The ends of the scale lie between summer sunset at about 8 hrs 30 mins, and winter sunset at 3 hrs 30 mins. These figures are a reasonable fit for the given latitude (because of the dial's condition more exact measurement is not possible). To the

4 I

West is a scale for the months of the year from 12 June to 12 December, divided into intervals of 5 days, alternately shaded. Thus the first appearance (or disappearance) of the gnomon's shadow gives the day of the month as well as the time of day.

There are two hour scales, that with large roman numerals is for a latitude of about 52°, while the small arabic numerals are for a latitude of about 42°. The arabic numerals are in the hand of Gemini. The values of latitude are from measurements made on the dial. The surviving gnomon is for *c.* 52°. The hour lines for 42° extend to the centre of the dial. To use such a 'double' dial two gnomons are needed. Consequently, the gnomon is held in a track formed by two rails fixed to the plate between which the gnomon slides, making a substitution easy. The present track is probably not the original one, as there are three pairs of small holes that could have been for attaching the original track. The present track may have been provided for the 52° gnomon when it was decided to locate the dial permanently in England.

The gnomon is thin (1.6–2.4), and as usual no allowance for what thickness it has was made when cutting the noon line. The shadow-casting edge is bevelled, and the back is cut into a scroll above an arc; the edges are fimbriated. At the upper end of the arc is a pear-shaped protrusion and the bottom of the arc is cut with a notch. These elements may be the hanging point for a plummet, and its fiducial marker. A dial intended to be used in two distinct locations 10° of latitude apart, and so needing two gnomons, could not be permanently fixed to a mounting. A plummet would therefore be required to level the dial.

Weathered, but still identifiable as the work of Gemini, it appears to be a companion piece for a quadrant in the British Museum (**2**). The quadrant has the same date, 1551, and its latitude is the same, 51° 34′; thus the present dial is associated with Edward VI whose name appears on the quadrant. Gemini is identified by his initials, T.G on the quadrant; these may be on the dial, but now obscured. What can

4 Dial plate.

be seen are the initials W B. These are also on the quadrant, **2**, and indicate William Buckley, Fellow of King's College, Cambridge, and tutor to Court Pages from 1551. In the space to the South of the gnomon is the motto, hard to read because of corrosion. The lines can be construed thus: ... *time can be redeemed with no cost, Therefore bestow it well and lett no howre be lost.* This is similar to part of the motto on both of Cole's 1575 compendia, **24** and **25**.

Provenance

This dial came on the market in 1984, and was purchased from a dealer by the Science Museum, London. It is said that it was in an auction sale at Chirk Castle, North Wales, during the Second World War.

Location

The Science Museum, London (1985–1389).

5 Astrolabe 1552

Signed and dated: *T. Gemini Anno 1552*

Brass, diameter 313

The throne matches closely that fitted to another Gemini astrolabe made for Edward VI of England (**1**). On the limb, the scale division into single degrees, alternately shaded, is shared between the hours in a day, twice I–XII in roman numerals, and four quarters to 90° (0–90°–0–90°–0), labelled every 5°. The mater is blank. One plate is provided, which is blank on one side. The face has on its edge a Zodiac scale, each sign with its full name and sigil, the degree divisions alternately shaded and labelled every 5°. Next is the calendar scale with names in Latin and the number of days in the months displayed. The divisions into units of a day match the Zodiac scale. Lastly is

Star list on the front

Engraved	Modern name	Designation
Οφθαλμος ♉	Aldebaran	α Tau
Αἴξ	Capella	α Aur
Κάνωβος	Canopus	α Car
κύον	Sirius	α CMa
κнφαλн ♊	Apollo/Castor	α Gem
Προκύων	Procyon	α CMi
ὕδρος	Alphard	α Hya
καρδία λεοντ.	Regulus	α Leo
Οὔρα λεοντς	Denebola	β Leo
Σαχυς	Spica	α Vir
Αρκτδρος	Arcturus	α Boo
Καρδία ♏	Antares	α Sco
Λύρα	Lyra/Vega	α Lyr
Αετος	Altair	α Aql
ὄυρα ♑	Deneb Algedi	γ Cap
Κυκνος	Deneb	α Cyg

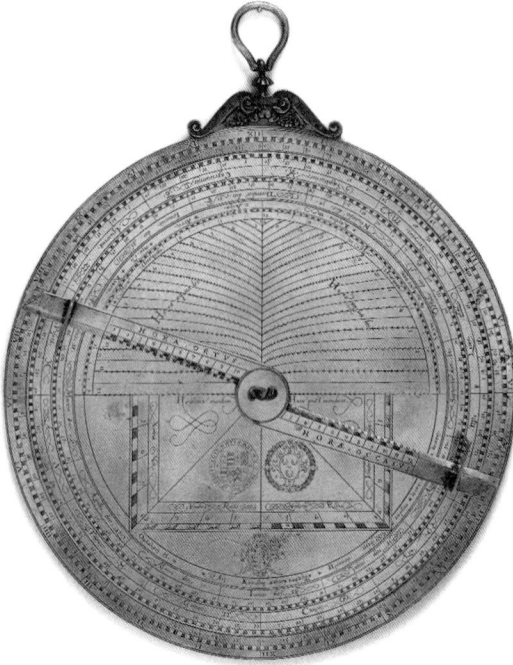

5 Front.

a ring of 16 star names in Greek with asterisms corresponding to their Right Ascensions. Since the disc can be rotated within an hour circle, it can be regarded as a 'mobile Zodiac' (Saunders 1984, p. 70; Michel 1947, p. 102). The exclusive use of Latin and particularly Greek is most likely due to Sir John Cheke, the young king's tutor, who was Professor of Greek at Cambridge University (see **2**).

NB. Gemini uses the same letter forms for his Greek as does Mercator on his celestial globe, which is dated April 1551. The lower case *eta* (η) is engraved identically by both as H. However, not all the stars named by Gemini are in Greek on the Mercator globe. On the back, the Greek names are the same except for the omission of Canopus and Lyra. The spelling of Sirius is correct on the back: κὐων. The spelling of Spica is incorrect on

both sides; should read: Στάχυς. The last two stars on the list are transposed.

The central area of the disc is in two parts. The upper is a diagram for conversions between equal and unequal hours. Two sets of curved lines are in half-hour divisions, the 12 full hour lines emphasized by arrow heads, and the half hours by plain lines. Across both sets of lines is engraved: *Horæ æquales*, and along the bottom lines; *Horæ ante meridiem*, *Horæ post meridiem*. The lines meet a semicircle divided in 12 hours with units of four minutes, and this scale is engraved: *Horæ Inæquales*. The alidade is attached to this side, and it is provided with hour scales on each arm labelled: HORÆ OCCASVS, and HORA ORTVS. The lower part is occupied by a shadow square divided in units to 12 at the apex, labelled in the normal way: *Vmbra Recta*, *Vmbra Versa*. Half the units are alternately shaded, the other half alternately with a prismatic ornament. The signature and date are in one of the plain divisions. Within the square are two coats of arms, each with cyphers. To the left, the arms of Dudley with seven quarterings flanked by the cypher I. N. in a Garter, for John Dudley (1502–53), created Earl of Warwick in 1547, and Duke of Northumberland in 1551. To the right, the arms of Sir John Cheke, *[Argent] three crescents [Gules] a crescent for cadency in the fess point*, the strapwork cartouche flanked by the cypher I. C., all in a wreath. Below the square are the Tudor royal arms in a strapwork cartouche set on a palm leaf (?) and olive branch, saltirewise, all between the cypher E. R.

The back of the instrument is engraved with the orthographic projection known as the Juan de Rojas projection (Michel 1947, pp. 103–9; Maddison 1966). This design was

with a cursor traversing it at right angles. The cursor is labelled in decorated capitals: LINEA AVRORA. The foot of the cursor has a four-fold foliated plate decorated with arabesques. Both sides are shown in Plates 2, 3.

Literature

Osley (1969, p. 97); Rasquin (1984, p. 40), no. 11; Boxmeer (1996, pp. 14–16); Madrid (1997, p. 174), no. A.3, plates pp. 117, 175.

Location

Observatoire Royale de Belge, Uccle. IC 450.

5 Back.

6 Astrolabe 1559

Signed and dated: *Thomas* ♊ 1559 ✠

Brass, gilded; diameter 355; thickness of limb 6

This astrolabe at Oxford is very like the one at Florence, **7**, so that the description of the one mostly serves both.

Throne

Attached at the top of the instrument is a curved bar with a roundel supported by a pair of S brackets simply decorated by lines resembling a roman one. Above the roundel are the remains of an attachment for a ring, now lost. To the front, the roundel is engraved:

Elizabeth Dei Gratia Angliæ Franciæ & Hiberniæ Regina

On the back of the roundel are the arms of England: France modern and England quarterly. At the sides are the initials: E R.

Limb

Divided into hours, twice 12, denoted by roman numerals. The single degree divisions are alternately shaded, and are numbered from the East point (centre right) running anti-clockwise 0–360°.

published at Paris in 1550, reissued 1551, and soon became popular as a change from the *Saphea*, a universal stereographic projection promoted by Gemma Frisius.

The outer edge is divided in single degrees, alternately shaded, and labelled from the top 0–90°–0–90°–0. The Rojas projection occupies the rest of the surface. It consists of two orthographic projections from infinity from either side of the sphere onto the plane of the solstitial colure. Azimuths are drawn every 3°, and the 15° lines (i.e. hour lines) are emphasized by arrow heads. The hours are numbered along the tropics, and are labelled: *Horæ ante Meridiem*, and *Horæ post Meridiem*. The 14 stars are marked by asterisms, and are named in Greek, as on the plate (see above). Measurements are made by using a regula

6 Throne, front.

6 Throne, back.

Mater

Filling the space is a QVADRATVM NAVTICVM, each side divided at the edge 90–0–90 in double units alternately shaded. Above the upper side is the description: *Longitudo minor siue Occidentalior* ✦/ *Longitudo maior seu Orientalior* ✦. At the left side the inscription reads: *Latitudo minor vel Australior* ✦ / *Latitudo maior aut borealior* ✦. Radiating from the centre are the 32 wind directions, or rhumbs, each marked with the name of the direction in

English. Beyond this set, 12 winds are named in both Latin and Greek, while four others are named in Italian. Below the lower edge of the square is the signature and date. The *Quadratum nauticum* of Gemma Frisius is in the many editions of his *Cosmographicus liber* (1529); see the Cole astrolabe, **23**.

Back

At the edge is the degree circle, divided as on the front, but labelled in quadrants anticlockwise from the top 90°–0, and so on. Within this circle is a universal projection, the *Saphea*. A large-scale illustration is in Gunther (1937), plate 13. The use of this projection was promoted by Gemma Frisius of Louvain, in his book *De astrolabio catholico et usu ejusdem* (Antwerp, 1556).

6 Front.

Fifteen star positions are named:

Venter Cæti	*Ceruix* ♌	*Wultur* [α Lyr]
Oculus ♉	*Cauda* ♌	*Aquila* [Altair]
Cing. Orio.	*Spica*	*Cauda Cygni*
Canis maior	*Alramech*	*Formahand*
Canis Minor	*Cap. Her.*	*Caput And.*

6 Mater.

In order to plot coordinates on the *Saphea* it is necessary to have a regula, cursor, and brachiolum; only the first is present. The regula (length 354) is mounted on a pin that fits in a hole at the centre of the astrolabe and is secured by a screw. Its edge is divided in 2° units, and labelled every 10°.

Rete

The rete is in the strapwork 'tulip' pattern typical of the Mercator and Arsenius workshops of Louvain. It was probably there that Gemini learnt his trade alongside Mercator during the 1530s. There is a close resemblance between the Gemini rete and the earliest of Mercator's astrolabes, datable to *c.* 1545. This topic is discussed in Chapter 2; see Fig. 2.5.

Each part of the strapwork is fimbriated. The ecliptic circle is divided into the 12 Signs of the Zodiac, which are identified by both sigil and name (Oxford) or sigil and vignette (Florence). The outer edge is divided into single degrees, alternately shaded. They are numbered in fives to thirty in every sign.

The rete is engraved with the names of 30 stars, most with pointers, but the positions of some are marked on the ecliptic circle or on part of the strapwork. A most unusual feature of the pointers on this astrolabe, and the one at Florence, lies in ending each pointer with a finely chiselled star shape. As far as is known, only Gemini affected this. The list of stars is in the table, alongside the stars on **7**.

The counterchanged alidade is used on the front of this astrolabe because sights are not possible when a regula is on the back. The centre is decorated by arabesques. From one arm the declination can be read; the scale units are 2°, labelled in tens, running North to 72° and South to 23°. The other arm is divided in hour intervals to 12 both ways, for use with the plate. The sight vanes have two holes, one about 0.5 mm, the other about 0.1 mm.

Plate

There is a single plate (diameter 321) that has on one side a tablet of horizons; on the other is

Star lists on **6** and **7**

Oxford IC 575, 6	Florence IC 489, 7	Designation	Common name
Pectus Casiop.	Pectus Casiopeiæ	α Cas	Schedar
Venter Cæti	Venter Cæti	ζ Cet	Baten Kaitos
Triangulus	Triangulus	β Tri	
Caput Gorgonis	Caput Gorgonis	β Per	Algol
Pleiades ★	Pleiades ★	η Tau	Alcyone
Oculus ♉	Oculus ♉	α Tau	Aldebaran
Venter Leporis ★	Venter leporis ★	α Lep	Arneb
Cingulum Ori.	*erased*	ε Ori	Alnilam
Canis ma. ★	Canis ma ★	α CMa	Sirius
Canis minor	Canis minor	α CMi	Procyon
Pollux ★	pollux ★	β Gem	Pollux
Pectus ★	Pectus ★	δ Gem	Wasat
Lucida Hydræ	Lucida Hydræ	α Hya	Alphard
Cor ♌ ★	Cor ♌	α Leo	Regulus
Dorsum Vrs. ma.	Dorsum Vr. maioris	α UMa	Dubhe
Dorsum	Dorsum ♌	γ Leo	Algieba
Cauda ♌	Cauda ♌	β Leo	Denebola
Algorab	Algorab	δ Crv	Algorab
Cavda Vr. ma.	Cavda Vrsæ	δ UMa	Megrez
Pes posterior Vr. ma.	Pes post. Vr. maioris	γ UMa	Phecda
Spica ♍ ★	Spicæ ♍ ★	α Vir	Spica
Alfeca Coro	Alph. Coronæ †	α CrB	Alphecca
Alramech	Alramech	β And	Mirach
absent	Lanx ♎ ★	α Lib	Zubenelgenubi
Lanx. Sep.	Lanx. Sep	β Lib	Zubeneschamali
Hasta Bootis	Hasta Bootis	η Boo	Mufrid
Cor ♏	Cor ♏	α Sco	Antares
Caput draconis	Caput draconis	γ Dra	Etamin
absent	Aquila	α Aql	Altair
Scheat	Scheat	β Peg	Scheat
Caput And.	Caput Andromedæ	α And	Alpheratz

★ The star position is on the strapwork or the ecliptic, so no extended pointer exists.

† The original scratched version of the name can still be seen, as the engraved name takes less space.

Note. The star names on the instruments are engraved in italic letters, but this has not been reproduced in this table.

a diagram for conversions between equal and unequal hours in the upper half, and a shadow square in the lower, divided to 12 on each side in tenths, alternately shaded. The plate locks in place in the mater by means of a small hole near the rim that fits over a pin protruding from the mater below the upper 12 o'clock position. There is room for only one plate to be installed at a time. This was also the practice of Gerard Mercator (G. Turner 1994).

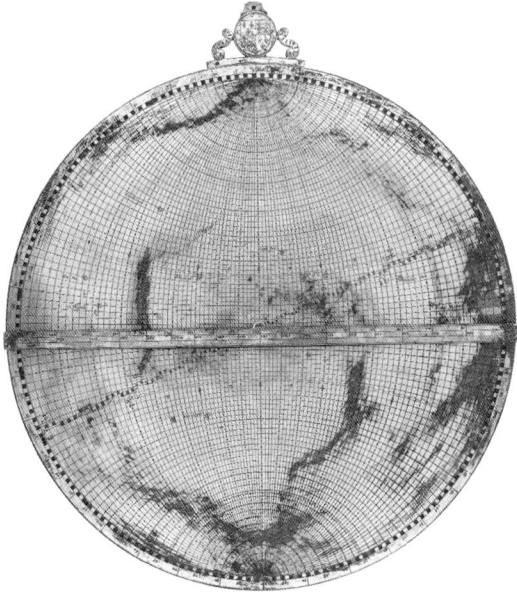

6 Back.

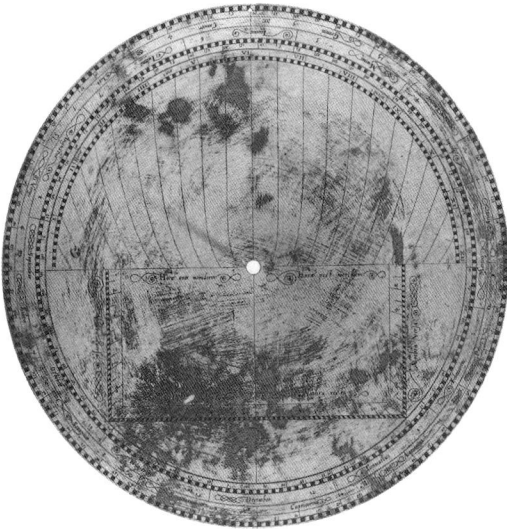

6 Plate.

engraver recorded, in 1659, the gift of the instrument to the University of Oxford.

Provenance

The gift of Nicolas Greaves to the University of Oxford in 1659. Previously in the possession of the elder brother of Nicolas, John Greaves, Savilian Professor of Astronomy from 1643 to 1648. For a possible earlier provenance, see Chapter 2, where it is proposed that it was purchased in 1559 for £10 by Robert Dudley, Earl of Leicester from 1564.

Literature

Gunther (1937, pp. 67–72), figs 1–3, plates 11–13.

Location

Museum of the History of Science, Oxford (36–6; 42223) IC 575.

The side with the tablet of horizons: *Horizontale Catholicum*, has curves every 2° from 0 to 90°. In some vacant space an

7 Astrolabe 155[?8]

Signed and dated: *Thomas* ♊ · *Anno* 155

Brass, gilded; diameter 352; thickness of limb 6; thickness of rete 3

This astrolabe at Florence is very like the one at Oxford, **6**, so that the description of the latter may be taken as serving both, with the exceptions noted below.

Fitted at the top of the instrument is a nearly round extension holding a ball attached to a shackle with ring, providing a ball joint that moves only in one sense. Arabesques decorate this extension.

In comparison with **6**, the limb and mater are the same. The rete is identical in design, but has a long pointer for *Cor Leonis*, and there are 30 stars not 29, with three variations in choice between the two (see Table); the back with its *Saphea* is the same except for a

7 Front.

7 Mater.

prominent division into degrees along the ecliptic line (degrees alternately shaded and staggered every five); the regula is the same, but the cursor and brachiolum are present; the 15 stars on the back are the same, but with small variations in spelling; the alidade is the same, but with two threaded holes, which were added later since they interfere with the original engraving. The sight vanes have two holes of about 0.5 mm and 1.5 mm in diameter. There is no plate with this instrument, but there is a hole to accept the lug on a plate.

On the lower part of the rete one may discern the rubbed-out star name: *Cingulum Orionis*. The pair of threaded holes in the alidade show signs of rubbing in the surrounding region. It seems that sights suitable for surveying could have been fitted.

The cursor is bevelled on both sides and marked out 0–90° in 2° intervals on one side. The opposite side is a scale of 0–100 by ones. Along the centre of the cursor is the inscription:

Cuncta Mathematices nunc instrumenta quiescant,
Quot sunt, quot'qæ fuere vnquam veterum atqæ recentum.

All mathematicians can now lay aside their implements,
however many they are, or were, and whether ancient or new.

The brachiolum is composed of two jointed arms, fimbriated; the second arm is tapered to a point.

The partial date engraved on this instrument, 155, obviously means it was made

intended for Mary, but she was succeeded on death by Elizabeth in November 1558, so he had to dedicate the volume to Elizabeth, but he left the engraved portrait of Mary (see Chapter 3, Fig. 3.1). A possible explanation may be suggested through the practice of map makers. A collection of maps required each to be dated on the publication of the group, so the last figure was not engraved but added in ink at the time of publication. Through the close similarity of this instrument to the astrolabe made for Elizabeth that is fully dated as 1559, it is likely that it was constructed in 1558, or earlier.

Provenance

One of the instruments brought to Florence in 1606 by Sir Robert Dudley.

Location

Museo di Storia della Scienza, Florence (1093)
IC 489.

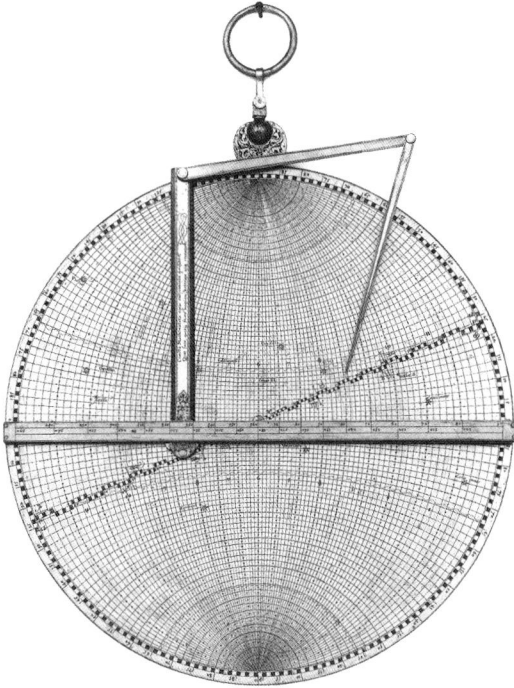

7 Back.

during the decade of the 1550s. However, it should be possible to propose a closer date. Gemini had already made astrolabes for Edward VI, **1, 5**, which are dated *c.* 1551 and 1552. Philip and Mary reigned from 1554 (Mary alone from 1553) to 1558, and in accordance with Gemini's obvious claims for royal patronage, he could well have made an astrolabe for the joint rulers of England. There is evidence that an edition of his *Anatomy* was

8 Compendium 1568

Signed and dated: · HVMFRAY · COOLE · MADE · THIS · BOKE · ✦ ANNO · 1568 ·

Brass, gilded; case 66 × 60, thickness 20

Comprising from the top:

a) engraved decoration

b) quadrant

---

7 The 15 stars positioned on the *Saphea* are:

Venter Cæti	*Ceruix* ♌	*Vultur*
Oculus ♉	*Cauda* ♌	*Aquila* [Altair]
Cing. Orio.	*Spica*	*Cauda Cig.*
Canis maior	*Alramech*	*Fomahand*
Canis minor	*Caput Her.*	*Cing. And.*

Note: *Caput And.* on **6** is replaced by *Cing. And.*, but both have the same coordinates.

c) compartments for drawing implements

d) magnetic compass

e) equinoctial sundial

f) table of latitudes of 33 European towns

g) calendar of fixed feast days

h) replacement base

a) At the centre of the lid is engraved, within a circle, a three-stemmed tree in leaf with flowers and a bird near the top. Above the tree is a scroll with the word repeated four times: IVGGE. This is the canting badge of the printer, Richard Jugge, who was Master of the Stationers' Company in 1568. The same badge is found on Cole's map of Canaan dated 1572 (see Chapter 3). Gunther (1927, p. 282) thought the bird was a nightingale, and quoted the *Oxford English Dictionary* for a 1598 reference. An alternative is the partridge, also in the *Dictionary* under 'Jug', with a reference of 1600. A partridge in a pear tree seems more appropriate. Around the circle is a strapwork frame, with two squirrels above, and two snails below. Squirrels are also found on Cole's compendium made in the following year, 1569 (**9**).

b) Cole names the quadrant: *A quadrant · to · take · yᵉ · hight · of · yᵉ · sōne · ☾ · ✶ · and · yᵉ ·*

8

8 a.

hight · of · toure9 · or · any · other · buildīge9.

The lower arc is labelled below:
Quadrans Astronomicus, and is divided in degrees, alternately shaded, marked from 10° to 90° both ways. The inner arc represents a

shadow square, and is labelled: *Quadrans Geometricus*. The division is in 24 parts, alternately shaded, and numbered every 3 from zero at each end to 12 at the centre point (45°). The parts are further divided in half, giving a total of 48 equal parts. The scale is labelled to the left: *Vmbræ Versæ*, and to the right: *Vmbræ Rectæ*. The quadrant is used with a pair of folding sights, with slit and pinhole. A peg at the centre of curvature of the quadrant supported a plummet, now lost.

c) The two compartments for drawing implements (now lost) have hinged lids each with a pair of catches in the form of a human hand. Decoration consists of four plain circles and stippling. The contents of each compartment is given in panels:

THE · COMPASE · SQUARE · & · RVLE ·
· THE · WRYT: · TINGE · PENNE ·

d) The magnetic compass, which can be removed, is surrounded by arabesque and stipple decoration. At the middle is the small

8 c, b.

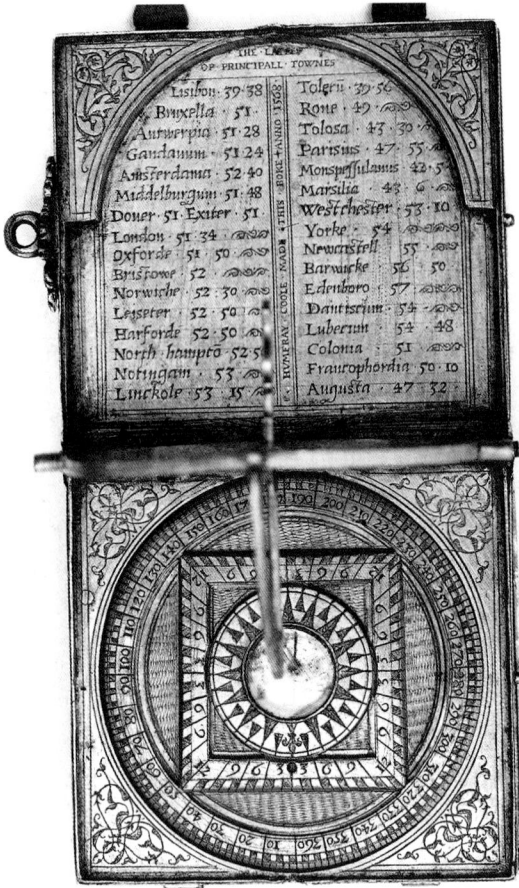

8 d, e, f.

compass box (diameter 13) surrounded by a rim on which are engraved the 32 wind directions, or points of the compass; a fleur-de-lis is at the North point. This is set inside a square, the Geometrical Quadrant, divided in units (alternately shaded) from 0 at the mid-points of the sides to 12 at the corners. The square is inside a circle (diameter 54.5) that is divided on its rim into 2° units, alternately shaded. Every 10° is marked to 360°. This complete unit can be released by pressing the split-ring on the underside; there is a location stud for alignment. The glass cover to the compass box

is domed, and is scratched in a way suggesting it is genuine; the needle is original. The only mark on the bottom is a line ending as a T and about 10° East of North. The blued needle has a point at the South and a † at the North.

e) The chapter ring is held in a semicircle cut into a plate fitted to the common hinge that runs across the width of the instrument. Opening the two halves of the instrument to lie flat, this plate is set vertical, and the chapter ring is positioned for latitude of use by setting a small quadrant in a notch in the plate. Divisions on the quadrant are to 2°, alternately shaded, and numbered every ten; subdivisions are to 1°. The quadrant and gnomon are on an axis pivoted inside the ring between the 12 o'clock positions. The plummet is missing. The hours on the upper side of the ring are marked in roman numerals to twice 12 hours, divided to quarter hours, alternately shaded. The hours on the lower (winter) side of the ring are from VI to VI. The chapter ring and its support plate pack elegantly into a recess in the second half of the instrument.

f) Formed like an archway to accommodate the equinoctial sundial when packed away, this side is engraved with what Cole called: THE · LATITV OF · PRINCIPALL · TOWNES.

For a commentary on this and on other latitude tables by Cole, see Chapter 4. It is shown that the latitudes of Continental towns were derived from Gemma Frisius, *Cosmographia*, various editions.

g) The calendar of fixed feasts, plus the entry of the Sun in the Zodiac, is taken from Leonard Digges, *A Prognostication everlasting*, Marshe editions (1564 or 1567). Originally

8   List of towns

Lisbon	39 · 38	Toletū	39 · 56
Bruxella	51 ·	Rone	49 ·
Antwerpia	51 · 28	Tolosa	43 · 30
Gandanum	51 · 24	Parisius	47 · 55
Amsterdama	52 · 40	Monspessulanus	42 · 5
Middelburgum	51 · 48	Marsilia	43 · 6
Douer	51 ·	Westchester	53 · 10
Exiter	51 ·	Yorke	54 ·
London	51 · 34	Newcastell	55 ·
Oxforde	51 · 50	Barwicke	56 · 50
Bistowe	52 ·	Edenboro	57 ·
Norwiche	52 · 30	Dantiscum	54 ·
Leiseter	52 · 50	Lubecum	54 · 48
Harforde	52 · 50	Colonia	51 ·
North · hamptō	52 · 50	Francophordia	50 · 10
Notingam	53 ·	Augusta	47 · 32
Linckole	53 · 15		

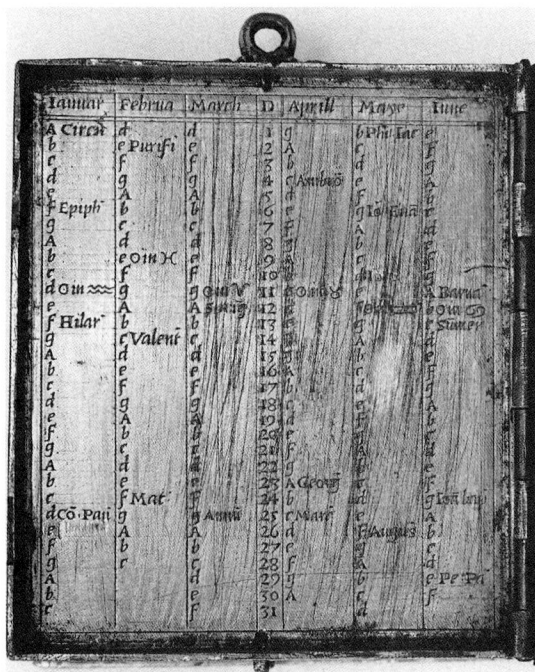

8 g.

published by Thomas Gemini in 1555, the several Marshe editions all print the same revised version of 'The generall Kalendar'. The table is over two pages, splitting the year. The compendium has the first half of the year, but the second half was on the bottom plate, now lost and replaced with a plain surface. The printed calendar is reproduced in Fig. 4.2, which shows what the missing half of the year will have been like. Compare the first half of the year with the photograph of the plate (**8** g). See Table 4.3 for a comparison of the Church calendar on seven Cole instruments.

h) The replacement base, with closure clips, is modern. The inner side is plain, the outer side has a foliate centre and C-scrolls at the sides.

The sides of the case are engraved with arabesque decoration. The small ring for attachment to a cord is fixed in one side of the lower half of the instrument, and is surrounded by an oblong collar cut in a leaf pattern.

Provenance

Bought by Lewis Evans for £168 from Sotheby, Wilkinson & Hodge, sale 19 April 1920, lot 316, 'The Property of a Lady'. The Lewis Evans Collection was presented to the University of Oxford in 1924.

Literature

Gunther (1926, pp. 292–3); Gunther (1927, pp. 281–5).

Location

Museum of the History of Science, Oxford (LE 1; 36313).

9 Compendium 1569

Signed and dated: ✦ *Humfray · Colle · made* [] *this · diall · anno · 1569* ✦

Brass, gilded; overall length 83, oval case 65 × 57, overall thickness 23

Comprising from the top:

a) engraved decoration

b) solar and lunar volvelles

c) perpetual calendar and table of fixed feast days

d) table of latitudes of 32 European towns

e) equinoctial sundial

9

f) miniature simple theodolite

g) establishment of the ports

h) tide computer

j) engraved decoration

a) The lid of the compendium has an engraving of Juno, with peacocks, holding a staff entwined by a snake. The figure is surrounded by grotesques.

b) · *The· course· of· the· Sonne· and · Mone · throughe 12 · Signes ·*

Pinned at the centre are two volvelles, rotating between calendar and Zodiac scales. The days of the year and the degrees of the Zodiac are divided into units of two, alternately shaded. Each Sign of the Zodiac begins with its number, name, and sigil, thus: I Aries ♈. The First Point of Aries is 11 March; of Libra 13 September. The solar

9 b.

volvelle has a pointer with a Sun's face to mark the Sun's position in the Zodiac, and round the edge is a scale of the Moon's age to $29\frac{1}{2}$ days to which the lunar volvelle is set by its pointer. There are a phase aspectarium, and astrological aspects: opposition, trine, quartile, sextile.

c) · *A kalender· with· y^e· saint'· daies · And· moueable feastes · for · euer ·*

The outer three circles list the feast days of the Church arranged in four quarters; the months are shown by their initial letters. This table follows in detail (with some omissions through lack of space) that printed by Leonard Digges, *A Prognostication*, second edition 1564. The entry of the Sun is on 11 January and 10 February on the instrument; 10 and 9 respectively in the 1555 edition, and 11 and 9 in the 1564 edition.

9 a.

Around the centre are seven circles for the dates of Easter. They are labelled: *East· D; Dō· Lt; Prime; Dō· L; Le· γ; Pri; Ep.* The circles provide the following information: Easter day, from 22 March to 25 April; Dominical Letter under each Easter date; Prime, an alternative denotation for the Golden Number, governing the date of Easter in a 19-year lunar cycle; Dominical Letter, also used in the determination of Easter, being the sequence of the days of the week following the paschal Moon; Leap Year, with two Dominical Letters, one before new year on 25 March, and the next in sequence after. The innermost two circles connect the Prime with the Epact for 19 years beginning with 1569 (*12/12*). At the centre are Cole's initials, H C, and the year 1569. The information for this table comes from Digges, *A Prognostication*, 1555 and later editions.

d) There are 32 European towns listed in two columns, giving their latitudes in degrees and minutes.

e) When the lid of the compendium is lifted, a lug at the hinge lifts up the arc holding the

9 e.

chapter ring. The arc is decorated on both sides with arabesques. The latitude quadrant, with its extension bar, is hinged to the ring to lie flat when not in use. For time-telling, the quadrant is turned through a right angle so bringing the gnomon into alignment with the

I	*Circuc̄* 1	*Epiph* 6	⊙ *in* ♒ 11	*Hil* 13	*Cō P̄* 25
F	*Purifi* 2	⊙ *in* ♓ 10	*Valen̄* 14	*Math* 24	
M	⊙ *in* ♈ 11	*Spring* 12	*Anunci* 25		
A	⊙ *in* ♉ 11	*George* 23	*Marcke* 25		
M	*Phillip Iacob* 1	⊙ *in* ♊ 12			
I	*Bam̄* 11	⊙ *in* ♋ 12	*Iō bā* 24	*Pē* 29	
I	*Doḡ beg* 6 [†]	⊙ *in* ♌ 14	*Mā Mā* 22	*Iā Apo* 25	
A	*Pē Vī* 1	⊙ *in* ♍ 14	*Dō eñ* 17 [†]	*Bā* 24	*D·Ī* 29
S	*Nā Mā* 8	⊙ *in* ♎ 14	*Mī* 29		
O	⊙ *in* ♏ 14	*Luke* 18	*Simon · Iude* 28		
N	*Om̄ Sā* 1	*Om̄ Añ* 2	⊙ *in* ♐ 13	*Andr̄* 30	
D	*Nī* 6	*Cō Mā* 8	⊙ *in* ♑ 12	*Thō · A* 21	*Na* 25

† Dog's beginning, Dog's end refers to the appearance of Sirius, the Dog Star. The helical rising of Sirius in the east at sunrise during the summer, after a period of 70 days' invisibility, marked the beginning of the Egyptian new year (Clagett, 1995, 57). The present-day solar conjunction is on 4 July (Bakich 1995, p. 162). Sirius has the brightest apparent magnitude of any star.

Douer	51 ·	Saulis	51 · 34
Exiter	51 ·	Cambrig	52 · 0
London	51 · 34	Baseingsto	51 · 34
Norwiche	52 · 30	Paris	47 · 55
Oxforde	51 · 50	Antvverp	51 · 28
Leiseter	52 · 50	Colonia	51 · 0
Harforde	52 · 50	Venetia	44 · 50
Bristowe	52 · 0	Basilea	48 · 0
North· hamptō	52 · 50	Compostellæ	45 · 0
W·Chester	53 · 10	Augusta	48 · 0
Linckolne	53 · 15	Lisbona	38 · 0
Notingam	53 · 0	Neapoli	39 · 0
Yorke	54 · 0	Lions	45 · 0
Newe· caste	55 · 0	Rouan	48 · 0
Barwic	56 · 50	Toledo	39 · 0
Edenbo	57 ·	Roma	42 ·

polar axis, which occurs when the quadrant moves through a slit and is held at the latitude of use. The scale is divided 0–90° in alternately shaded 2° units. The upper side of the chapter ring is marked out in twice 24 hours, for use during the summer half of the year, while the under side is marked from VI to VI for use during the winter. It is this side that bears the full signature.

f) A miniature simple theodolite (diameter 50) is contained in the next section of the compendium. It matches quite closely Cole's simple theodolite of 1574, **15**. The edge is graduated to 360° in single degrees, and within is a geometrical square labelled at each of its sides with calligraphic elaboration: *The Geometricall Square.* At the middle is a magnetic compass, the base divided into the 32 points, alternately shaded. The glass cover contains air bubbles. Around the housing rotates an alidade with a pair of edge and pinhole sight vanes that are hinged for

packing. Underneath the compass box is a circular protrusion with a shallow waist. This waist is gripped by a pair of sprung claws resembling those of a crab (Fig. 2.6). This mechanism is reached by lifting up the hinged

9 f.

panel that crosses the next leaf in the compendium. It is easy to press the claws apart and so release the miniature theodolite.

g) THE NAMES OF PRĪCIPAL PORTES *&c* HAVENS OF VROPE WHAT MONE MAKETH A FVLL SEA. The time of high tide is important when entering or leaving port. The highest tides occur when the Moon is new or full, and the time of such a tide varies according to the geographical location. For ease of memory, seamen noted the compass bearing of the Moon at the time of high water ('full sea'). A port has its own particular bearing, and so the establishment of the port is found in a table listing ports under 18 wind directions. The table has 18 rather than 32 because the bearing of new moons is diametrically opposite that of full moons (Waters, 1958, pp. 31–3). The rule is given by William Bourne, *An Almanacke and*

Prognostication for … 1571 and 1572, & 1573 (London [1567]), sig. Cviii, 'The fourth Rule teacheth howe to knowe by the age of the moone when it doeth flowe at any place, where you doe knowe what moone maketh a ful sea'. Sig. Diiii has a print of a wind rose. Note that although the title of the edition has the years from 1571, at the end of a chapter Bourne writes 'Nowe this yeare 1567, …'. Most of this edition is a reprinting.

h) AN INSTRUMENT TO KNOWE THE EBE AND FLVDDES. The outer circle is divided into 360° in 2° intervals alternately shaded; the middle into twice XII hours; the inner into the 32 wind directions. At the centre is the solar volvelle with a pointer, the rim divided to 30 days, approximately the age of the Moon (actually $29\frac{1}{2}$); subdivisions are to 6 hours. Finally comes the lunar phase aspectarium, with a pointer marked by a crescent.

9 g.

9 h.

9 j.

Arabesques complete the decoration at the centre and edges. This is the earliest known tide computer of English workmanship according to Waters (1958, caption to plate XLIV). A closely similar computer by Cole, but not dated, is in the British Museum, **12**. It is backed by a nocturnal.

j) The underside of the compendium is engraved with the figure of Jupiter standing over an eagle and holding a shaft. Among the grotesques are two squirrels below, and above are two monks with rabbit's ears, reading books. The monk on the left is wearing nose spectacles; see Winkler (1988, pp. 32–52).

The compendium may be hung on a chain by a ring attached to a casting fixed to the top opposite the hinge. The urn finial is supported by two scrolls. The side of the casing is deco-

rated with a beaded rope edging. The hinge to one leaf has been repaired by the addition of a small brass plate fixed by five rivets.

Provenance

This compendium was presented by the Right Honourable Philip Stanhope to the Reverend Philip Bigsby, who had married Stanhope's widowed stepmother, Frances, in 1783. Bigsby gave it to his brother, Robert, in 1812, who presented it to King William IV. The King gave it to Greenwich Naval Hospital in 1833, and it was later lent to the National Maritime Museum. The compendium has, not infrequently, been referred to as 'Sir Francis Drake's dial'; however, there is no supporting evidence for this (*Armada* 1988, p. 212).

Literature

Waters (1958, pp. 517–18), plate XLIII; *Armada* (1988, p. 212); Ackermann (1998, pp. 55–9).

Location

National Maritime Museum, Greenwich (AST0172; old no. D. 318).

10 Gunner's folding rule

c. 1570

Signed: *Humfray · Coolle · Mad · This · ;* undated

Brass; length 320; width of arms 28

This complicated instrument is used either closed, to read tables that cover both arms, or opened out completely to form a geometrical quadrant. The joint has a relatively large central hole (diameter 15) probably to hold a magnetic compass. The hole is surrounded by arabesque decoration. Side 2 has Cole's signature.

Side 1

On this side are scales for the calculation of timber and board measure on the left arm. From measuring the width of a given board, the intention is to find the length in inches that produces a square foot. Similarly for timber, knowing the area of a balk, the length to make a cubic foot is to be found. The left arm begins 144 / 12, which is engraved on the point; this is a table to find the square from a measure of the width. Thus the area of a board of side 12 is 144, or of side 1ft 6in is $2\frac{1}{4}$ square feet, shown on the instrument as 1· 6 / 2· 3. These scales are labelled on the outer edge (left): *Borde measure* and (right) *Tim measure*. The right arm has an inch scale 0–11, divided into thirty-seconds of an inch, numbered sequentially 0–312. See Chapter 4, 'The folding rule', for the explanation of timber measure.

Side 2

On this side the upper part has a table for gunnery running over both arms when closed together. The six columns are headed with their names at right angles to the table; the first five beginning with H, meaning 'How'. The punctuation between words is omitted here.

H· *Many poūde euery shote wayeth*
H· *many lli of pouder euery pece shot*
H· *Many Inches hie the bullett is*
H· *Hie the peece is in the mouthe*
H· *Much euery pece wayeth*
The Names of Ordenan

This table is discussed with other similar tables in Chapter 4. See **11**, **22**, for comparable tables.

Below the table are nine empty rows. There follows on the left arm a table named: *The*

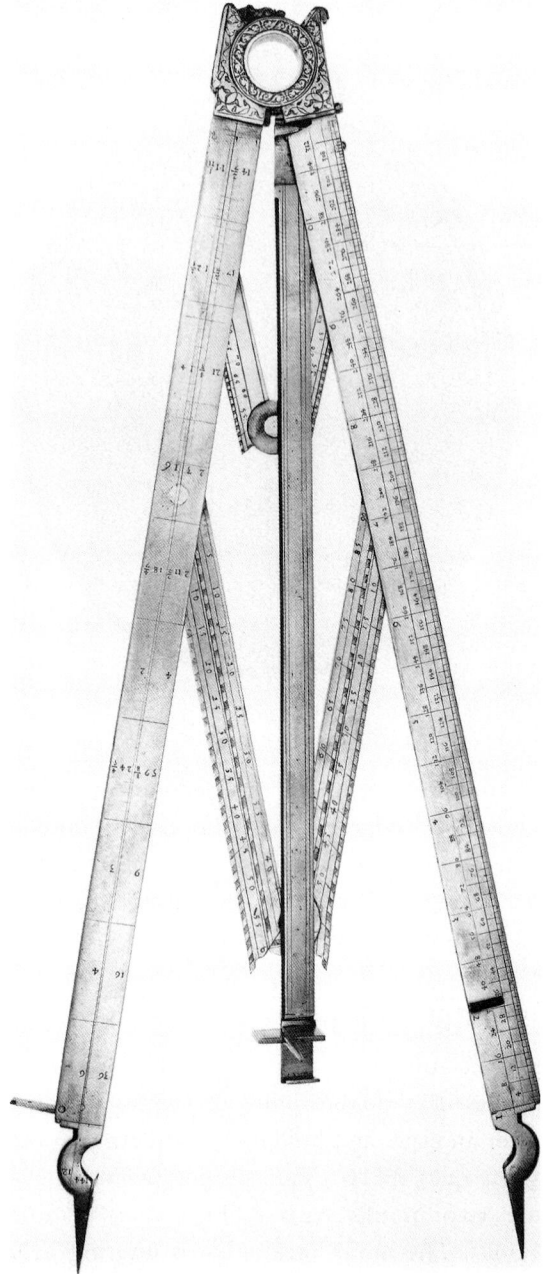

10 Side 1.

Name of gun	Gun weight	Gun ø	Shot ø	Powder weight	Shot weight
Robinet	200	$1\frac{1}{4}$	1	$\frac{1}{2}$	1
Fawconet	500	2	$1\frac{3}{4}$	2	2
Fawcon	800	$2\frac{1}{2}$	$2\frac{1}{4}$	$2\frac{1}{2}$	$2\frac{1}{2}$
Minion	1100	$3\frac{1}{4}$	3	$4\frac{1}{2}$	$4\frac{1}{2}$
Sacar	1500	$3\frac{1}{2}$	$3\frac{1}{4}$	5	5
D Cull	3000	$4\frac{1}{4}$	4	9	9
Culluer	4000	$5\frac{1}{2}$	$5\frac{1}{4}$	18	18
D Canō	6000	$6\frac{1}{2}$	$6\frac{1}{4}$	28	30
Canon	7000	8	$7\frac{3}{4}$	44	60
E Canō	8000	7	$6\frac{3}{4}$	20	42

table of timber measure. The three columns are headed: *Foot, Inches, partes*. A further column is on the side edge, which runs from 1 adjacent to 144 on the face, to 9, adjacent to 0 on the face. Also engraved on the edge is: SQVARE YNCHES OF THE TYMBRE, and *Tim measure*. On the face, the table begins at 144, and runs down to 36.

On the other arm is a table of linear measure written in Latin. The heading reads: DE PARTIBVS MENSURÆ SEV SPECIEBVS GEOMETRIÆ PRACTICÆ. The data is in four groups.

Granū igitur hordie est minima mēsura

Digitus habet	*4 grana*
per latera contigue disposita	
Vnica habet	*3 digitos*
Palmus habet	*4 digitos*
Dichas habet	*2 palmos*
Spithania habet	*3 palmos*
Pes habet	*4 palmos*
Sesquipes habet	*6 palmos*
Gradus habet	*2 pedes*
Passus simplex	*2 pedes*
	cum dimidio

Passus Geometricus quo vtitus

Cosmometra habet	*5 pedes*
Pertica habet	*10 pedes*
Cubitus habet	*6 palmos*
Stadium habet	*125 passus*
Leuca habet	*1500 passus*
Miliare Italicū habet	*1000 passus*
Miliare Italicum habet	*8 stadia*
Miliare Germ cont	*4000 passus*
Miliare Ger magnum	*5000 passus*
Miliare Germ cōmune	*32 stadia*

For explanations of lengths, see Doursther (1840) and Zupko (1981).

The length of each arm, measured from the pointed tip to the apex of the triangle that is formed when the arms are set at right angles, is 12 inches. The right arm is calibrated from the end of the first inch to just short of the end of the eleventh inch in thirty-seconds.

When the two arms are fully opened to form a right angle, two pairs of joined arms pull out from slots in the outer arms to form two geometrical squares, or shadow squares. The larger is engraved *Latus Vmbræ Rectæ*, and *Latus Vmbræ Versæ*, with arms divided in two scales, one in degrees to 45° at the centre, the

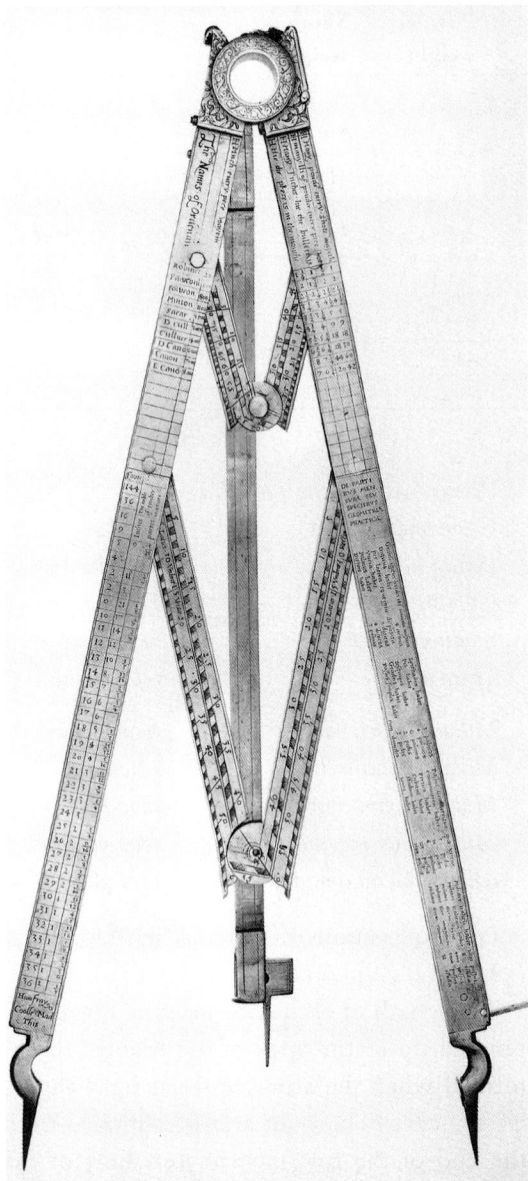

10 Side 2.

other to 60 at the centre. The units are alternately shaded. The smaller pair has divisions that run continuously from one end through the centre to the other end, 0–50, and 90°–0. These scales can be read from the fiducial

edge of a central bar that also packs into the cavity in one arm. It has a sight at the joint now broken off and another at the opposite end that folds and fits into a slot in one arm when the instrument is closed. Another arm has provision for two folding sights, the one by the joint now broken off.

The spelling of Cole's name (see Table 2.2) with double *oo*, double *ll*, and the *a* in *Humfray*, suggest an early date of around 1570. A similar rule is described under **22**.

Provenance

Donated to the British Museum in 1878 by Major General Augustus Meyrick.

Literature

Gunther (1927, pp. 294–6), plate 73; Ward (1981, p. 105), no. 311; Ackermann (1998, pp. 80–2).

Location

The British Museum, London (1878, 11–1.114).

11   Gunner's folding rule

c. 1570–75

Signed: HC; undated

Brass; overall length opened out 308; width of arms 16; thickness of arms 4.5

It is not an easy matter to describe all the parts of this rule because within each arm are two pull-out leaves with tables on both sides. Taking the side with the inch scales at the top, and with the curved joint uppermost, the arm on the left is A1, its underside A2; the arm on the right is D1, and the underside D2. Within the A side are the leaves B1 and C1, their reverses B2 and C2. In the D arm are the leaves E1 and F1, with their reverses E2 and F2. The outer edge of both arms is engraved, and at each end are sockets for pull-out sights, now lost.

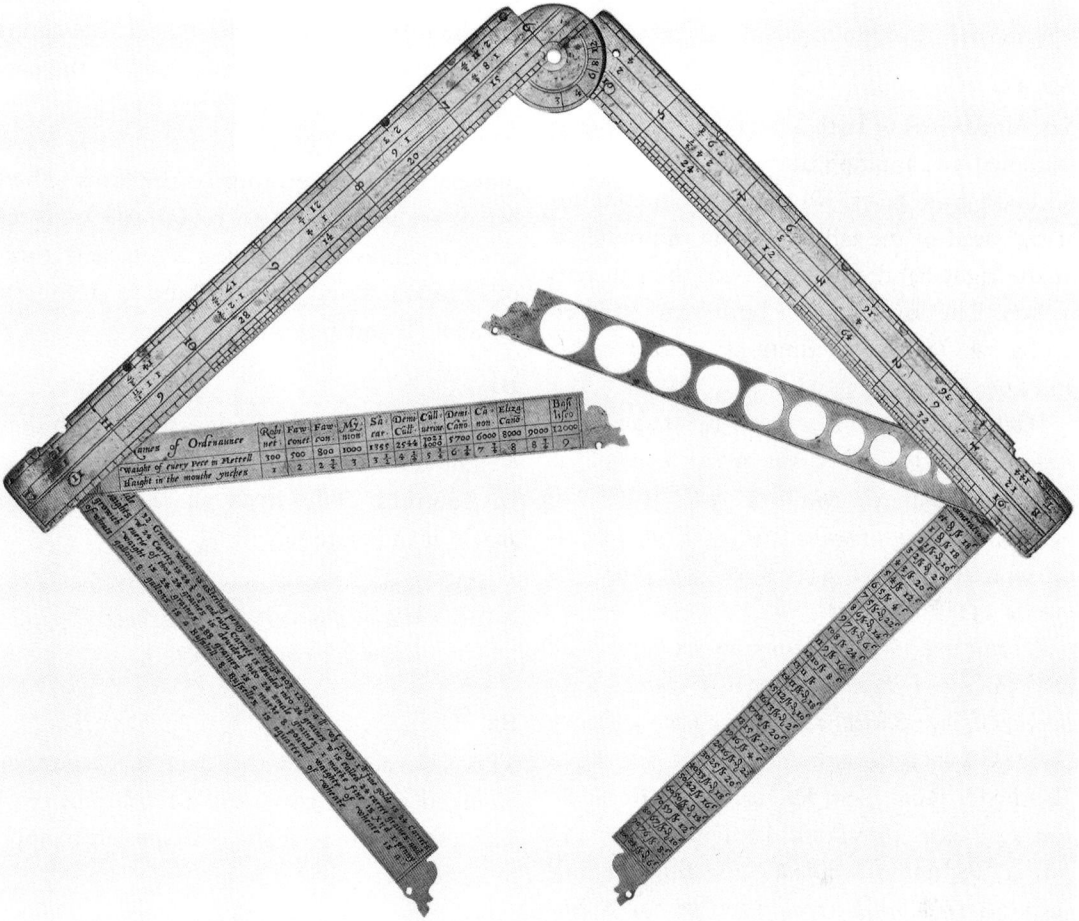

11 Side 1, leaves A1, B1, C1, D1, E1, F1.

A1, D1

Opened out, a 12-inch rule is formed, with the ends of the measure being lines close to the ends of the arms, which is why the overall length is 308 mm and not 305 mm. The 12 inches are divided in ⅛ths along the outer rim, while along the inner rim 12 separate inches are divided into scale parts: 11, 9, 28, 14, 20, 15, 10, 24, 12, 64, 32, 16. The other scales on this surface are marked **T** (timber measure), and **B** (board measure); the figures are engraved sideways with respect to the inch scale. On the face, the timber scale runs down from 144 to 12, while the board scale runs down from 12 to 1. Part of both these scales is engraved on the outside edge of the rule, which means the numbers are cramped. Beginning at the left, from the lost sight at the end of D1, the upper label reads: (on the lost sight) [timber] MEASVRE; the lower label: BORDE· MEASVRE. At the joint, the semicircle is marked for polygons: 3, 4, 6, 8, 12. These mark the positions to set the arms for drawing figures of 3–12 sides.

A2, D2

The upper half of both arms taken together is occupied by a multiplication table. The multiplicand figures 9, 17, 18, 19, 20 are in columns at the head of the table, and the multiplier is in the right-hand column, 1–20, then in tens to 100. The lower half of both arms is taken up by two tables of definitions: wool weights and linear measure.

Woll waightes · 7^{ll} is a claue· 14^{ll} a stone· 26· claue is $^1/_2$· a sacke· 13· stone is $^1/_2$· a sacke· 52· claue is a sacke· 26· stone is a sacke· 364^{ll} · is a sacke· 3· sackes is a surplus·104^c· is ·2· sa; 208^c ·is ·4· sa; 416· is ·8· sa ·832· is ·16· sa; 1664^c · is ·32· sa 3328^c· is ·64· sa·

A barlie corne is the least measur yet frσ it do all other Procedes 3· Barlie cornes cont 1· ynche · 4· in thickne' a finge Bredeth 12· ynchs a foote 3· foote a yarde ·5· foote a Geometricall pace· 15· fote ·$^1/_2$· a Pearch ·20· Pearc ·$^1/_2$· a Ro ·40· Pear a Roode 80 Pea $^1/_2$: a Acre ·160· Pea a Acre 660· foote or 132 pace a Eng furlog 8 furlongs is a Englishe mile ·1500 pace is a french league· 1000 pace or ·8· furlongs an Itallian mile· 32 furlongs is a comō Germain mile·253440· Barlie corns in length is an Englishe mile ·

At the bottom, running sideways, is a table marked: *Acre / Bre:deth · Len:gth ·*

This gives the length and breadth, in perches, which multiplied together make 1 acre.

At the joint is a *Geometrical Quadrant*, also labelled CONTRA S / RYGHT SHADO. The sides of the shadow square are divided in units, alternately shaded, to 12 at the corner (45°). A line from this corner to the edge of the joint is named: *Line of leauel*. The line and the geo-metrical quadrant are used with a plumb-line attached to a peg from the hole at the centre of the joint. With the arms set at 90°, the ends of the arms form a base for the level.

Within each arm are two thin brass leaves pinned at the lower ends of the arms. These leaves are swung out to reveal one with ten circular holes for gauging shot, and three others with tables. Leaves B and C are in arm A, while E and F are in arm D.

B1

Twelve sizes of gun, eleven named, with their weight and calibre, are set out in three rows of 12 columns. The rows share the definite article at the extreme left:

> The   *Names of Ordenaunce*
>   *Waight of euery Pece in Mettell*
>   *Haight in the mouthe ynches*

B2

This side continues the data for a gunner, arranged in four rows and 12 columns, with the definite article at the extreme left to apply to each row.

> The   *Haight of the Bullett ynches*
>   *Powder that euery Pece showeth ll*
>   *Scores at Pointe blancke*
>   *Waight of euery shote ll*

C1

A table of weights for use by goldsmiths and jewellers, named: *Golde waights growith of wheate:*

31 Grains wheate is a sterling penny· 20· sterlings a oz· 12· oz a ll w^{t}z Troy· 1^{ll} fine golde is 24 Carrets w^c· 24 Carrtts is· 24$^1/_2$ oz· and euey Carrett is deuided into 24 graines w^c is called· 24· carret graines: and euerie of thos· 24· graines is diuided into 24 smale graines ·1· marke fine is xij d

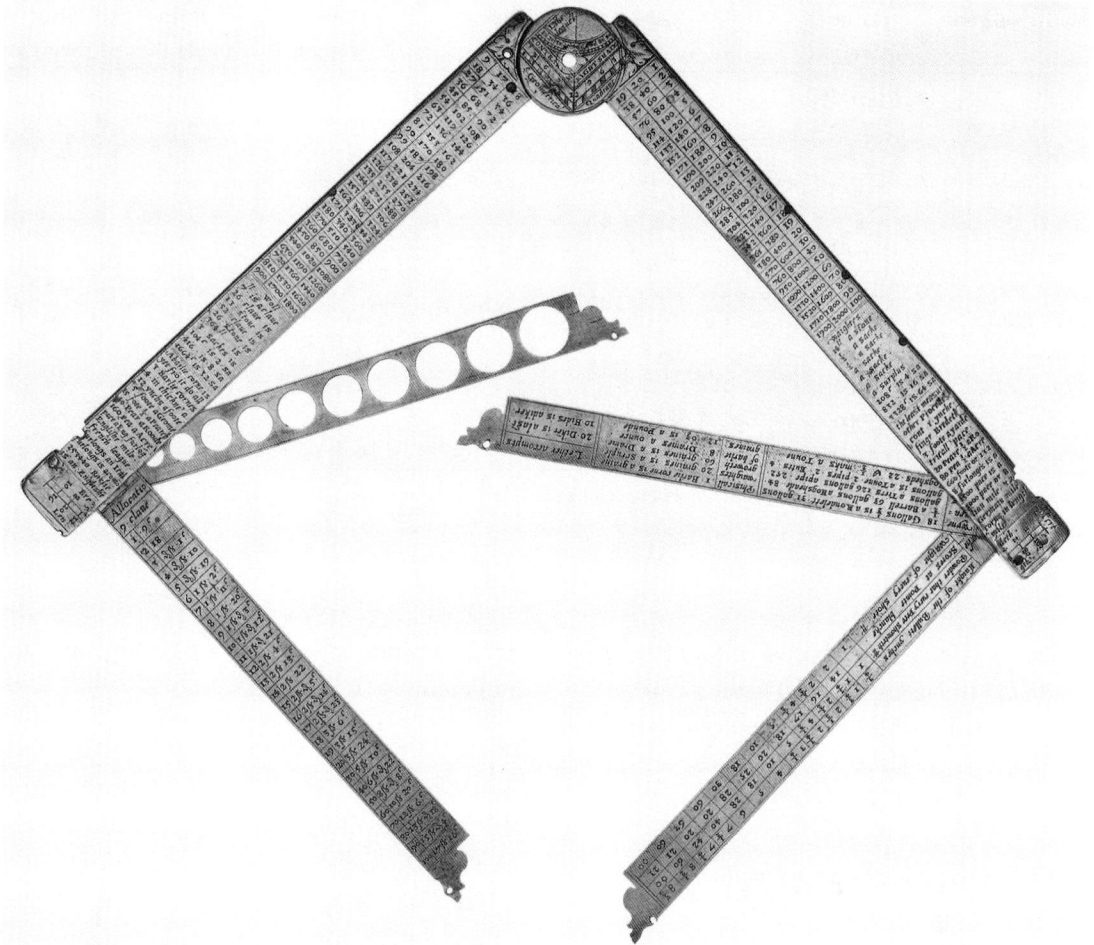

11 Side 2, leaves A2, B2, C2, D2, E2, F2.

·1· *penny waight is· 24· graines 288 graines is a marke ·8· pounde waight of wheate is a gallon ·8· gallons is a Bulshell ·8· Bulsells is a quarter of wheate.*

C2

Three tables are on this leaf, measures for wine merchants, apothecaries, and leather sellers. The very old established wine measures are given by Cole, and he seems to include the tierce introduced by Henry VIII in 1536 (Connor, 1987, p. 171). But Cole writes '84 gallons a Tiers', whereas the firkin was 84 gallons, and the tierce 42 gallons (a sixth of a tun). One may suppose Cole jumped a line, for the firkin (a third of a tun), would have been familiar to him.

The first table is named: *Wyne mea:sures*: 18 *Gallons* ½ *is a Rondelett ·31 gallons* ½ *a Barrell 63 gallons a Hogsheade ·84· gallons a Tiers* 126

Name of gun	Gun weight	Gun ø	Shot ø	Powder weight	Point blank	Shot weight
Robinet	300	1	$\frac{3}{4}$	1		1
Fawconet	500	2	$1\frac{3}{4}$	$1\frac{1}{4}$	14	2
Fawcon	800	$2\frac{3}{4}$	$2\frac{1}{2}$	$2\frac{1}{2}$	16	$2\frac{1}{2}$
Mynion	1000	3	$2\frac{3}{4}$	$4\frac{1}{2}$	17	$4\frac{1}{2}$
Sacar	1355	$3\frac{1}{2}$	$3\frac{1}{4}$	5	18	5
Demi Cull	2544	$4\frac{1}{4}$	4	10	20	10
Culluerine	3021 / 4000	$5\frac{1}{4}$	5	18	25	18
Demi Canō	5700	$6\frac{1}{4}$	6	28	28	30
Canon	6000	$7\frac{3}{4}$	$7\frac{1}{2}$	40	20	60
Eliza Canō	8000	8	$7\frac{3}{4}$	42	20	63
	9000	$8\frac{3}{4}$	$8\frac{1}{2}$	60	21	60
Basilisco	12000	9	$8\frac{3}{4}$	60	21	90

Note that the last entry, for the biggest gun, is not in Cole's hand.

This table is similar to **22**, but with several variations to the figures. See Chapter 4.

gallons a pipe ·252· gallons is a Toune ·2· pipes ·2· Butts ·4 Hogsheds ·22 & $\frac{1}{2}$ maks a Toune:

The middle table: *Phisicall waightes groweth of barlie graines: 1· Barlie corne is a graine 20· graines is a scruple ·60· graines is a Drame ·8· Drames a ounce ·12· oz is a Pounde:*

The third table: *Lether accompts: 20· Diker is a last ·10· Hides is a diker:*

E1, E2

This leaf is a ring gauge, pierced with ten holes. The diameters in millimetres are (approximately): 4.6, 5.8, 7, 7.9, 8.5, 9, 9.5, 9.85, 10.5, 11. Millburn (1995) has published a description of a set of folding shot and calibre gauges for small arms by William Deane of London. From 1723 to 1748, he was Mathematical Instrument Maker to his Majesty's Office of Ordnance. His gauges are named as 'Coll. Borgards Gauges for Small Armes'. Albert Borgard (1659–1751) became colonel-commandant of the Royal Artillery in 1722. The diameters of the 24 holes range from 11.4

mm to 29 mm. It seems that Cole's gauges are the precursor of Borgard's.

F1

Numeratio de et· ϑ $\int8$ 18^c. A multiplication table in units of 44 cloves expressed in sacks (52 cloves), half sacks, and cloves. A clove is an English weight for wool and cheese (Connor 1987, pp. 135–7; fig. 29), equivalent in Cole's day to 7 pounds avoirdupois, as noted by Robert Recorde, *The Ground of Artes* (1543, p. 203). The multiplier is in the left-hand column, 1–20, then in tens to 100. The multiplicand begins with half a sack plus 18 cloves, denoted by symbols: ϑ (half), $\int8$ (sack), ^{c} (clove). Thus multiplied by 100 the result is 84 $\int8$ + ϑ + 6^c; reducing to cloves, this becomes: 4368 + 26 + 6 = 4400 cloves.

F2

Allocatio 9 claue. A multiplication table of units of nine cloves (63 lb). The arrangement is the same as on face F1, and employs the same

symbols. Thus multiplied by ten the result is
1∫8 + ϑ +12ᶜ; reducing to 52 + 26
+ 12 = 90 cloves.

When this instrument is compared with the
four 2-foot jointed rules **17–20**, it is seen that
the type of information is the same and displayed in the same manner, but in this case
remarkably condensed and showing Cole's
engraving skill.

Literature
Ackermann (1998, pp. 82–5).

Location
Museum of Artillery, Woolwich (Class XXIV 200).

12 Nocturnal *c.* 1570

Signed: · *H · Cole* · ; undated

Brass, gilded; diameter 95; length including handle
119; thickness 1

Side 1

The rim is engraved with the calendar, the
outer edge carrying the days, and the following ring the names of the months, in Latin,
with their number in the year. On this instrument the months are numbered from January.
The last date in December is given as 31¼, and
an extra dividing line allows for the extra
quarter. Rotating round a rivet with a central
hole is the solar volvelle, with a pointer with a
Sun's face running over the calendar. The rim
is divided to twice 12 hours with notches,
longer at the hour and shorter at the half hour.
These notches enable the user, in the dark, to
feel the number of hours or half hours
between the pointer at 12 and the position of
the index arm. Also rotating around the central
rivet is an arm labelled: ⊤ INDEX ⊤ HORA ⊤
NOCTIS. The root of the arm is inscribed:

12 Side 1.

STELLA POLARIS. The handle, which is engraved
with the initials WF, is in line with 23
October; on the opposite side is 21 April. The
nocturnal is set for use by putting the solar
pointer to the date, holding the instrument by
the handle, vertically downwards, sighting the
Pole Star through the central hole, and aligning
the guards of the Little Bear (*Ursa minor*), in
effect *Kochab* (β UMi), along the side of the
index arm. The hour is then felt along the
toothed edge of the volvelle. The volvelle is
decorated with arabesques.

Side 2

This side is a tide computer. The outer circle
is divided into 360° in 1° intervals; the middle
into twice 12 hours; the inner into the 32
wind directions. There follows the solar
volvelle with a pointer, the rim divided to 30

12 Side 2.

days, approximately the age of the Moon (actually 29½ days). Finally comes the lunar volvelle with aspectarium, and a pointer marked by a crescent. The face is completed with the aspects: opposition, trine, quartile, sextile. According to Waters (1958, plate xliv caption), the earliest known English tide computer is that on Cole's compendium of 1569, **9**, and he considers the one described above can be dated to the period shortly after 1570 because the Moon's age is given as 30 days and not 29½ days. However, 30 is more practical for seamen when working out the date of the New Moon; see Chapter 4.

Provenance

Purchased by the British Museum in 1857.

Literature

Gunther (1927, pp. 293–4); Ward (1981, pp. 74–5), no. 208; Ackermann (1998, p. 47), plate III, fig. 16.

Location

The British Museum, London (1857, 11–16.2).

13 Altitude sundial *c.* 1574

Signed: *humfray Colle*; undated; probably *c.* 1574 by comparison with similar dated dial, **14**
Silver; horizontal plate 73 × 78

Side 1

The rectangular plate is engraved with a quadrant divided 0–90° in degree units, alternately shaded, numbered in tens. The arc is labelled: *The heighte · of · the · sonne*, and below is the latitude:

<u>G</u> <u>M</u>

51 30

Outside the quadrant is a double circle with the seven planetary sigils in 12 segments to show the planet governing each hour of the

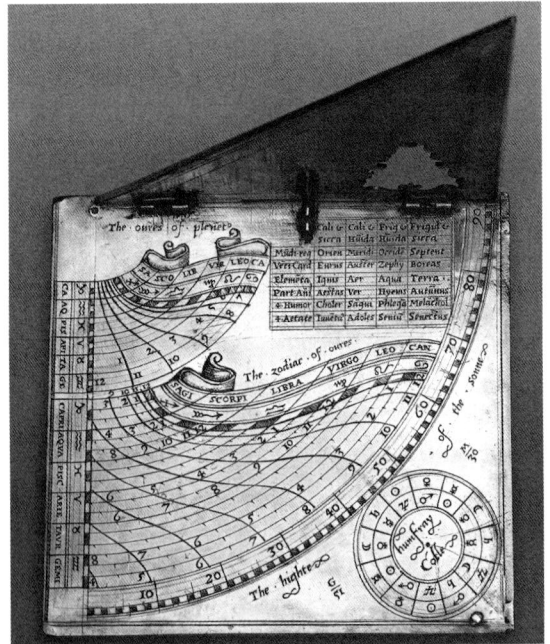

13 Side 1.

130

day and night. Within the circle is the signature. Cole began to write *Coole*, and finally engraved an *l* over the second *o*.

Inside the quadrant are two sets of hour lines, the outer giving the equal hours, and the inner the unequal, or planetary, hours. The inner is labelled: *The · ours · of · plenet⁹*. At the sides of both scales are the divisions of the Zodiac, which act as a calendar: *The · zodiac · of · oures ·*. Each Sign is marked in 10° arcs, alternately shaded, and the curved hour lines are numbered, with half hours indicated by dots. In the space between the sets of hour lines is a table, in Latin, of regions of the world, the four cardinal winds, the elements, the seasons, the four humours, and the four ages of man, arranged under the headings: hot & dry, hot & wet, cold & wet, cold & dry.

Along the side above the scales is hinged a right-angled isosceles triangular gnomon, with a catch to hold it upright. A scallop-edged aperture is intended to hold a plummet (missing). The time of day is found by aligning the dial plate so that the vertical edge of the gnomon casts the shadow of the sun parallel to the edge of the plate, when the shadow of the hypotenuse cuts the hour lines in the manner of the usual plumb-bob. The hour line is that which corresponds to the Sun's place in the Zodiac. See Plate 4.

Side 2

A quadrant is engraved with a degree scale as on the upper side, while within is a shadow square in units to 12 at the 45° position. The two parts are labelled: *Latus Vmbræ Versæ / Latus Vmbræ Rectæ*. This scale is used with a counter-changed alidade attached to the apex of the quadrant to act as a plumb-line, and a

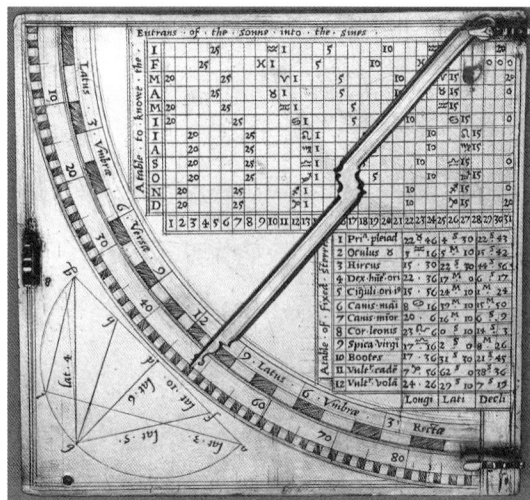

13 Side 2.

pair of sights fixed to the edge. A third foot is attached to the opposite edge, the feet being necessary when the upper side is used. Outside the quadrant arc is a semicircle inscribed with triangles marked: *lat·* 3; (4, 5, 6) *lat·* 10. These represent the lengths of the sides of regular polygons drawn inside a circle of the same radius as the semicircle.

Inside the arc of the quadrant are two tables. One is a calendrical table to find the date of the Sun's entry into each Sign of the Zodiac, titled: *A table · to · knowe · the · Entrans · of · the · sonne · into · the · sines ·*. The second table, titled: *A table · of · fixed · sterres*, gives the ecliptic longitude and latitude coordinates, and declination North or South of the celestial equator for 12 stars (S = septentrio; M = meridies); see Table below.

Case

The case is probably original, and is made of wood covered with blind-tooled leather and lined with red baize. The tooled motif on the under side is a fleur-de-lis in squares, and

No.	Cole's name	Longitude	Latitude	Declination	Modern name	Designation
1	*Pria^a · pleiad*	22.♉ 46	4.S 30	22.S 43	Alcyone	η Tau
2	*Oculus ♉*	3. ♊ 16	5.M 10	15.S 42	Aldebaran	α Tau
3	*Hircus*	15. 30	33.S 30	44.S 56	Capella	α Aur
4	*Dex· hūe^r· ori*	22. 36	17.M 0	6.S 17	Betelgeuse	α Ori
5	*Cigūli· ori· i^o*	15. 56	24.M 10	1.M 24	Alnilam	ε Ori
6	*Canis· maī*	8. ♋ 16	39.M 10	15.M 50	Sirius	α CMa
7	*Canis· nπor*	20. 6	16.M 10	6.S 9	Procyon	α CMi
8	*Cor· leonis*	23.♌ 6	0.S 10	14.S 3	Regulus	α Leo
9	*Spica· virgı̄*	17.♎ 16	2.S 0 †	8.M 26	Spica	α Vir
10	*Bootes*	17. 36	31.S 30	21.S 45	Arcturus	α Boo
11	*Vultr^r cade̅*	7.♑ 56	62.S 0	38.S 36	Vega	α Lyr
12	*Vultr^r vola̅*	24. 26	29.S 10	7.S 19	Altair	α Aql

† Should read: 2.M 0. Cole was probably thinking of S for South.

on the top is a Tudor rose surrounded by fleurs-de-lis and other motifs.

Provenance

The Cottrell-Dormer family, Rousham Park, Oxfordshire, by descent. Bought by the Science Museum, London, 1984.

Literature

Higgins (1950); Christie's South Kensington, sale 13 December 1984, lot 219; Vaughan (1993); Ackermann (1998, pp. 42–3).

Location

The Science Museum, London (1985–100).

14 Altitude sundial 1574

Signed: *Humfridus Côle Anglus*; dated: 1574

Gilt brass; horizontal plate 75 × 75

This dial is similar in form to **13**, but is two dials in one, being reversible. It is designed for the latitudes 52° and 55°. The descriptive wording is in Latin, not English, and Cole has

13 Case.

Latinized his name and appended his nationality by adding Anglus. One may presume that this dial was made for a Continental customer, and it would be appropriate to sites in North Germany or Denmark. See also Cole's astrolabe of 1582, **23**.

Side 1

The rectangular plate is engraved with a large and small set of hour lines as described before (**13**). In a corner the latitude is noted:

G	M
52	0

The labels read: *Altitudo Solis*; *Horæ Diei*; *Horæ planetarum*. The roundel for the governors of the hours is labelled: *Dominia planetarum*. Outside the hour lines is a table giving the entry of the Sun into the Zodiac: *Introitus · Solis · in · 12 · Signa*. The isosceles right-angled triangular gnomon hinged to this side bears Cole's signature and the date. The plummet is missing from its scallop-edged aperture.

14 Side 2.

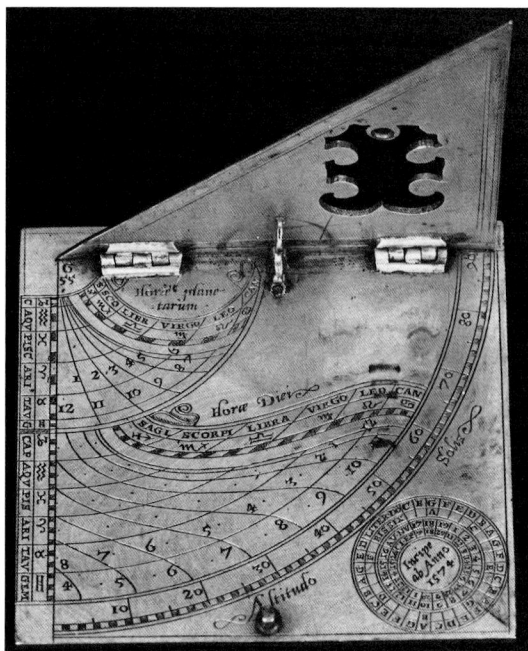

14 Side 1.

Side 2

The plate on the reverse is engraved in the same manner as before (**13**). The differences are: the latitude (G/55); the absence of a table of the Sun's entry in the Zodiac; a perpetual calendar substituted for the governors of the hours. This last is named: *Incipe ab Anno 1574·*, and it begins: LITER· DŌ / BIS SIX / AV· NV / CI· EP. (Dominical Letter, leap year, Golden Number, Epact). A similar circular table beginning with 1575 is on the ring, or poke, dial, **21**.

Literature

Hausmann (1979, p. 54), no. 5; Vaughan (1993, pp. 191–200); Ackermann (1998, p. 41).

Location

Staatliche Museen Preußischer Kulturbesitz, Kunstgewerbmuseum, Berlin (K 4670).

15 Simple theodolite 1574

Signed and dated: ✦ H ✦ Côle · 1574 ✦

Brass; diameter 170; thickness 2

Although this instrument has appeared many times in the literature as an illustration of an altazimuth theodolite, it was originally intended to be a simple theodolite. The altitude semicircle and its supporting brackets over the compass are a modern reconstruction. Cole produced a simple theodolite, and in this form it will be described. For an earlier simple theodolite by Cole, albeit small, see the compendium dated 1569, **9**.

The azimuth circle is made of a solid piece of brass. Around the rim is engraved a 360° scale divided in units alternately shaded, each further divided to half a degree. At 10° intervals are two sequences of numbers, one from 0 to 360°, the other 0 to 90° in four quarters. There follows a geometric square (in appearance a double shadow square), which Cole has

15 Azimuth circle.

labelled at the four sides: ✦ *The* ✦ *Geometricall* ✦ *Square* ✦; the capitals have elaborate flourishes. The squares are divided from centres to corners in two ways, one to 12, in units, the other to 60, in units alternately shaded. Inside the square is the circle of 32 wind directions, each direction marked by initials.

Pivoted at the centre is the alidade, with ogee curved ends. Each arm has at one edge a set of marks to provide five divisions subdivided in threes, from the mid-point of the lines forming the shadow squares and their corners. The alidade is provided with two turn-buttons to secure the pair of lugs attached to a band that fits over the compass. This band appears to be part of the modern reconstruction that has been added. Originally, such a band could have held a sighting bar similar to a plain table alidade. The magnetic compass in its circular box is a push-fit into a keyed hole at the centre. No degrees are marked, nor are the four cardinal points. The declination, at 12° E, is indicated by an engraved pointer exactly resembling the magnetic needle. The glass cover appears to be original.

Underneath the instrument is a brass tube on a trilobed bracket that attaches to the plate by three winged screws. This enables the instrument to be fitted on a pole or tripod. Engraved on the underside is the name: ✦ *Edward* ✦ *Williams* ✦. The writing is in Cole's hand; it is likely that this is the owner's name.

The modern altitude semicircle is anachronistic; the decoration of rays is on one side only, and the scrolled brackets and the shape of the sights are atypical. As is bound to be the case, the divisions and numbers are not a good match to those of Cole. This reconstruction was made to the order of G.H.

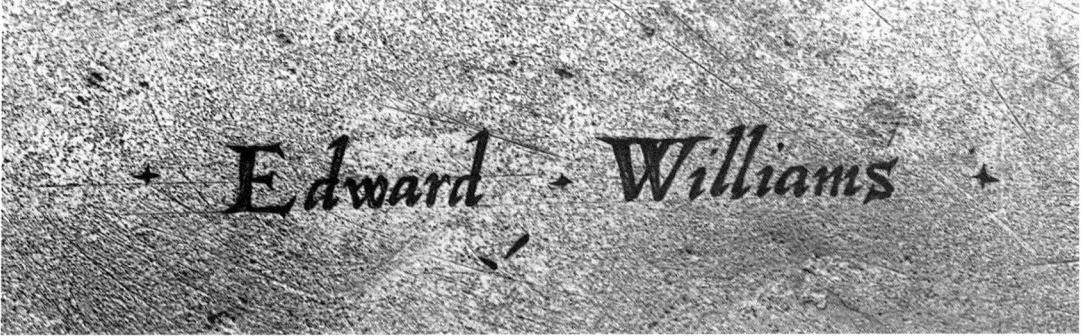

15 Name on the underside.

Gabb, from whom the National Maritime Museum acquired the instrument. The records say: 'When discovered the semi-circle was missing. This was restored and the instrument as it now stands represents the earliest known theodolite.'

Provenance

F.3 in the G.H. Gabb Collection, purchased by the Museum in 1937.

Literature

Waters (1958, p. xxii), plate LVIa.

Location

National Maritime Museum, Greenwich (NAV1448; old no. SI/T.1).

16 Astrolabe 1574

Signed and dated: *Made by Humfrey Cole 1574*

Also engraved (after 1610): A D / *Henr: Princ: Magn Brittan*

Brass, gilded; diameter 88; thickness of limb 6

This very small instrument was called by Gunther (1927) a 'Three and a half inch astrolabe', and is a finely made device suitable for a gift. The front shows the limb,

rete, and a choice of plates. On the back is a universal stereographic projection, the *Saphea*, complete with rule, cursor, and brachiolum.

16 Front.

Throne and limb

The astrolabe is suspended from a ring over a cylinder supported by a pair of scrolls. The cylinder is hollowed out (diameter 12) to take a magnetic compass covered by a mica disc held in place by a split ring, all protected by a swing-out disc that is engraved with an armillary sphere, as is the reverse of the cylinder. The limb is engraved with two circles, the outer of 24 hours, marked in roman numerals I–XII twice, and the inner divided into single degrees, shaded in blocks of two, and numbered every ten. The numbering is from the East, 0°–360°.

Mater

The mater is occupied by a nautical quadrant labelled at the top: *QVADRATV NAVTICV*. This has the 32 wind directions marked with initials

(e.g. N.W.B.W.). There is also a scale description along the top and left side, while the signature is to the right. At the bottom is the additional engraving, as above. This is not in the hand of Cole, who was dead before the prince was born in 1594 (see below). In view of the care Cole took with making his nautical and geometrical squares symmetrical, it is clear that whatever was in the bottom segment (descriptive words or a dedication) has been rubbed down to provide space for the dedication. This was engraved between guide lines, but the craftsman did not allow for the width, and the last word is cramped.

Back

On the back of the astrolabe is the universal projection known as the *Saphea*, favoured and promoted by Gemma Frisius. It is a horizontal projection from the equinoctial points onto the

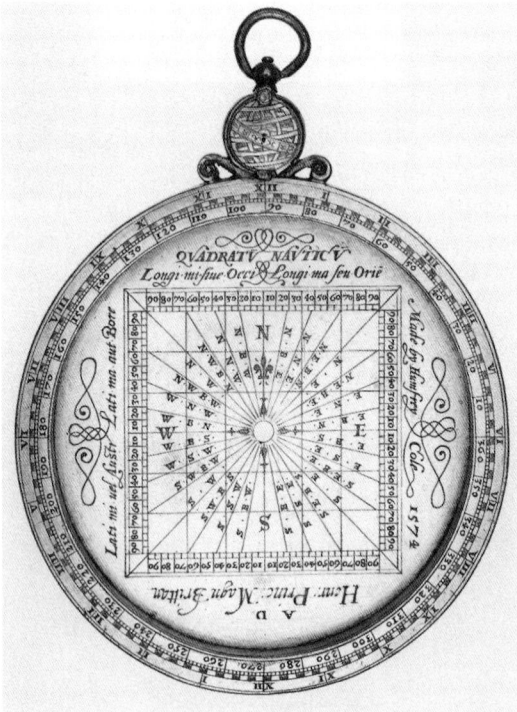

16 Mater.

16 Back.

solstitial colure, or great circle passing through the Poles and the solstitial points. The centre of the projection is crossed by the ecliptic, which is divided into the Zodiac. With this projection, the first half of the year is on the same line as the second half, which is why each section of 30° has two Zodiac sigils. The line is divided in 2° intervals, alternately shaded. The outer edge is in 360°, double units alternately shaded, numbered every 10° in four quadrants. The tropics are each divided into 12 hours.

Because of the lack of space, the seven stars on the universal projection are indicated by asterisms each with a letter, A–G. These stars are identified, with others not on the projection (H–M), by the star names engraved around the outer rim of the instrument. Front view, beginning to the right of the suspension:

A. *Aquila.* ♂. ♃
B. *Lanceatr.* ♃. ♂
C. *Cor.* ♌. ♂. ♃
D. *Oculus.* ♉. ♂
E. *Cani⁹. mar.* ♃. ♂
F. *Spica.* ♍. ♀. ♂
G. *Cor.* ♏. ♃. ♂
H. *Lucida.hyd.* ♄. ♀
I. *Arcturu⁹.* ♃. ♂
K. *Vmbilic⁹. Andr.* ♀
L. *Orion.* ♂. ☿
M. *Crus.* ♒. ☿. ♄

Note: B, Lanceator, and I, Arcturus, refer to the same star, α Boötis.

Rule and cursor

For taking coordinates, a rule pivots on a central pin, and along the rule moves the cursor: *Cursor · Brachiolum*, to the tip of which is attached a brachiolum, or jointed arm. For

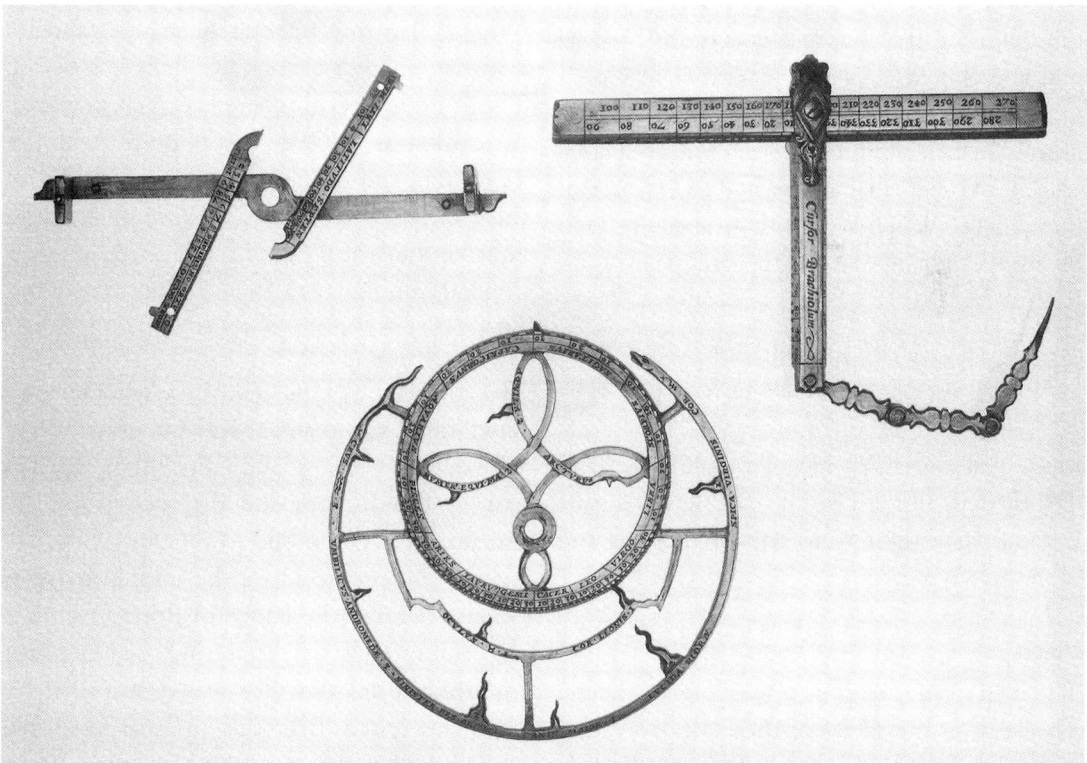

16 Rete, alidade, cursor.

the use of this projection, see Saunders (1984). The rule is graduated from the centre 0°–90°, and then in reverse to the other end, 90°–270°; then 270°–360° back at the centre. The cursor is divided to the left 0–90°, and to the right 0–100 units; with both scales the division is in double units, alternately shaded.

Rete

The rete is cut for 12 stars, named with their magnitudes:

VMBILICVS · ANDROMEDÆ · 3	SPICA · VIRGINIS 1
OCVLVS · ♉ · 1	ARCTVRV⁹
SINISTER · PES · ORIONIS 1	COR · ♏
CANIS · MAIOR · 1	AQVILA · 2
LVCIDA · HIDRÆ 2	CRVS · ♒ · 3
COR · LEONIS · 1	HVMER⁹ · EQVI · MA 2

Alidade

On the front, the alidade pivots on the central pin, and is held in place by another of Cole's ingenious devices. The pin has a circular groove, which is gripped by the pincer-shaped ends of two bars that are pegged to the base of the alidade. When closed these bars are held by studs, and the alidade appears to be solid. Each is divided:

a) *LATITVDO · SEPTEN*, above a scale 0–90°; and (inverted) *LATI · MERI*, below a scale 0–20°.

b) *HO · ORTV⁹ · ☉ ·*, left of a scale 1–12; and (inverted) *HORÆ · OCCA · ☉ ·*, left of a scale 1–12. The sight vanes are of the peg and pinhole type.

Plates

The mater can hold four plates below the rete. The lugs on the edges of the plates are offset to the left, as is the locating slit in the limb at the top of the mater. The offset lug is easier to position accurately than the old symmetrical lug. The plates are cut as follows:

a) latitudes 51° 30′ / 52° 30′

b) latitudes 53° 40′ / 55° 00′

c) tablet of horizons / calendar–Zodiac, equal–unequal hour, and shadow square diagrams

d) plate plain on both sides, roughly finished

The latitude plates have almucantars every 5°, numbered every ten; azimuths are every 10°, all numbered. On the horizon plate, *Horizontale · catholicum*, the circular lines are every 5° numbered in tens. With this plate the latitude can be found from the time of sunrise or sunset; see Saunders (1984, p. 15). The reverse of the horizon plate is engraved with the typical back of a conventional astrolabe. The outer, concentric scales, are the Zodiac, with 2° divisions alternately shaded, and the calendar, in 2-day intervals, shaded as before. The months are numbered as well as named (December is 12). The first point of Aries is at 11 March, and of Libra 14 September. Within these circles is a conversion diagram between equal and unequal hours, and a shadow square divided in units to 12.

Case

The astrolabe is contained in a wooden case, lined in red velvet and covered in green velvet, with silver fastenings and embellishments. The case is fitted, with the protrusions on the instrument accommodated on the inside, and hidden and protected on the outside by silver covers. Because the badge of Prince Henry at the centre, and the lettering of the Latin inscriptions are inexpert in execution, it has been suggested that the original case was embellished in the nineteenth century. There is, however, nothing anachronistic about the

16 Case.

of a case for it if it had none, or its decoration if one existed, are reasonable. It is possible that the decorative features on the case were the work of one of his teachers, or even of the Prince himself.

Provenance

Part of the collection of Ralph Bernal (*c.* 1785–1854), art collector, and a Member of Parliament from 1818 to 1852, which was sold in 1855 in 4294 lots realizing £71,000 (*DNB*). Bought by the British Museum in 1855.

Literature

Gunther (1927, pp. 278–80); Gunther (1932, pp. 484–7), no. 306; Ward (1981, p. 116), no. 336; Saunders (1984, pp. 16, 24, 98); Ackermann (1998, pp. 32–3).

Location

The British Museum, London (1855, 12–1.223) IC 306.

case, which could well be original, or have been provided at the time the astrolabe came into the Prince's possession. The instrument was made in 1574, 20 years before Prince Henry was born. He was designated Prince of Wales in 1610, and died in 1612. He is known to have been interested in natural philosophy, under the influence of his tutor, Sir Thomas Chaloner, and the acquisition of a small astrolabe for educational purposes, and the provision

17 Surveyor's folding rule 1574

Signed: ✦ *Humfrey · Côle · 1574* ✦

Brass; radius (excluding decorative dolphin heads) 305; width of arms 35; thickness *c.* 5

The rule is pierced with four threaded holes, marked: A, B on the underside of one arm, and C, D on the upper side of the other, for fixing sights. The hole at the joint can attach the rule to a tripod.

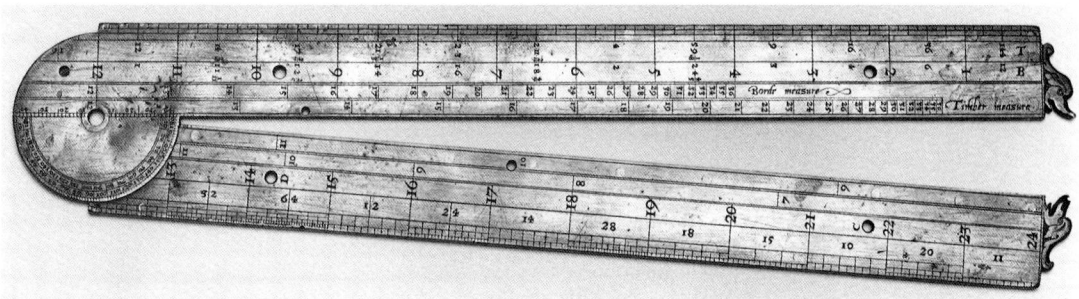

17 Side 1.

Side 1

a) *Timber measure*, for estimating the volume of a tree or balk of timber given the cross-section area;

b) *Borde measure*, for estimating the area of a board or plank given the width;

c) an inch scale, numbered from 1 to 24, the inches divided on one arm to eighths and on the other arm into the following parts: 16, 32, 64, 12, 24, 14, 28, 18, 15, 10, 20, 11. This choice of scales is remarked on further in the description of the alidade, **32**;

d) at the joint is a degree scale on the upper surface, divided to 2° intervals and marked 0–180° both ways. The joint is locked at 90° or 180° by inserting a pin in holes.

The inside edge of each arm is divided 500–0 (at centre of joint) to 500 in two-unit divisions, so making 1000 parts in 24 inches. This 1000-part scale is used with the arms locked at 180°, forming a plane table alidade as mentioned by Digges (**32**).

Side 2

This bears the signature and date, with a quadrant, or clinometer, to 65° by ones, and a geometrical quadrant, both scales for use with the joint locked at right angles. These scales are labelled: *Quadrans; Quadratum Geometricum: Latus Vmbra Versa; Latus Vmbra Recta*. The quadrant, used as a clinometer, requires a plumb-line attached to a peg that fits the hole at the edge of the arm by the bottom of the solar altitude scale. Beyond the clinometer scale are two tables, for lengths and areas. The first reads:

> *3 Barly cornes, to an ynche · 12· ynches a foote · 3· foote a yarde · 5· yards and ·1/2· a pearche: 40· pearches in lengthe and · 4· in breadthe, an acre. So an acre contayneth · 160· pearches: the halfe acre · 80· pearches a roode · 40· pearches* ✚

The next table gives the length and breadth, in perches, which multiplied together come to one acre in area. (A perch measures 16.5 feet, or 5 metres.)

One arm has a solar altitude dial described as: *The houres of the daye and what Signe the sonne is in at all tymes*. The altitude hour scale resembles a cylinder dial or horary quadrant, and the rod gnomon is to be inserted into a small hole 67 mm from the centre of the hinge. The vertical height of the engraved dial to the noon mark at the summer solstice is

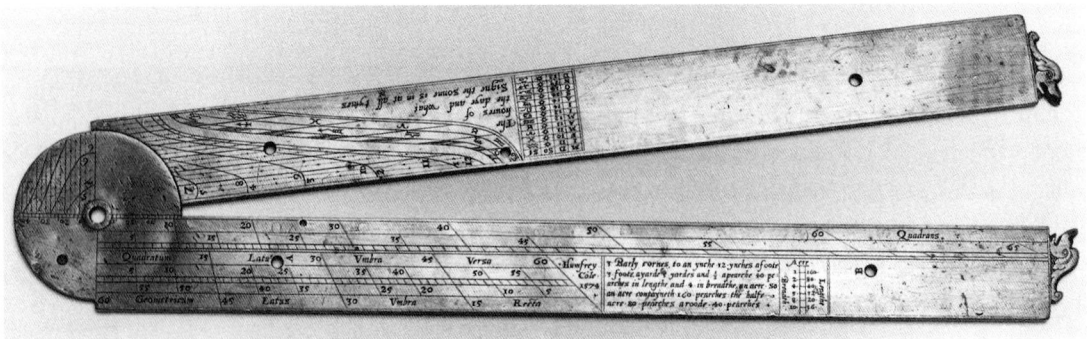

17 Side 2.

125 mm, which shows, geometrically, that the dial was made for a latitude of $51\frac{1}{2}°$. Below is a table giving the date of the Sun's entry in the Signs of the Zodiac.

M	D	So	Si
I	11	☉	♒
F	10	☉	♓
M	11	☉	♈
A	11	☉	♉
M	12	☉	♊
I	12	☉	♌
A	14	☉	♍
S	14	☉	♎
O	14	☉	♏
N	13	☉	♐
D	12	☉	♑

Provenance

Sotheby's, New Bond Street, London, sale 12 June 1984, lot 329.

Literature

Vaughan (1993); Ackermann (1998, pp. 74–5).

Location

The Science Museum, London (1984–742).

18 Surveyor's folding rule 1574

Signed: *Humfrey · Côle · 1574 ·*

Brass; radius (excluding scrolls) 305; width of arms 33; thickness *c.* 4.5

The arms are made from three brass bars riveted together. About 30 rivets are used in each arm. The ends are decorated with small scrolls. The rule is pierced with four threaded holes, marked: A, B, •, ••, on side 1, and ••••, •••, C, D, on side 2, for fixing sights (none

present). The hole at the joint can attach the rule to a tripod.

Side 1

This rule is the same as that described under **17**, with only very minor variations in the sequence of subdivisions of the inch scale. The parts are: 16, 32, 64, 6, 12, 24, 7, 14, 28, 10, 15, 20. The inside edges of the arms comprise a 1000-part rule as before.

Side 2

This bears the signature and date, with a quadrant, or clinometer, to 63° by ones, and a shadow square, both scales for use with the joint locked at right angles. These scales are labelled as on **17**, and similarly there is a table of length definitions and area measure in acres. At the joint is a diagram to show the angles of polygons with various numbers of sides: 3, 4, 5, 6, 10, arranged inside a protractor.

Provenance

Purchased from a private collector in 1986.

Literature

Ackermann (1998, pp. 72–3), fig. 29 shows side 2.

Location

Whipple Museum of the History of Science, Cambridge (Wh: 3199).

19 Surveyor's folding rule 1575

Signed: ✦ *Humfrey · Côle · 1575* ✦

Brass; radius 305; width of each arm 31.5; thickness 4.9 to 5.5

The arms are made from four brass bars riveted together. About 30 rivets are used in each arm. The rule is pierced with four straight holes, marked: A, B, C, D, for attaching sights; the hole at the joint can fix the rule to a tripod.

Side 1

On this upper side the four scales are as described under **17**, with a minor variation into the divisions of the inch: 16, 32, 64, 36, 24, 42, 28, 18, 15, 30, 22, 26. The inside edges form the 1000-part scale.

Side 2

This bears the signature and date, with a quadrant, or clinometer, to 66° by ones, and a shadow square, both scales for use with the joint locked at right angles, as described in **17**.

Cole has made a few errors. On the timber scale at 27, the 7 is over a rubbed-out 6, while on the plotting scale at 36, the 6 is over a 2.

Literature

G. Turner (1983, p. 103), plate 51, both sides; Ackermann (1998, p. 76).

Location

Museum of the History of Science, Oxford (50–41; 49631).

20 Surveyor's folding rule 1575

Signed: ✳ *Humfrey · Côle · 1575* ✳

Brass; overall length fully opened 633; radius (excluding decorative dolphin heads) 305; width of arms 35; thickness *c.* 5

The rule is composed of four sheets of brass giving a total thickness of about 5 mm. The construction uses some 60 rivets to hold the sheets together. Decorative dolphin heads are at the ends of the arms, as on **17**.

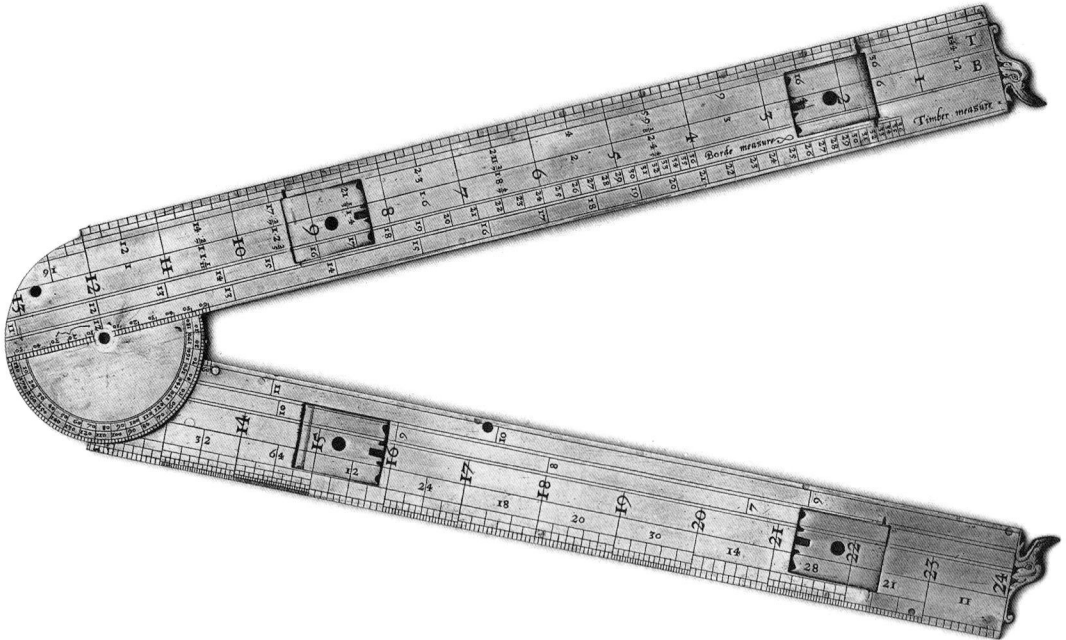

20 Side 1.

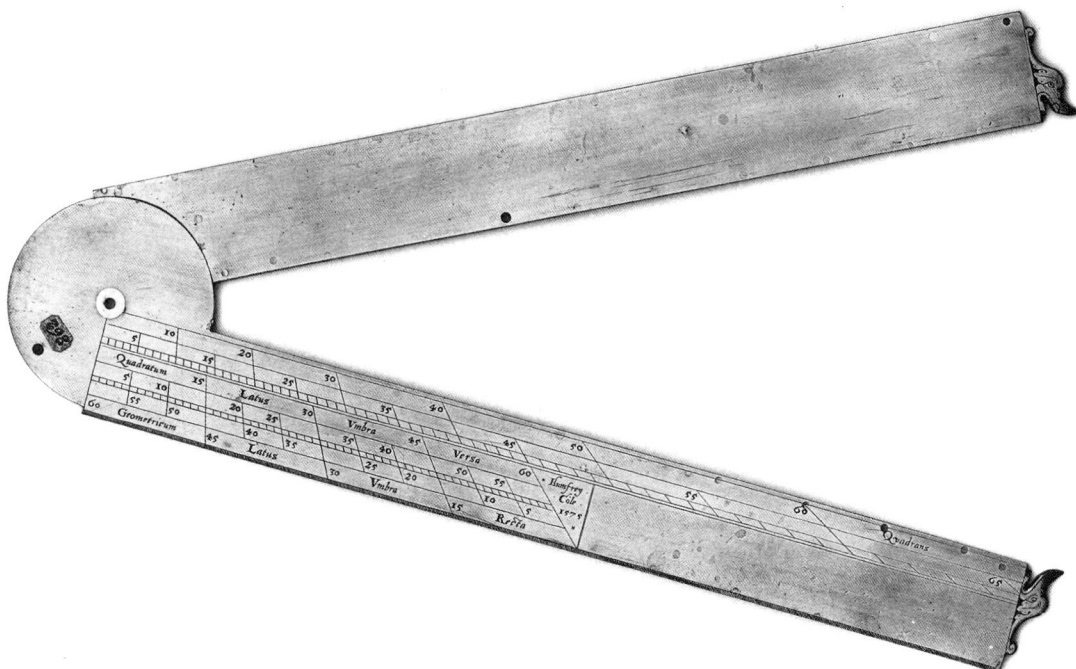

20 Side 2.

The scales on side 1 are the same as those described on the rule **17**, with just two changes in the parts of an inch: 30 and 21 substituted for 15 and 10. The scales on the inside edges and on side 2 are identical with those on **19**.

This particular rule is superior to others through having hinged, inset sights, two in each arm, that are flush with the surface when closed. Each sight is provided with a slit and a hole.

Provenance

One of the instruments taken to Florence by Sir Robert Dudley in 1606.

Literature

G. Turner (1991, p. 202), colour plate; Ackermann (1998, p. 77).

Location

Museo di Storia della Scienza, Florence (2527).

21 **Ring dial** 1575

Signed: *H. Côle*; dated at beginning of perpetual calendar 1575

Brass, gilded; diameter 74, width 22.5, thickness 4

The dial is supported by a stirrup with a triangular knife-edge to locate in a groove inside the ring. Beside this groove are two circular tables with calendrical information.

The first table reads: THIS TABLE · BEGINETH · AT A · 1575 · AND · SO · FOR · EVER. The surrounding two circles are labelled: G· NVBE, and EPACT. These tables begin with 18 | 19 | 1 …, and 18 | 29 | 11 …, both correct for the starting year of 1575. There are the required 19 sections before the sequence repeats in accordance with the Julian, or Old Style, calendar. Cole failed to engrave the 1 for the Epact 15, so it now reads 5.

21

site curve is read with respect to the latitude of use by a scale 50°–56°, and for solar declination by a Zodiac scale, the divisions marked by sigils.

Between these two scales on the inner curve, opposite the calendrical circles, is a table for the Sun's entry into the 12 Signs of the Zodiac. The four columns are headed: M [month; D [day]; SO [Sun]; Si [Sign]. This table is identical with the one on Cole's surveyor's folding rule, **17**.

On the outside of the ring are four fixed bands and three movable that alternate with the fixed; the movable are for planetary aspects, Moon's age, and for setting the pin-hole, with which the time and Sun's altitude can be found. The information on each band is as follows:

a) Fixed. Holy days.

b) Movable. Planetary aspects; crescent boss as a 'handle'.

c) Fixed. Zodiac, each 30°.

d) Movable. Moon's age, to 29^1/$_2$; full moon boss as 'handle'.

e) Fixed. Calendar months.

f) Movable. Pin-hole; towns and their latitudes.

g) Fixed. Thin rim to keep the bands in place. Outer edge continues the towns and latitudes.

a) The holy days of the Church are arranged on the band in 12 sections. The months are read from the third fixed band (given here by their initial letters); the day, and then the holy day in an abbreviated form, as on Cole's compendia.

The second circular table has the same wording at the centre as the previous one. The surrounding two circles give the Dominical Letters, the inner circle for the leap years, but there is no label. There are 28 sections, beginning with B, correct for 1575, and the years to 1602 Old Style.

These circles are followed on one side of the inner curve by hour lines for the seven latitudes from 50° to 56°, which cover the whole of England from Plymouth to Berwick (56° 50′ according to Cole). It is these lines that receive the spot of sunlight from the pin-hole in the opposite curve. This side has a slit over which moves a band containing the pin-hole. The spot of light received on the oppo-

a

[I] 1· CIRCV·	6 EPIPH	13 HILA	25 C̄O· P̄A
[F] 2· PVRIFI MARI	14 VALE	25 MAT	
[M] 12· GREGORI	25 ANVN: MAR		
[A] 23· GEORGE·	25· MARCVS·		
[M] 1· PHILLI· IACO·	6· IOAN· EVAN·		
[I] 11· BARN	24· IO· BAP	29· PETER· PAVL	
[I] 6· DOGE· BEGIN	22· MAG·	25· IĀ· APO	
[A] 1· PE· VIN·	17 DOG· EN	24· BART	29· DEC· IO
[S] 8· NATI· MARI	21 MAT·	29· MICH	
[O] 13· EDWARD CON	18· LVC	28· SI· IVD	
[N] 1· OMNI· SAN	2· OM· AN·	30· ANDR	
[D] 6· NICO	8· CON· MA	21: THO·	25· NAT· D̄O

f

Douer 51· 0 ✳}	*Exiter* 51· 0 ✳	*London* 51· 34 ✳	*Oxforde* 51· 50 ✳
Bristo 51· 20 ✳	*Northampton* 52 · 50 ✳	*Norwiche* 52 · 30 ✳	*Harford* 52 · 50 ✳
Leiseter 52· 50 ✳	*Notingame* 53 · ✳		

g

West Chester 53 · 10 ✳	*Linckolne* 53 · 15 ✳	*Yorke* 54 · 0 ✳
Newe Castell 55 · 0 ✳	*Barwick* 56 · 0 ✳	*Saulisburie* 51 · 15 ✳
Winchester 51 · 0 ✳	*Antwerpia* 51 · 28 ✳	*Louanium* 50 · 58 ✳

Note: The month initials, in brackets, are not in the band of holy days, but on the month band, e.

b) Beginning at the crescent boss, the conjunctions are named, with their sigils.

SEXTILE ✳ | QVADRINE ☐ | TRINALL ◁ | OPPOSITION ∞ | TRINALL ◁ | QUADRINE ☐ | SEXTILE ✳ | CONIVNCTION ♀ |

f) Beginning at the pin-hole: see above.

g) Continued on edge of band: see above.

Provenance

Presented to the British Museum in 1905 by C.J. Wertheimer.

Literature

Gunther (1927, pp. 291–3); Ward (1981, pp. 125–6), no. 161; Ackermann (1998, pp. 45–6).

Location

The British Museum, London (1905, 6–8.1).

22 Gunner's folding rule 1575

Signed: ✚ *H Côle* ✚ 1575

Brass; overall length of arms 187; width of arms 16; thickness 4.5

This instrument is similar in many ways to **10**, and some of the information engraved on it is virtually the same. It can be used closed to

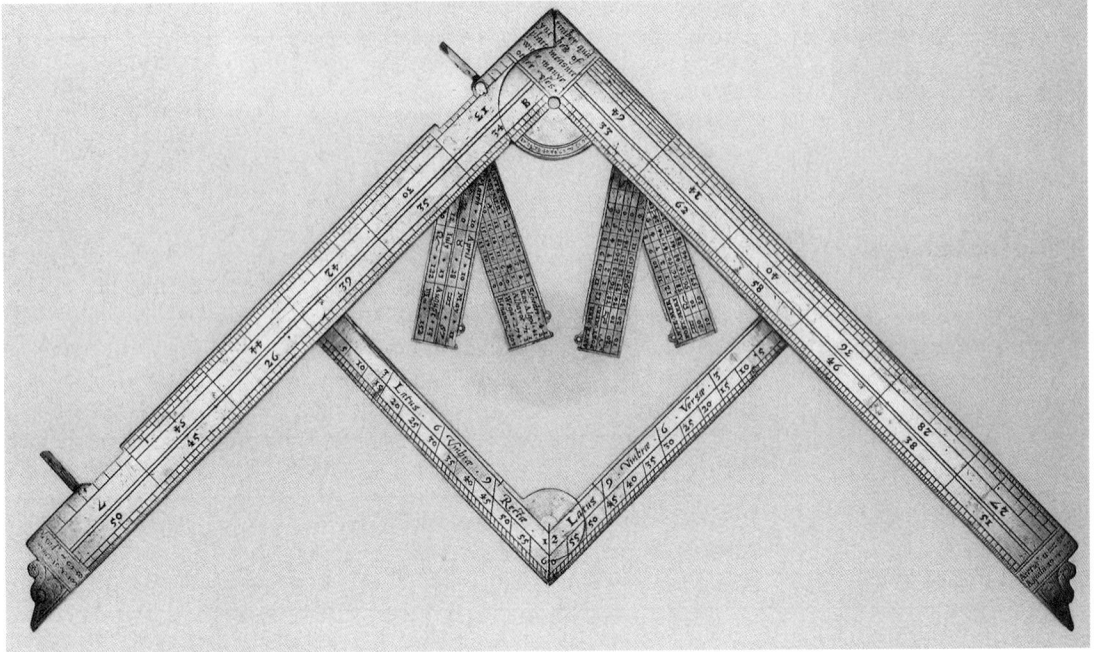

22 Side 1, leaves A1, B1, C1, D1.

read tables that cross both arms, and can be opened out to form a geometrical square. In the upper part of both arms are slits that contain pull-out leaves that carry a variety of information. The signature is along an outer edge, as is Cole's explanation of the use of the device: *This Instrument is a rule a square a peare of Compasses A quadrant to knowe ye howres heightes and distances of any thinge ye shotinge of Ordenaunce to measure.*

Side 1

At the joint of the two arms is a rectangle containing the words: *Timber and ynches of platt measure withe manye other vses.* A quarter-circle part of the joint has a degree scale 0–90° in units of two, and numbered in tens. Both arms have two runs of 6 inches divided into a variety of parts for drawing maps to a desired

scale. For this usage, see the plane-table alidade **32**. Beginning at the left arm, the divisions of an inch are into: 7, 54, 44, 42, 30, 13, 50, 45, 26, 39, 35, 34; right arm, 27, 28, 36, 40, 24, 64, 51, 38, 46, 58, 62, 33. At the joint end of the left arm is the letter B, close to a small hole at the centre of the joint. Two sight vanes, each with a pin hole, are hinged in the edge and lift up for use.

When the arms are fully open they form a right angle, and a pair of jointed arms pull out from slots in the main arms to form a shadow square. These arms are labelled: *Latus Vmbræ Rectæ*, and *Latus Vmbræ Versæ*. The scales on each arm are in units to 60 at the apex, numbered in fives. The divisions are also numbered in threes to 12 at the apex. On the reverse, the scale division is to 90° both ways in degree units numbered in tens. This square

146

is to be read by a plummet (missing) pegged in the hole at the centre joint, the sighting of the target being through a pair of folding vanes let into the edge of the left arm.

The ends of the arms finish with triangular points, immediately above which are two lines giving the Right Ascension for four stars. These numbers are reasonable, but the choice of Zodiac sigil is curious, as is the preceding number. Modern designations added in **22**a.

Side 2

At the joint is a roundel with four concentric circles labelled at the centre: *The entraunc of γᵉ soñe int the 12 signes.* The 12 segments give the month, day, ☉, and Zodiac sigil. The dates differ somewhat from some other similar tables by Cole, and especially with a similar table elsewhere on this instrument, see **22**b.

The gunnery table is announced on the edge after the signature: *The haightes and waightes of Ordenance.* It spreads across the upper part of both arms when closed. This table is closely similar to that on **10**, but with one extra row (scores at point blank, i.e., score (20) of paces, a pace being 5 ft), and one column (the heaviest cannon, Basilisco). The table on the present rule is virtually identical to that published in Holinshed (1587); **22**c.

> Howe Maney pounde euery Shote wayeth
> Maney Scores at pointe blancke
> Maney lli of powder euery pece shot
> Many Inches hye the bullet is
> Hie γᵉ pece is in the mouthe
> Moche euery pece wayethe
> The Names of Ordenaunce

For similar tables, see **11** and Chapter 4, 'Gunnery tables'.

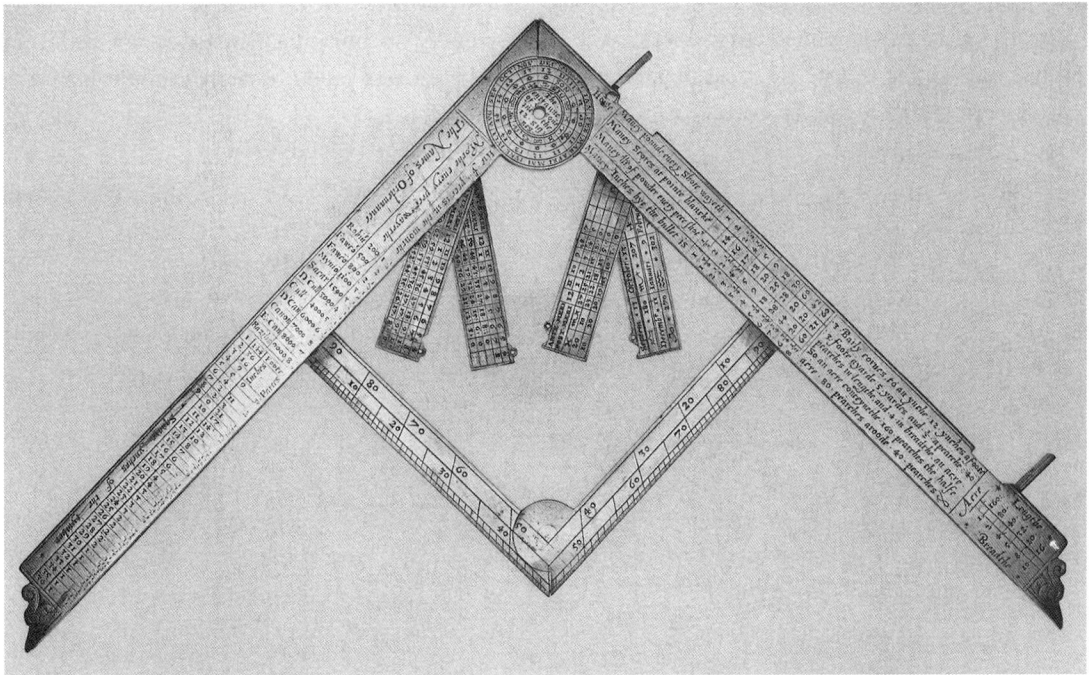

22 Side 2, leaves D2, C2, B2, A2.

22a.

*Pleyad*ˢ· 7 · ♋ · 50 [η Tau]	*Arctur*⁹· 11· ♊ · 200 [α Boo]
Algomey· 15· ♈ · 109 [β CMi]	*Aquila*· 19· ♈ · 290 [α Aql]

22b.

IAN 11	FEB 10	MAR 11	APRI 11	MAI 12	IVNE 12
IVLI 14	AUG 14	SEP 14	OCT 14	NOV 13	DEC 12

The lower parts contain on the left a timber scale, labelled: ✳ *Square ynches of the tymber* ✳, that is the dimensions to give a square foot (144 square inches). On the left is a table defining lengths from three barley corns on; this is identical to tables on the surveyor's folding rules (for example **17**). There follows the small table for the dimensions in perches to make an acre, which is also on Cole's surveyor's rules.

The four leaves

With the arms fully opened, the leaves can be pulled out of their slots by a small hole in a raised part of the edge of each leaf. Viewed from side 1, the leaves will be designated A1, B1, C1, D1, and viewed from side 2 the reverse of these leaves will be designated A2, B2, C2, D2.

A1, A2

The Sun's entry in the Zodiac, with the month, day, and Right Ascension are in **22**d. Cole often includes a table of this kind on his instruments, and it is usual to find an inconsistency in one or two of the dates. Here, however, all but one (June 12) are different from the dates given in the roundel on side 2, see **22**a.

22c.

Name of gun	Gun weight	Gun ø	Shot ø	Powder weight	Point blank	Shot weight
*Robine*ᵗ	200	$1\frac{1}{4}$	1	$\frac{1}{2}$		1
*Fawcō*ᵗ	500	2	$1\frac{3}{4}$	2	14	2
Fawcō	800	$2\frac{1}{2}$	$2\frac{1}{4}$	$2\frac{1}{2}$	16	$2\frac{1}{2}$
Myniō	1100	$3\frac{1}{4}$	3	$4\frac{1}{2}$	17	$4\frac{1}{2}$
Sacar	1500	$3\frac{1}{2}$	$3\frac{1}{4}$	5	18	5
D Cull	3000	$4\frac{1}{4}$	4	9	20	9
Cull	4000	$5\frac{1}{2}$	$5\frac{1}{4}$	18	25	18
D Cañ	6000	$6\frac{1}{2}$	$6\frac{1}{4}$	28	28	30
Canon	7000	8	$7\frac{3}{4}$	44	20	60
E Can	8000	7	$6\frac{3}{4}$	20	20	42
Bazilis	9000	$8\frac{3}{4}$	$8\frac{1}{2}$	60	21	60

22d.

March ✻ 10	April ✻ 10	May ✻ 11	Iune ✻ 12	Iuly ✻ 13	August ✻ 12
♈ ✻ 0	♉ ✻ 28	♊ ✻ 57	♋ ✻ 90	♌ ✻ 122	♍ ✻ 152
Septem ✻ 13	October ✻ 13	Novemb ✻ 12	Decembr ✻ 11	Ianuari ✻ 9	Februar ✻ 8
♎ ✻ 180	♏ ✻ 207	♐ ✻ [2]37	♑ ✻ 270	♒ ✻ 302	♓ ✻ 332

22e.

ANTE	MER	12	11	10	9	8	7	6
POST	MERI		1	2	3	4	5	6
♈	♎	38 :	36 :	32 :	26 '	18 '	9 "	0
20	10	34 :	33 '	28 :	22 :	14 :	6 '	0
10	20	30 :	29 :	25 ·	19 :	11 :	3	0

22f.

Cole's name	Cole's RA	Modern name	Designation
Scheder	4½	Schedar	α Cas
Ras· Algol	41	Algol	β Per
Alhayot	71½	Capella (Alhajoth)	α Aur
Wega	274	Vega	α Lyr
Benanaz	203	Alkaid (Benetnash)	η UMa

The other three leaves contain, over both their sides, tables of the solar altitude every 10 days and for every hour of the day. The time of year is indicated by sigils of the Zodiac, in appropriate pairs as required. Capricorn is on its own line below Aquarius/ Sagittarius. These tables are in the form of **22**e.

Calculations show that the symbols after some numerals represent tens of minutes of arc.

Minutes	0–10	'
	10–20	"
	20–30	·
	30–40	:
	40–50	:'
	50–60	:"

N.B. The symbol " may occur as one mark above the other.

The end of leaf B1 is finished with a table giving the Right Ascension for five stars. **22**f. Allowing for precession, and his approximations to half a degree, Cole's values are good.

Provenance
Bought by the British Museum in 1912.

Literature
Gunther (1927, pp. 296–9); Ward (1981, p. 105, no. 310); Ackermann (1998, pp. 78–80).

Location
The British Museum, London (1912, 11–1.1).

23 Astrolabe 1575

Signed and dated: *Humfridus· Côle· Londinensis· hoc· instrumentum fabricauit· 21 die Maÿ· Aº· Dñi· 1575* ✻

Brass; diameter 610; thickness of limb 9.5

The largest and finest of English astrolabes, this is in many ways Cole's masterpiece. It has been in the possession of the University of St Andrews for over two centuries, although it is not known how or when it arrived there. There is space for three plates, with the rete and alidade above. At present there are two plates with the instrument, one by Cole and one by Marke. The plate for 52° is engraved by Cole; it is backed by a tablet of horizons. The other plate was made some 100 years later. It is one sided, the projection is for 56° 25′, and it is signed: *Iohn Marke fecit*. John Marke was a mathematical instrument maker whose premises were in the Strand, near Somerset House, and he is known to have been working between 1665 and 1673 (Clifton 1995, p. 179). By contemporary standards the latitude of the substitute plate is correct for St Andrews; it may, therefore, be supposed that the astrolabe had reached Scotland *c.* 1670. Cole's signature and descriptions are in Latin, which suggests a Continental customer, and this supposition is given weight when comparison is made with the two azimuth sundials of 1574. One, **14**, in silver, has English for the signature and descriptions: it is cut for the latitude 51° 30′. The other, **13**, gilded, uses Latin for descriptions and the signature: the two latitudes provided are 52° and 55°; Cole here uniquely adds 'Anglus' after his name. On the astrolabe he makes the same point through 'Londinensis'. In view of Cole's practice, it is possible that both the astrolabe and the sundial (dated 1574) were made for Continental customers.

Throne

Attached at the top is a semicircular extension with a shackle and ring, which is of square section except for the rounded inner surface.

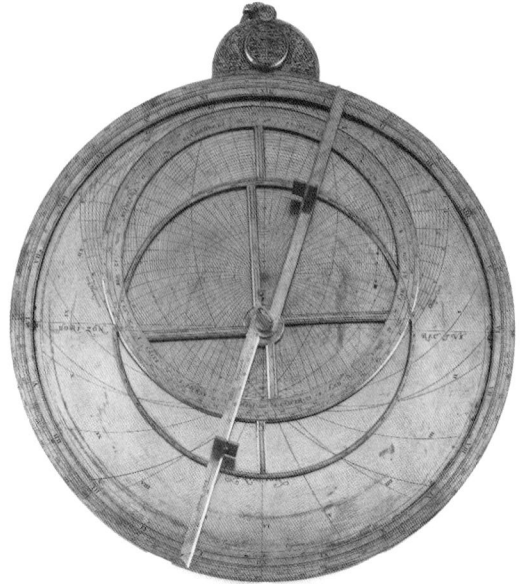

23 Front.

Set into the brass is a small compass box (inner diameter 48) with a line (length 43) to mark the direction of magnetic north, simulating a compass needle with an arrow at one end and a double S at the other. The needle and cover glass are missing. A hinged cover protects the

23 Throne.

compass, and the outside is engraved with a complete armillary sphere. The space around the compass has a foliate decoration. The same motifs are on the reverse of the throne.

Limb

The brass annulus forming the limb is made up of four curved strips of brass in three sections spliced in the regions of 80°, 200°, and 320°. It is fixed to the back plate by 24 rivets. There are two carefully divided scales. The first is for 0–360° (from the East) subdivided to 10 minutes of arc, labelled in two bands one for the tens the other for the fives. The second scale is for time, twice XII hours in roman numerals, subdivisions to one minute.

Mater

The entire space is occupied by a *QVADRATVM NAVTICVM*, each side divided at the edge 90–0–90 in single units. Above the upper side is the description:

Longitudo minor siue Occidentalior / Longitudo maior seu Orientalior. At the left side the inscription reads: *Latitudo minor uel Australior / Latitudo maior aut Borealior*. At the right side are the signature and date (transcribed above). From the centre radiate the 32 wind directions, or rhumbs, each marked with the initials of the English names. Near the centre the cardinal directions are named: *Septentrio, Oriens, Meridies, Occidens*. Towards the sides 12 winds are named in Latin, Greek, and Italian, and the four on the diagonals in Latin and Italian. For transcriptions, see Ackermann (1998, p. 35). The design is that found in the many Gemma Frisius editions of *Cosmographia*. See also the Thomas Gemini astrolabes, **6**, **7**.

Back

The edge is divided to 10 minutes of arc, and the degrees are marked every five in quadrants, 0–90°. Within this band is a universal

23 Mater.

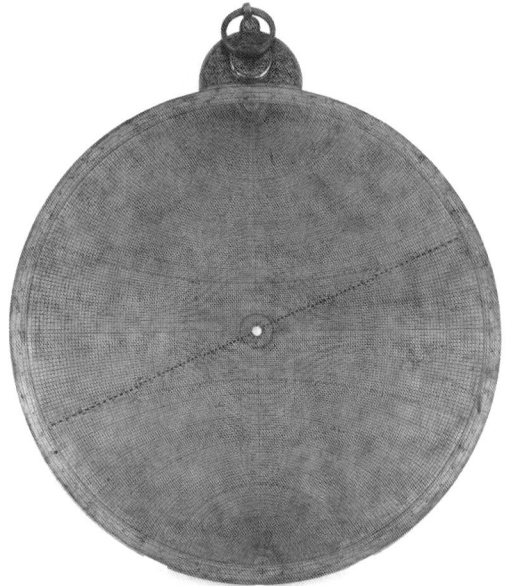

23 Back.

Star positions on the back of the astrolabe

As engraved	Designation	RA		Declination		Dekker calculation	
		° '		° '		°	°
Cap Medusæ	β Per	40 30		39 50		40.3	39.3
Extre Eridan	θ Eri	43 00		−40 40			
Hircus	α Aur	71 40		45 10		71.4	45.1
Sinis pes Orio	α Ori	73 10		−09 20		73.5	−09.1
Dex hu Ori	α Ori	83 20		06 20		83.2	06.3
Canopus	α Car	93 30		−51 40			
Canis Ma	α CMa	97 00		−15 50		96.9	−5.9
Apollo	α Gem	106 50		32 20			
Canis Mi	α CMi	109 30		06 10		109.4	06.0
Hercules	β Gem	110 10		28 30			
Cor ♌	α Leo	146 00		13 40		145.7	14.0
Cauda ♌	β Leo	171 20		16 40		171.3	16.6
Spica ♍	α Vir	196 00		−08 50		195.40	−08.8
Benan	η UMa	202 10		51 10		202.1	51.2
Arcturus	α Boo	209 10		22 00		209.0	22.0
Luci Coronæ	α CrB	228 50		28 30		228.7	28.5
Cor ♏	α Sco	240 50		−24 50		240.7	−24.9
Lu Lyræ	α Lyr	275 00		38 40		275.0	38.6
Vult Volans	α Aqu	291 40		07 30		291.6	07.5
Fomahand	α PsA	338 30		−33 30			

The star names underlined are engraved upside down.

Right Ascension and declination scales are divided to 1°, and estimated here to 10'. I am grateful to Dr A.D.C. Simpson for recording the star positions on the astrolabe. The values for epoch 1550 were calculated by Dr Elly Dekker and published in G. Turner (1994, p. 353). They have here been rounded to one tenth of a degree. These values are sufficiently close to show that Cole was using the same epoch.

projection, the *Saphea*. This is two stereographic projections from the equinoctial points onto the North–South plane known as the solstitial colure (see Saunders, 1984, p. 56). The use of this old projection was furthered by Gemma Frisius of Louvain, in his book *De Astrolabio catholico et usu ejusdem* (Antwerp, 1556). The coordinate lines are drawn with considerable care in 1° intervals. Every 10° the lines are emphasized by arrow heads, and smaller marks show the 5° lines. The tropics and polar circles are emphasized by wavy lines. The line of the ecliptic across the centre is divided in degrees alternately shaded, and numbered in fives. Each group of 30° is counterchanged to show the beginning and end of each Sign of the Zodiac. Twenty stars are pinpointed by asterisms and are named. Because of the projection of the heavenly sphere onto a plane, to facilitate interpretation, the star names are inverted in quadrants 2 and 3 (90°–270°). Both the regula and cursor are missing.

Rete

The rete has no star pointers. It comprises the ecliptic circle (diameter 385) and the equatorial circle (diameter 350). The latter is formed from two arcs soldered to the underside of the ecliptic circle. A cross frame, counterchanged, holds the parts together, including the central pivot point. Additional splicing and insets can be seen from the underside. The ecliptic circle is divided into the Signs, with full names and sigils, each numbered from 1 at Aries. The bevelled edge is divided to half a degree, the degrees numbered in fives to 30°. The equatorial circle is also divided to half a degree, numbered in tens from the First Point of Aries, 0–360°.

The counterchanged alidade (length 610, width 16, thickness 3) is used on the front of this astrolabe because sights are not possible when a regula is on the back. The sights are of the slit and pin-hole type. One arm is blank; from the other the declination can be read. The scale units are to 1°, labelled in tens, running North to 80° and South to $23\frac{1}{2}°$. The sections are named: *Latitudo Septentrionalis* and *Latitudo Meridionalis*. The pin is a replacement.

Plate

Of Cole's manufacture, there is only the single plate (diameter 553; thickness 1.5). On one side is a stereographic projection for *Latitudo* $\cdot 52 \cdot | \cdot o \cdot$, while on the reverse is a tablet of horizons. The azimuths are every 5°, and the almucantars every 1°. Below *Tropicus Cancri* are the 12 divisions into unequal, or planetary, hours, labelled in roman numerals. Across the whole face are the 12 divisions into the Houses of Heaven, labelled in arabic numerals. Other lines are: *ÆQVATOR, Tropicvs Capricorni, HORIZON RECTVS*. There is no twilight line. The tablet of horizons is engraved with 90 arcs at 1° intervals from 0–90°, labelled every

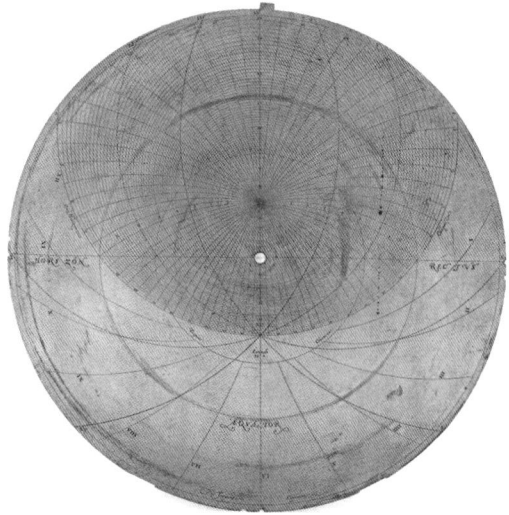

23 Plate for latitude 52°.

10°, the lines hatched every 5°. The diameters are labelled: *HORIZON RECTVS* and *Linea Meridiana*; at the bottom is: *Tropicus Capricorni*. The 90° circle is marked in two places: *Equinoctialis*. With this plate it is possible to find the time of sunrise and sunset at all latitudes. Alternatively, the latitude can be determined

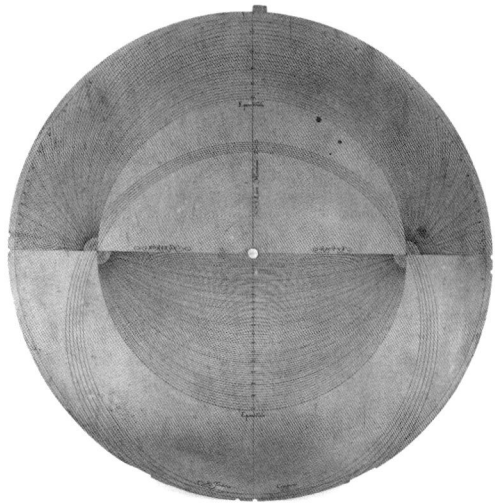

23 Plate with horizons.

from the time of the Sun's rising or setting on a given day (Saunders 1984, pp. 15–17). There are six other circles with centres on the meridian line between 47° and 52°, which are not labelled, and whose use is not known.

The plate has a lug offset to the right, which is common for the large Renaissance astrolabes, giving better accuracy. To keep the plate securely in position, at the bottom and both sides, are circular catches (diameter 8). These are inset into the limb and are pinned so they may be pulled out to press down on the plate. A very small extension, with a hole, is to the side of each catch, and the plate is notched so it may be easily inserted over the extensions, which are used to extract the catches from their sockets (Fig. 2.7). There are two notches at each position because of the offset lug at the top of the plate, the left side of which is always in line with the meridian line on each side of the plate. Gunther (1927, p. 277) made this comment:

> To secure the position of the plates their margins are notched for engaging in three hinged clips on the inside of the rim of the mother. This arrangement is believed to be unique: it is one of several examples of the ingenuity of Humphrey Cole in introducing into his instruments some useful constructional detail, for which a modern maker would take out a patent.

Provenance

In Gunther's paper, read on 17 June 1926 to the Society of Antiquaries (Gunther 1927), he referred to: 'The recent discovery of the finest known example of an English Astrolabe of the largest size'. He continued: 'It seems to have been in the possession of the University of St. Andrews for the past two or three centuries.' The existence of a new plate for the latitude of

56° 25′ by John Marke, who was working between 1665 and 1673, suggests that the astrolabe was in Scotland at that time, acquired by James Gregory during his 1673 stay in London. Marke is known to have supplied various instruments for Gregory. It is obvious that Marke made his plate specifically to fit Cole's astrolabe, and he has done his best to imitate Cole's style of lettering. Gregory may have refined the latitude by the time he wrote in a letter dated St Andrews 30 April 1674: 'the latitude here is 56°: 22′ (Turnbull 1939, p. 281).

On Gregory's life, and the instruments at St Andrews, see Turnbull (1939, *passim*). James Gregory (1638–75) was elected the first Professor of Mathematics at St Andrews in 1668, and he was provided with a room in the library that also served as an observatory. A university commission on 10 June 1673 permitted Gregory to build an observatory and equip it with instruments, and he went on a purchasing trip to London later in 1673. A professorship at Edinburgh University persuaded Gregory to leave in 1674. His new observatory was not properly maintained, and it was pulled down for road widening in 1846. A few items from Gregory's period in the library building remain, for example three Joseph Knibb clocks. It could be that the Cole instruments were given to Gregory and were neglected by those who came after him, until about 1926 when Gunther described them. Nothing further can be said about the instrument's origins. See also Cole's nautical hemisphere, **30**.

Literature

Gunther (1926, p. 293); Gunther (1927, pp. 273–8); Gunther (1932, pp. 488–92); Wray (1984); Macdonald and Morrison-Low (1994, p. 33); Ackermann (1998, pp. 33–5)

Location

University Collections, University of St Andrews, St Andrews, Scotland. IC 307.

24 Compendium 1575

Signed and dated: ✱ *Humfrey* ✚ *Côle 1575* ✱

Brass, gilded; diameter 55, thickness 22

Comprising from the top:

a) nocturnal, calendar, and Zodiac scales

b) table of latitudes of 40 European towns

c) equinoctial sundial

d) magnetic compass

e) perpetual calendar and table of fixed feast days

f) engraved motto on outer rims

g) suspension bracket

h) leather case

a) Around the central hole of the nocturnal are the words: ✱ THE ✚ POLE ✚ STERRE ✱, and through the hole the observer lines the instrument on the Pole and turns an arm to line with *Kochab* (β UMi) in the Little Bear. Below is a disc divided into hours, from 3 am to 9 pm, each hour division notched for identification by the fingers during the hours of darkness. At the 12 o'clock position is a short index, with a Sun's face, to set the disc according to the day of the year. The disc is engraved: *The· nighte· houres·* Finally, there is a small volvelle with a replacement index bearing a crescent Moon. The scales that follow are the calendar, with 20 April at the top and 22 October at the bottom, and the names of the months written out; the Zodiac, with the ordinal number, name, and sigil for the Sign. The scales for both days and degrees

24 a.

24 b.

are divided to intervals of two, alternately shaded, but the divisions vary in width. The Sun enters Aries on 11 March.

b) The inside of the lid has an innermost circle engraved, between clear guide lines: *The latitudes of all the principall townes and cites of evrope.* The 40 towns are arranged in eight panels, beginning at the hinge:

W· Chester	53	10	Francopho	50	12
Linckoll	53	15	Cracouia	51	0
Yorke	54	0	Dantisc̄	55	0
N· Cast	55	0	Lubecn̄	54	48
Barwi	56	50	Emdc̄	53	42
Edenboro	57	0	Constanti	43	0
Cōpostel	44	20	Florentia	43	4
Lisbonᵃ	39	38	Neapolis	41	0
Barsal	41	35	Venetia	44	50
Nātes	48	12	Roma	41	50
Parisius	47	55	Douer	51	0
Marsilia	43	6	Exiter	51	0
Tolosa	43	30	London	51	34
Louani	50	58	Oxford	51	50
Rone	49	0	Bristo	51	0
Antwerpia	51	28	Northampt	52	50
Bruxella	51	0	Norwiche	52	30
Brugæ	51	30	Harford	52	50
Ganda	51	24	Leisetr	52	50
Tořac̄	51	40	Noting	53	0

c) Attached to the same hinge as the lid is the arc supporting the hour circle on which is engraved the signature and date. This circle is divided on the upper side to twice 12 hours in roman numbers, with half-hour divisions, alternately shaded, while the lower side is divided into one set of 12 hours for the winter half of the year. On the meridian line a bar projects towards the centre on which is hinged the polar gnomon, with a quadrant attached to the lower end; this is divided in 2° of latitude. The dial is set at the correct angle for use by running the edge of the quadrant through a slot in the vertical mounting past a pair of knife-edges against which is set the appropriate degree of latitude. The small hole at the apex of the quadrant is for hanging a plummet (missing).

d) The bottom section of the instrument contains a magnetic compass, with what appears to be the original magnetic needle (cross and point ends). The cover is a disc of mica. The compass card (diameter 51) is of printed paper, and is type A-1 (see Chapter 4). The 32 directions are indicated by pointed ribbons that are hand-painted in burgundy red, apple green,

I	1 Circn̄	6 Epiph	13 Hill	25 Co Pau
F	2 Puri Ma	16 Valē	25 Mat Ap	
M	12 Grego	25 Annū Ma		
A	4 Ambro	23 George	25 Marcus	
M	1 Phi Iac	6 Iō Euā	26 Augu	
I	11 Barn	24 Iō Ba	29 Pe Pa	
I	2 Visi Ma	22 Ma Mag	23 Iā Ap	
A	1 Pe Vī	10 Lau	24 Bar	29 De Io
S	8 Nat Ma	21 Mat	29 Mic	
O	13 Edwa Cō	18 Lu	28 Sī·Iude	
N	1 Om Sā	2 Om Anī	30 Andr	
D	6 Ni	8 Cō Ma	21 Thō	25 Na

and deep blue. Some of the lettering and edging is in gold paint. The edge of the card has been trimmed. The declination is shown by marks on the brass rim above the cover, but this is now loose and rotates.

e) The outer three circles on the base list feast days, and the inner four provide a perpetual calendar. At the centre the starting date is given: THIS TABLE BEGIÑETH AT 1572 & SO FOR EVER. The circles are named: DO LE | LE Y (Dominical Letter, leap year); PRIM | EPA (Prime [or Golden Number], Epact). They begin with E/F (1572 is a leap year), with 28 sections, the Prime with 15, and Epact with 15, both with 19 sections.

The fixed feast days of the church are arranged in three circles in four quarters. The months are shown by their initial letters, followed by the day and then the feast day in an abbreviated form.

f) Engraved around the outer rim is the motto:

✷ AS TIME AND HOVRES PASITH AWAYE ✦ SO DOETH THE LIFE OF MAN DECAY ✷ AS TIME CAN BE REDEMED WITH NO COSTE ✦ BESTOW IT WELL AND LET NO HOWR BE LOST ✷

This motto is the same as inscribed on **25**, also dated 1575 (Gatty, all editions from 1872). It may be compared with the Cole compendia dated 1579, **26, 27**, where the mottos are in Latin.

g) The suspension bracket is in the form of a sphere like an orb, held by a pair of S-shaped supports. A small ring fixed to the top of the sphere has a free ring passing through it. The same construction is used on **25**.

h) A brown leather fitted case (length 75, width 69, depth 30), the interior lined with burgundy velvet, contains the instrument. Top and bottom are gold tooled using a punch with a very small square floral motif, and the rims are edged by gold tooling with a pomegranate motif.

Provenance

Octavius Morgan Bequest to the British Museum, 1888. Charles Octavius Swinnerton Morgan (1803–88), was a Member of Parliament from 1841 to 1874, and an antiquary.

Literature

Gunther (1927, pp. 289–91), figs 34, 35; Ward (1981, pp. 125–6), no. 360; Ackermann (1998, pp. 59–61).

Location

The British Museum, London (1888, 12–1.293).

24 e.

24 h.

25 Compendium 1575

Signed and dated: ✳ *Humfrey* ✦ *Côle* ✦ 1575 ✳

Brass, gilded; diameter 54, thickness 20

Comprising from the top:

a) nocturnal, calendar, and Zodiac scales

b) table of latitudes of 40 European towns

c) equinoctial sundial

d) magnetic compass

e) perpetual calendar and table of fixed feast days

f) engraved motto on outer rims

g) suspension bracket

This compendium has many parts that are identical to the one made in the same year, **24**. Consequently, the description of that one serves this. Some minor variations will be noted.

25

25 Compass rose.

Plate 14(a). The 32 points have ribbons hand-painted in pale grey, grey-blue, and burgundy. Some lettering and lining is in gold paint. The edge of the card is divided and numbered in ink every 10° in quadrants 0–90°.

e) The outer three circles on the base list feast days, and the inner four provide a perpetual calendar. All are identical to the engraving on **24**, with only very minor variations in the spelling of two feast days.

f) Engraved around the outer rim is the motto:

✳ AS TIME AND HOVRES PASITH AWAYE [HINGE] SO DOETH THE LIFE OF MAN DECAYE ✳ AS TIME CAN BE REDEMED WITH NO COST [] BESTOW IT WELL AND LET NO HOWR BE LOST.

The motto is the same as that on **24**. This dial is the only source stated for the motto in Gatty (1872), and all later editions.

g) The suspension bracket is a match to that on **24**.

a) The lunar index that would have had a crescent on it has broken off; its stub is visible.

b) The inside of the lid has an innermost ring engraved: ✳ *The latitudes of all the principall townes and cities of evrope* ✳. The 40 towns are arranged in eight panels in the same order and sequence as on **24**. There are only very minor variations in the spelling of six towns; the latitude figures are the same.

c) The equinoctial dial, with signature and date, is the same as that on **24**.

d) The bottom section of the instrument contains a magnetic compass, with what appear to be the original magnetic needle (arrow head and double S ends) and original glass cover, which has numerous air-bubbles, some elongated to 1 mm in length. Such bubbles are to be expected in glass of this date. The compass card is of paper, and is type A-1. See Chapter 4, and

Provenance

Presented to the Society of Antiquaries of Scotland on 4 November 1783 by Joseph Edmondson, Esq., Moubray Herald. *Synopsis of the Museum of the Society of Antiquaries of Scotland* (Edinburgh, 1849), p. 113, no. 31. *Catalogue of Antiquities in the Museum of the Society of Antiquaries of Scotland* (Edinburgh, 1863), p. 76, K77.

Literature

Gatty (1872, p. 6), no. 16; Gunther (1927, pp. 289–91), fig. 33.

Location

National Museums of Scotland (NL 18).

26 Compendium 1579

Signed and dated: ✦ *Humfrey* ✦ *Côle* ✦ *1579* ✦

Brass, gilded; diameter 56, overall thickness 20

26 c, a, f. (See also Plate 8.)

Comprising from the top:

a) planisphere

b) table of latitudes of 32 European towns

c) hinged plate: nocturnal; the Sun's entry in the Signs

d) equinoctial sundial

e) magnetic compass

f) perpetual calendar and table of fixed feast days

g) motto around rim

a) The top is inscribed: *Hora· planitarū.* Around the rim is a degree scale, divided in 1° intervals from 0 to 90° in four quarters. The upper semicircle has an unequal (planetary) hour diagram, and below is a shadow square, labelled: *Vmbra / Versa*; *Vmbra* ✦ *Recta*. This scale is divided in units to 12 at the corners. An alidade pivots at the centre. It was originally fitted with a pair of folding sights, but these have been removed and the spaces filled by pieces of brass. At one end is a cylindrical stub with a central hole piercing the arm.

b) The purpose of this table is engraved in a circle at the centre: ✳ *The · names · of · sites · and · townes · of · vrope* [flourish]. In four circles are the names of 32 towns.

Douer	51	0	*Norwiche*	52	30
Exiter	51	0	*Harforde*	52	50
Londō	51	30	*Lesetur*	52	50
Oxfor	51	50	*Bristo*	52	0
Northhamtū	52	50	*Nucastell*	55	0
Notingā	53	0	*Barwick*	56	50
W.chestr	53	10	*Edenbor*	57	0
Linco	53	15	*Yorke*	54	0
Composte	44	20	*Parisius*	47	55
Lisbona	39	38	*Marsili*	43	6
Barsal	41	35	*Louani*	50	58

Rone	49 0	*Tolosa*	43 30
Anwarpe	51 28	*Dantis*	55 0
Bruge	51 30	*Emdem*	53 42
Cracoui	51 0	*Veniti*	44 50
Praga	50 4	*Colon*	51 0

c) Hinged to the lower part of the compendium is a nocturnal, which is swung out for use. Round the central hole for sighting *Polaris* is the identification: ✦ *The • pole ⊾ star* ✦. This is on a rotating disc from which extends the arm marked: INDEX POLARIS. The alignment is with the star *Kochab* (β UMi) in the Little Bear. Below this is a volvelle divided into two 12-hour periods, each hour division notched for identification by the fingers during the hours of darkness. At a 12 o'clock mark is a short index to set the volvelle to the day of the year. Finally, there is a calendar scale, with 20 April at the top and 20 October at the bottom.

On the reverse of the nocturnal is a table connecting the date of the entry of the Sun in the Zodiac. At the centre is the explanation and another signature: *A table· for· the· soñes· entraun in· to· anye* [of] *the· 12· Signes ~ · H* ✦ *Côle* ✦. This is surrounded by four circles giving the initial letter of each month, the day of the month on which the Sun enters a division of the Zodiac, the symbol for the Sun (☉), and the Zodiac denoted by its conventional sigils. The dates are the same as those tabulated under **17**.

d) When the lid of the compendium is lifted the semicircular bracket holding the chapter ring of the equinoctial dial is raised. The latitude quadrant, with its extension bar, is hinged to the ring to lie flat when not in use. To tell the time the quadrant is turned, so bringing the gnomon into alignment with the polar axis. The latitude marked on the quadrant is set by the quadrant's position in a slit. The scale is in 2° units, from 0 to 90°. The upper side of the chapter ring is divided to twice 12 hours, subdivided in half hours, alternately shaded. The under side is marked from VI to VI for use during the winter. This side bears the full signature. The plummet is missing, as in all the compendia.

e) The compass has an elaborate card of the A-1 type (see Chapter 4). The rather short magnetic needle has a point at one end and a double cross at the other. The glass cover is present.

I	1 *Circn*	6 *Epiph*	13 *Hill*		25 *Cōn· Pau*
F	2 *Puri· Ma*	16 *Vali*	24 *Mat· Apo*		
M	12 *Grigo·*	25 *Annũ· Mary*			
A	4 *Ambro·*	23 *George*	25 *Marcus*		
M	1 *Phi· Iac*	6 *Iō· Euā*	26 *Augus*		
I	11 *Barn*	24 *Iō· Ba*	29 *Pē· Pa*		
I	2 *Visi· Ma·*	22 *Ma· Mag·*	25 *Iā· Apo·*		
A	1 *Pe· V· in·*	24 *Barth·*	29 *Dec· Ione*		
S	8 *Nat· Ma·*	21 *Mat·*	29 *Mich*		
O	18 *Lucke· Euā·*	28 *Simon· Iude*			
N	1 *Al saintes*	30 *Andro· Apos·*			
D	6 *Ni*	8 *Cō ma*	21 *Thō· ap·*		25 *Na*

f) The base has engraved at the centre: THIS TABLE BEGINITH AT A 1579 AN SO FOR EVER. Above this are four circles in two pairs. They are labelled: DOMI L [Dominical Letter], LEP Y [leap year], PRIM [Prime, the same as the Golden Number], EPAC [Epact]. This lower pair of circles occupy a lunar cycle of 19 years, while the pair of circles above cover a complete Solar Cycle of 28 years from 1579 to 1606 inclusive. The Dominical Letter for 1579 is D, and the Prime Number and Epact are both 3, as shown on this plate. The outermost three circles, each divided into four parts, list the initials of the months and the days of the month for 35 holy days. This listing is similar, but not identical, to those on **24** and **25**.

g) The outer edge of each half of the case bears two mottos:

27

1) TEMPORA ✦ MVTANTVR ✦ [HINGE] ✦ & ✦ NOS [†] ✦ MVTAMVR ✦ IN ✦ ILLIS

(Times change and we change with them.)

2) SICVT ✦ VMBRA ✦ DIES [HINGE] ✦ NOSTRI ✦ SVPER ✦ TERRAM ✦

(As a shadow are our days on Earth. I Chr. 29:15.)

† The form given here ('& nos') is in Raphael Holinshed, *The Chronicles of England* (1577), fol. 99b. The proper form has 'nos et', and is from the compilation of Matthew Borbonius, a sixteenth-century poet, *c.* 1540 (Benham 1924, col. 669b, note; Gatty 1889, p. 303).

The central outer rim of the case has an acanthus leaf decoration. The suspension is a ring held in a small fixed ring that is supported by two plant leaves.

Provenance
Christie's South Kensington, sale 29 March 1990, lot 231.

Literature
G. Turner (1990, pp. 126–7), colour plates; Ackermann (1998, pp. 64–6).

Location
Private collection.

27 Compendium 1579

Signed and dated on the equinoctial dial:
✳ *Humfrey* ✦ *Côle* ✦ *1579* ✳

Brass, with modern gilding; includes modern parts. Diameter 53; overall thickness 23

Comprising from the top:

a) modern lid

b) planisphere

c) table of latitudes of 32 European towns

d) hinged plate: nocturnal

e) hinged plate: governors of the hours

f) equinoctial sundial

g) missing magnetic compass replaced by modern disc

h) perpetual calendar and table of fixed feast days

j) motto around rim

a) The modern outer disc to the lid is engraved with flowers; the inner side has anachronistic grotesque decoration.

b) The lid covers a circular plate, which resembles the back of an astrolabe. Round the edge is a degree scale divided to 1° and numbered 0–90° in four quarters. The upper semicircle has an unequal 12-hour diagram, and in the middle is engraved: *Horæ planetarū*. The lower semicircle has a shadow square: *Vumbra · / · Versa ·* | *· Vmbra ✦ Recta ·*. Pivoted at the centre is a modern replacement alidade, with two hinged sighting vanes. The engraving on the arm is purely decorative.

c) The next plate, backing the planisphere, is named on the innermost circle: *The · names · of · cites · and · tounes · of · vrope.* In four circles are the names of 32 towns.

Londō	51	34	Cōpostel	44	20
Exiter	51	0	Lisbōa	39	38
Douer	51	0	Brsalo	41	35
Oxfo	51	50	Rone	49	0
Norwich	52	30	Parisius	47	55
Harfor	52	50	Marsil	43	6
Leset	52	50	Louāi	50	58
Bris	52	0	Tolos	43	30
Norhātū	52	50	Anwarp	51	28
Noting	53	0	Bruge	51	30

W· ches	53	10	Craco	51	0
Lincō	53	15	Prag	50	4
N· castell	55	0	Dantis	55	0
Barwic	56	50	Emdem	53	42
Edēbo	57	0	Venit	44	50
York	54	0	Colō	51	0

d) On a supplementary hinge is the nocturnal that is swung out for use. The central sighting hole now has a screw inserted. Around the centre is the label: THE ✦ POVL ✦ ANTAR. This is on a disc with an index arm engraved: · *Index · Polaris ·*. The volvelle below is divided into two 12-hour periods, each hour division with a notch for identification by the fingers during hours of darkness. A longer pointer is at one 12-hour position to set the volvelle to the time of year against the calendar scale. At the top is 20 April, and at the bottom, the handle position on a nocturnal, is 20 October, near the hinge.

e) On the reverse of the nocturnal is a circular table showing the planets that govern each hour of the day. An index arm (a replacement) on a small central disc points to the planets, the sigils for which are engraved in two circles. The third, inner circle is in seven sections, one for each day of the week, denoted by the word *Dies* and the sigils for the governing planet for the first hour of that day, thus ☉ indicates Sunday, and so on. The purpose of this table is defined on another circle: *Tabula gubernationis planitarum.*

f) The equinoctial sundial is the same as that on the other Cole compendium, **26**, with the same signature and date, 1579.

g) The magnetic compass has been stripped out, and the space is taken by one modern brass disc. It is heavily chiselled in an attempt to produce a decoration resembling the grotesque.

h) On the base, the fixed feast days of the Church are arranged in three circles in four quarters. The months are shown by their initial letters, followed by the day and then the feast day in an abbreviated form. Note that compared with the other 1579 compendium, **26**, 14 February is correct, 13 October and 2 November are added, and 6 December is omitted.

The table to find Easter is the same as that on Cole's other compendium of 1579, **26**. The centre has the label: THIS BEGINES AT A 1579 & SO FOR EVER.

I	1 *Circū*	6 *Epiph*	13 *Hill*	25 *Cō· Pa*
F	2 *Pur· Ma*	14 *Valē*	24 *Ma· Ap*	
M	12 *Grigo·*	25 *Ann̄· Ma*		
A	4 *Ambro·*	23 *Georg*	25 *Marcu'*	
M	1 *Phi· Iac*	6 *Iō· Euā*	26 *Aug*	
I	11 *Bar*	24 *Iō· ba*	29 *Pe· Pa*	
I	2 *Visi· Ma·*	22 *Ma· Mag·*	25 *Iac· Ap·*	
A	1 *Pe· V· in·*	24 *Bart·*	29 *De· Iō*	
S	8 *Na· Ma·*	21 *Mat·*	29 *Mi*	
O	13 *Edwa*	18 *Luc*	28 *Simō· Iud*	
N	1 *Om sa*	2 *Om an*	30 *Andr·*	
D	8 *Cō ma*	21 *Tho·*	25 *Nat*	

j) The outer edge of each half of the case is inscribed with a motto:

1) ✦ NON ✦ QVAM ✦ DIV [HINGE] ✦ SED ✦ QVAM ✦ BENE ✦

(NOT HOW LONG BUT HOW WELL. SENECA, EP. 101.)

2) ✦ MORA ✦ NVLLA ✦ [HINGE] FVGACIBVS ✦ HORIS ✦

(There is no delay for the fleeting hours.)

The suspension bracket is in the form of a sphere like an orb, held by a pair of S-shaped supports. A small ring fixed to the top of the sphere has a free ring passing through. This suspension resembles those on the two compendia of 1575, **24, 25**.

Provenance

Bequeathed by Max Elskamp to the city of Liège in 1932.

Literature

Michel (1966, p. 62), plate 26; Michel (1974, pp. 47–8).

Location

Musée de la Vie Wallonne, Liège (505).

28 Horizontal sundial 1579

Signed and dated: ✦ *H* ✦ *Côle* ✦ *1579* ✦

Also engraved: ✦ Sur ✦ Henry ✦ darcy ✦

Brass; 180 × 181; thickness 1.3 to 1.5

The rectangular plate is now black with age. On the chapter ring the roman hour numbers are contained in a circular band. At the corners, around the fixing holes, are the ends of a rectangular frame marked with a double line. The hours are numbered IIII to VIII, and are divided to half-hour intervals, alternately shaded.

At the centre is a circular band from which radiate eight triangular pointers to the cardinal points, at the ends of which are engraved the initial letters of the wind directions. The gnomon angle is 52°, and the hour angles engraved on the plate show that they were calculated for the same angle. The corners of the plate show many graffiti that are clearly old: for example, HK, TW, HKA.

The thin, gnomon has lugs that fit into a pair of slots in the plate, and hold by bending the lugs flush with the underside. The gnomon is supported upright by a triangle of

28

The rectangular plate, dark with age, has the hour circle contained between bands; the edges are engraved with a double line. At the corners are the fixing holes. The hours are numbered IIII–VIII, and are divided to half-hour intervals. The gnomon angle is about 52°–53°, and the hour angles engraved on the plate show that they were calculated for the same angle. The gnomon is thin with a bevelled edge, and the meridian hour line makes no allowance for its thickness. Smaller and plainer than Cole's dial of three years earlier, **28**.

The gnomon (vertical height 66) has lugs that fit into a pair of slots in the plate, and hold by bending the lugs flush with the underside, which is rough finished. The northern side of the gnomon is decorated by ogee curves.

Provenance

The Gabb Collection was purchased from G.H. Gabb by Sir James Caird for the

metal that slots into the lower edge with struts to support the plate vertically. The northern edge is cut with ogee curves. The underside of the plate is rough finished.

Sir Henry Darcy was knighted by the Earl of Leicester in 1566 at Fotheringhay, according to Metcalfe (1885, p. 119). But Shaw (1906, II, 71) has an earlier year: '1565, Aug. 21. Henry Darcey (at Kenilworth), by the earl of Leicester'.

Provenance

Bought by Lewis Evans for £8 in October 1909 from Lewis & Simmons, 77 Knightsbridge, London.

Location

Museum of the History of Science, Oxford (LE 232; 51646).

29 Horizontal sundial 1582

Signed and dated: *Humfræ ✛ Colle · 1582*

Bronze, 129 × 128; thickness 2

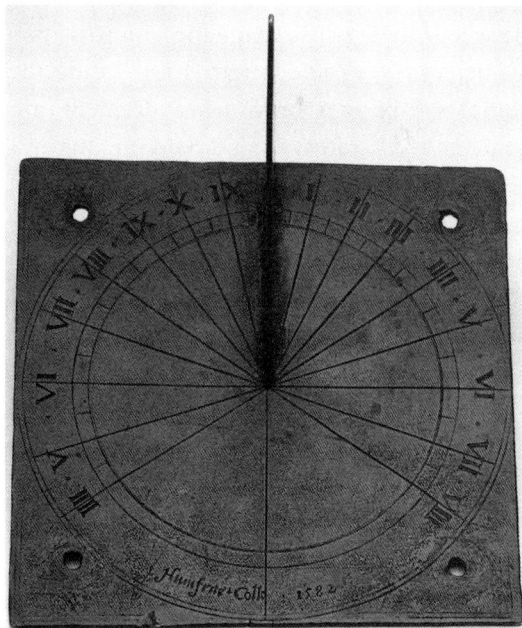

29

National Maritime Museum in February 1937. Gabb had spent some 50 years collecting instruments. This dial is No. 1 in section G of the Gabb Collection.

Location

National Maritime Museum, Greenwich (AST0390; old no. D.249/37–10C).

30 Nautical hemisphere 1582

Signed and dated: · *Humfrey · Côle · fecit · 1582 ·*

Brass; height 445; diameter of base 220; diameter of meridian ring 360

The horizon ring is supported by two semi-circular arcs that are integral with the bowed legs which screw to the thick base. Each toe is marked by straight cuts, one to four, for identification with its screw. The base is pierced by four holes, roughly cut (diameter about 15 mm), one of which has on the underside an area around it chiselled away. Between the legs is a magnetic compass (box diameter 118, needle length 104), Inside the compass box is a brass band divided into degrees at 2° intervals, marked in quarters 0–90°. The compass card is of white paper, with meridian and horizontal lines only. The fleur-de-lis, E, S, W, are in ink. Above the compass is a structure of four curved bars to locate a plumb-bob and the fiducial point (the bob is a replacement).

Within the framework is pivoted an equinoctial ring with a meridian ring spliced into it. The pivot points are at the E and W positions on the horizon ring. The meridian ring has two bifurcated lugs that slide over the supporting semicircle, which is engraved with degrees of latitude, 0–90°, in quadrants. The variable meridian, a brass disc, is pivoted from

30

the meridian ring at the positions of the Poles. The disc is composed of a plate riveted at the blank back (rough finished) to an annulus with 48 rivets to provide rigidity. Within the annulus is a plain disc which is slightly dished; it is engraved with a circle at the edge and two diameters at right angles. It is attached by the pin holding an alidade on the front side by an ingenious small circular button (diameter 22) with a central hole surrounded by four others. The central hole has a radial slot to give the form of a key-hole. This pushes onto the pin, and the slot engages in a cut made in the shaft of the pin. The button is attached over a small domed brass disc, or washer (diameter 26; see Fig. 2.8). The large and small discs ensure that the alidade is free to rotate, but is held firmly in the position to which it has been set. The counterchanged alidade (length 265, width 11, thickness *c.* 3) carries a pair of sight vanes, each pierced with two holes.

Tercera   fo.lxxxvij.

Esta es la demonstracion.

30 Engraving from Cortés, *Compendio de la Sphera* (1556). BL Oxford.

The variable meridian disc is engraved with an orthographic projection of the celestial sphere onto the plane passing through the solstitial colure. This derives from the *Planisphærium* of Ptolemy of Alexandria, and is known as the *organum Ptolemei*. It is to be found in a rotatable form in the fifteenth century, and in a fixed form in the sixteenth century, commonly under the name of the Rojas astrolabe projection (King and Turner 1994, pp. 178–9). It was used for reckoning time accurately from solar altitude in equinoctial hours. In the present case, the equatorial

ring is set to the latitude of the observer, the instrument aligned N–S by the magnetic compass, and the alidade set to the time of year by the declination scale (labelled by sigils of the Zodiac). The disc is then rotated to receive the Sun's rays through the sights, the time then being given by a bifurcated pointer running over the hour scale engraved on the equinoctial ring. One side is engraved in 24 hours, XII–XII (for summer), the other side in 12 hours, VI–VI (for winter). Both sides have a degree scale, in units labelled every 10°. The distance between *Polus Arcticus* and *Stella polaris* is $4\frac{1}{2}°$, the rather high figure put forward by Martin Cortés (see **96**).

The engravings on the variable meridian, in addition to the signature, are:

Polus · Arcticus:	*Polus · Zodiaci:*	*Stella · polaris:*
Polus · Antarctus:	*Tropicus Cancri*	*Tropicus Capricorni*
Æquino | ctialis	*Linea Æcliptica*	*Axis* [of the plate]

Some markings, and the fixing, of the horizon ring seem to suggest that an alteration occurred during construction. The scale is in single degrees, labelled every 10°, and the 32 points of the compass are engraved. The guide lines used for positioning the initial letters for the winds are noticeable. The four supports to the instrument are located at the cardinal points of the horizon ring, and, to accommodate the mortice and tenon joint, semicircular extensions have been incorporated into the ring. At the West, the tenon protrudes into the engraved W, and similarly at the other side with E. On the degree scale the labelling is immediately adjacent to the tenth division mark, except for 90° and 270°, which are engraved lower down the scale to accommodate the fixture by which the variable meridian plate is pivoted. The same shift in labelling the degree scale occurs at 180°

and 360°, to allow space for the cut–away section that permits the passage of the lugs attached to the meridian ring. These displaced degree markings show that the horizon was made for the instrument. Nevertheless, the initials N, S, E, W have been cut by the insertion of the tenons; this is probably fortuitous. On the other hand, at the North and South positions there are open slots, which will have been for an overarching altitude semicircle that has been lost. Where the meridian circle joins the equinoctial circle is a bifurcated lug that would have engaged on the now missing semicircle. For a semicircle that is present, see the nautical hemisphere (**47**) made by Charles Whitwell some 15 years later.

For the design of such an instrument, see Martin Cortés, *Breve Compendio de la Sphera y de la Arte de Navegar* (Seville, 1556, fol. 87r). This was translated by Richard Eden and published by Richard Jugge at London as *The Arte of Navigation* (1561). Jugge issued it again in 1572, the same year that he published Cole's map of Canaan in Parker's *Bible*. (See Fig. 3.3 for the map, and **8** for Cole's compendium made for Jugge). A more advanced version of the hemisphere was designed by Michiel Coignet, published in his *Nieuwe Onderwijsinghe* (1580), and in his *Instructions nouvelle* (1581). Cole's nautical hemisphere resembles the Cortés instrument more than that of Coignet. The Whitwell, **47**, follows Coignet.

Provenance

Gunther (1927), in his note, says that this was 'part of the ancient scientific equipment of the University of St. Andrews'. Allen (1928) gives reasons for believing that some of the St Andrew's instruments were acquired when James Gregory was empowered, in June 1673, to buy 'such instruments'. This is more fully discussed in Turnbull (1939, *passim*). See also **23**.

Literature

Gunther (1927, pp. 300–1); Wray (1984); Macdonald and Morrison-Low (1994, p. 29); Ackermann (1998, pp. 36–8).

Location

University Collections, University of St Andrews, St Andrews, Scotland.

31 Altazimuth theodolite 1586

Signed and dated on the vertical semicircle: ⋎ H ⋎ *Cole* ⋎ *1586* ⋎

Signed and dated on the base: · H · *Cole* · *1586* ·

Brass; azimuth circle diameter 200, thickness 2.5–3; altitude semicircle diameter 123, thickness *c.* 2

The azimuth circle is divided 0–360° in 1° units, and within this scale are marked the 32 points of the compass. The rectangular square is named on each of the four sides, above the scales: *Quadratum Geometricum*. It is attached to the circle by eight brass straps, connections being made by strips on the underside soldered in place. The straps are cut to resemble Tuscan pilasters; on one of these, at the North, is the second signature. The geometrical quadrant is divided as on the simple theodolite of 1574, **15**. An alidade, with ogee curved ends, rotates about the centre, and each arm has a tapered brass dovetail strip on which to slide sights, now missing. The strips are numbered 2 and 4 to identify the appropriate sight. Further locations for sights are at the North and South positions; these are numbered 1 and 3, and only the no. 2 sight remains (height 129, width 25). This is tall,

31 Side view.

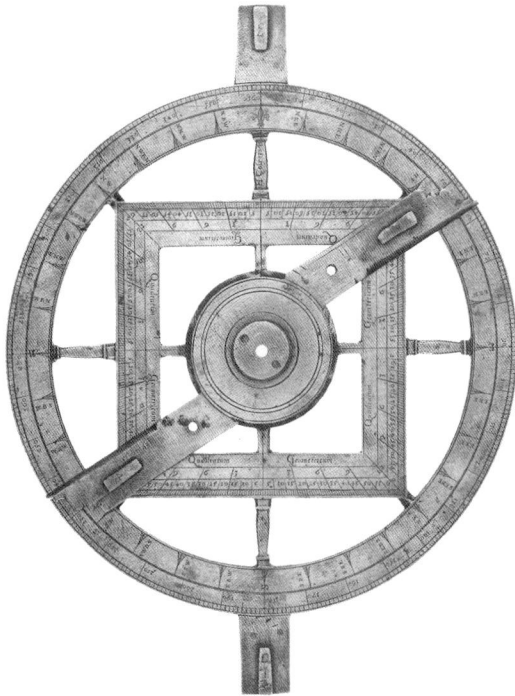

31 Azimuth circle.

and contains a slit to one side. At the centre is a band with two lugs for attachment by wing nuts to the alidade. This band rotates around the compass box, and supports the brackets that hold the vertical semicircle. The compass needle and glass cover are missing. When this superstructure is removed, the instrument becomes a simple theodolite.

Above the brackets the vertical support is divided so that the semicircle and shadow square pass between two bars with a clamping screw at the top. Below, just above the compass, is a plummet. The semicircle is divided in single degrees from 0 at the middle to 90° at either end. The shadow square has each part labelled: ⱱ *Vmbra* ⱱ *Recta* ⱱ ; ⱱ *Vmbra* ⱱ *Versa* ⱱ. The front vertical bar has a scale from 1 to 25, for reading against the edge of the shadow square. Similar scales are on the arms of the alidade.

Above the semicircle is the square-sectioned sighting bar (length 231; section 5). Each sight is different. The backsight has a pinhole and an edge above, while the foresight has a peg in a wide aperture with an edge above. The peg sight resembles those used on modern rifles.

Provenance

The upper part, including the compass, was discovered in the library of St John's College, Oxford, in 1924 (Gunther 1927, p. 301). The lower part was also found in the College in about 1950. St John's College was founded in 1555, and owned much land. This theodolite was most probably purchased for use on the estates.

Literature

Gunther (1927, pp. 301–2), figs 45–7; G. Turner (1983, p. 102), plate 48; Bennett and Johnston (1996, p. 56), no. 46.

Location

Museum of the History of Science, Oxford (24–37 and 51–54; 55130).

32 Plane table alidade 1588

Signed and dated: *H ✦ Côle ✦ 1588 ✦*

Brass; length 445; width 25; thickness 3.5

The sides are bevelled, with the exception of one end, which is scalloped. The English inch is marked out from the top of the bevel at one end, extending to 17 inches divided into $\frac{1}{8}$ths. There is a further run of 17 inches each divided into a different number of parts denoted by engraved numbers, which are: 7, 20, 21, 22, 24, 26, 28, 30, 32, 36, 40, 48, 64, 56, 18. This gives a choice of 16 scales for use in drawing maps or plans. The two holes near the middle are threaded, and were for fixing a sight for use with a plane table.

The theodolite was described in Leonard Digges, *A Geometrical Practise Named Pantometria* (1571). His son, Thomas Digges, added to the 1591 edition an appendix at the end of the First Book called *Longimetra*:

> A note touching a Platting Instrument for such as are ignorant of Arithmeticall Calculations. Some take the Semicircle of my Topological Instrument [theodolite], setting the Perpendiculare thereof vpon a straighte long Ruler deuided into a thousande or more equall parts, and instead of the Horizontall Circle vse only a plaine Table or boarde: whereon a large Sheet of Parchement or Paper may be fastened... . This being an Instrument onely for the ignorante and vnlearned, that haue no knowledge of Noumbers, and not to be practized but in fayre weather, and where yee may haue time sufficient euen in the Fielde to pricke and set downe the Charte.

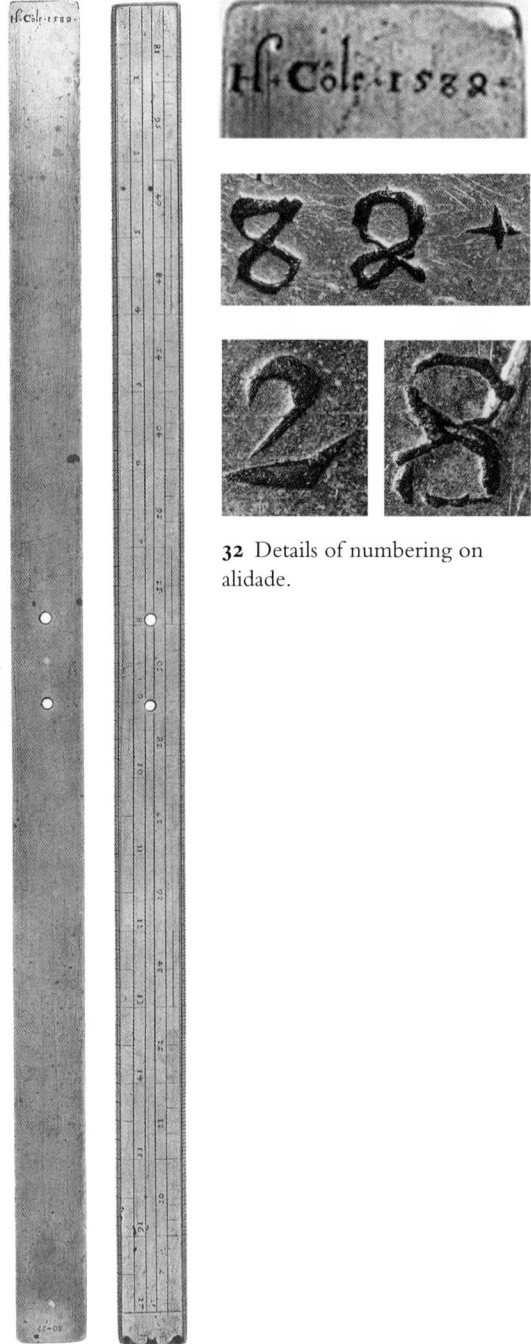

32 Details of numbering on alidade.

32 Back and front of alidade.

The divisions of an inch require an explanation. Plane tables are not particularly large, and can measure 15 × 12 inches, which can take a full sheet of foolscap paper measuring 17 × 13½ inches. A surveyor would wish his plan to fill the sheet of paper, so he would, accordingly, need to choose a convenient scale. For example:

110 yards	in 15 inches		can use a scale of		1:7
300 yards	"	"	"	"	1:20
400 perches	"	"	"	"	1:27
440 yards	"	"	"	"	1:30
600 yards	"	"	"	"	1:40
960 feet	"	"	"	"	1:64

This explains the series of divisions of the inch engraved on this alidade, and a comparison can be made with the surveyor's folding rules, which have 12 choices of scales (**17, 18, 19, 20**), and a gunner's compasses with a choice of 24, **10**. Comparison may also be made with a modern plastic ruler, which has inches in these parts: 8, 10, 16, 32, 64.

To the eye, the date seems to be 1582, and it is this that is in all earlier literature. However, a microscope reveals that the 8 is over an original 7, and the final numeral is 8, and there is evidence that Cole sometimes squashed the lower part of an 8. See the figures above.

33 b, c, d.

Provenance

Bought at Sotheby's, New Bond Street, London, sale 13 June 1980, lot 27.

Literature

G. Turner (1983, pp. 102–3), plate 50; Ackermann (1998, p. 89), fig. 42.

Location

Museum of the History of Science, Oxford (80–22; 47108).

33 Compendium 1590

Signed and dated: · *Humfrey* · *Cole* · *1590* ·

Brass, gilded; diameter 52

Comprising from the top:

a) nocturnal

b) table of latitudes of 16 European towns

c) horizontal sundial

d) magnetic compass well

e) decorated base with motto

f) side and suspension bracket

33 a.

This compendium, dated in the year before Cole's death, is smaller than all his earlier compendia, and some of his normal information is restricted or absent. There are two novelties. One is the slit type of nocturnal, where both *Polaris* (α Ursa minor) and *Kochab* (β Ursa minor) are viewed together in the slit, and not by using a central hole and index arm. This may be the first example of the use of this method. (See compendia by Whitwell and Allen.) The other variation is in the provision of a horizontal dial with hour lines curved to allow for five latitudes. Wear has removed much of the gilding on the outside.

a) On the nocturnal, the calendar is inscribed round the outer edge, and the date is in 2-day

33 e.

units, numbered in tens, ending at 28, 30, or 31. The volvelle that follows is edged with a tooth at each hour for ease of identification in the dark, the range of hours being from 3 am to 9 pm. At the 12 o'clock position is a raised knob in the form of a six-pointed star. Inset in the middle of the volvelle is the disc with a slit from the centre to the edge for aligning with the Little Bear. Engraved in a label is the use for this part: STELLA · POLARIS · OR · VRSA · MINOR. Just offset from the centre, in line with the slit, is an asterism to represent the Pole Star, and running just under the slit are six other asterisms to complete the representation of the constellation *Ursa minor*, ending with *Kochab* (β UMi).

b) There are 16 towns listed in eight panels; only Antwerp is not British. The list reads, starting to the right of the hinge:

Exceter	50	45	Oxford	51	47
Couentre	52	28	Newcastel	55	
Canterburi	51	18	Coulchester	52	
Notingame	53		Barwick	56	
Bath	51	26	N· hampton	52	15
Norwich	52	44	Edenborongh	57	
London	51	33	Cambridge	52	14
Yorke	54		Antwerp	51	28

c) Unusually for Cole, and other makers of compendia, the sundial is horizontal, to be used with a folding gnomon, which is unfortunately missing. The attachment points are at XII and at the centre. Although the face is cut for five latitudes: 50°, 51°, 53°, 55°, 57°, we do not know whether the gnomon could be adjusted for latitude. The hours are marked in roman numbers from 3 am to 9 pm. The hour lines are curved to provide for the changes in latitude; half hours are indicated by dots. At the centre is Cole's description of the dial: *This dial* [inserted *is*]

ieuerall [overall] *for Ingland and Scotland and for all contres yt lie east and west from them.*

d) A circular well is in the lower half of the plate, and was occupied by the magnetic compass; only the remains of the pivot point are visible.

e) The base is inscribed with foliage, flowers, and bunches of grapes. Prominent is a standing bird with a crested head and raised wings. In a band surrounding this decoration is the motto: VIGILATE · QVIA · NESCITIS · QVA · HORA · DOMINVS · VESTER · VENTVRVS · SIT (Watch therefore: for ye know not what hour your Lord doth come. Matt. 24:42.)

f) The outer rim is well engraved with an acanthus leaf motif. The suspension is very simple, with a small fixed ring that holds another larger ring.

Literature

Holbrook (1974, p. 8), no. 31; Holbrook (1992, p. 157); Ackermann (1998, pp. 68–9).

Location

Horniman Museum, Forest Hill, London (31–183 A).

34 Compendium 1588

Signed and dated: ✶ *ARyther* ✶ *Fecit* ✶ 1588 ✶

Brass, gilded; diameter 55, thickness when closed 16

Comprising from the top:

a) plain top

b) the names and latitudes of 30 European towns

c) equinoctial sundial

d) magnetic compass

e) plain bottom

Plimmouth	51	1	London	51	32	Norwiche	52	10
Southãpto	51	12	Bristowe	51	42	Ipswiche	52	0
Exciter	51	0	Hereford	52	2	Worcester	52	30
Douer	51	26	Cambridge	52	0	Nottinghã	53	0
Rye	51	5	Oxford	51	50	Yorke	54	1
Lincolne	53	6	Antuerpia	51	0	Compostella	44	20
Newcastell	55	0	Amstredem	52	0	Lisbona	39	40
Barwicke	56	23	Brugæ	51	0	Parisius	47	55
Edenburg	57	0	Colonia	50	0	Noruegia	63	0
Carlyl	55	2	Roma	42	0	Venetiæ	45	0

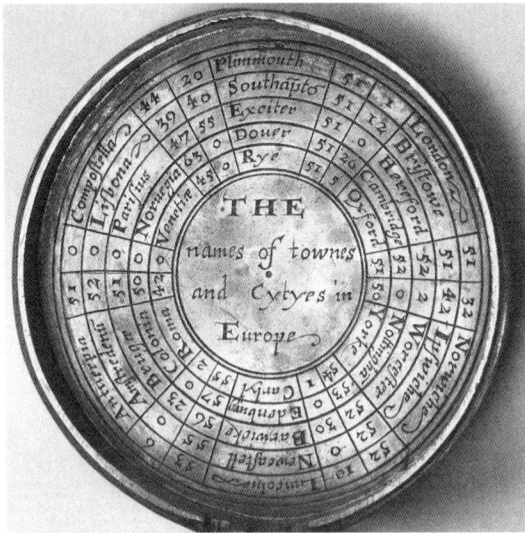

34 b.

a) The lid is simply decorated with three concentric circles near the edge, the inner two close together.

b) The names and latitudes, in degrees and minutes, are engraved in five concentric rows and six columns. At the centre is the description:

THE *names of townes and Cytyes in* Europe.

This table has some similarities with the latitude list on the Whitwell compendium dated 1600 (**58**). Both have the rare entry for

Norway, given as 63° (modern value for Oslo is 60°).

c) The equinoctial dial is erected when the lid is fully open, in the usual manner (see Plate 9). The gnomon with latitude quadrant is hinged as a unit, the quadrant passing through

34 c.

a slit to set the latitude. The scale is divided in 2° units, numbered every 10°. The upper chapter ring is marked in roman numerals in two 12-hour periods, subdivided in half hours. The lower side is marked out for 12 hours, 6–6, for the winter half of the year, and carries the signature and date. The plummet is missing.

d) The hand-drawn paper compass card (diameter 50) has 16 coloured points, blue, red, gold, and 16 shown by dots. For another Ryther hand-drawn card, see the theodolite, **35**. The magnetic needle is missing; the cover is of mica. The brass cover rim is engraved: [flower head] WILLIAM PAWLEY OWETH THIS SAME IN ANNO DOMINI 1588 ✶ The rim is not fixed, but under the A in SAME is a mark to show the magnetic declination, and the rim is correctly set by a line above the flower head matching a line cut on the hinge.

e) The bottom is plain, merely decorated with lines as on the top.

Provenance
Sotheby's, New Bond Street, London, sale 23 October 1985, lot 331.

Literature
Armada (1988, p. 213), no. 12.12.

Location
The Science Museum, London (1985–2021).

35 Altazimuth theodolite 1590

Signed and dated on the vertical semicircle:
✶ *ARyther* ✦ *fecit* ✦ 1590 ✶

Brass; azimuth circle diameter 214, thickness 1.6; altitude semicircle diameter 109; sighting bar 235

The azimuth circle is divided 0–360° in 1° units, labelled in tens. The inner square (sides

35 a

139) is marked out as a geometrical quadrant, all four sides divided in the same way, both 0–12 and 0–60 by ones. The diameter of the central circle is 113. All these parts are cut from a single cast brass plate, the underside left rough, as beaten and filed.

The alidade (length 219, width 21, thickness 1.5) is counterchanged, and at each end are hinged sights, one with a slit, the other with a wire (missing). There is a central hole for the post of the compass box, and on each arm is a turn-catch to fix the collar, which is placed over the compass, and which supports the vertical semicircle. Attached across the collar (diameter 88) is an arc from which rises the support for the plummet (missing) and the semicircle, which is secured by a wing bolt. The semicircle is divided in units 90°–0–90°; the shadow square in units to 12 at the corners.

Attached to the semicircle is the square-sectioned sighting bar, grooved to accept the semicircle. At each end sights (29 × 26) are fitted into grooves and soldered. Both sights

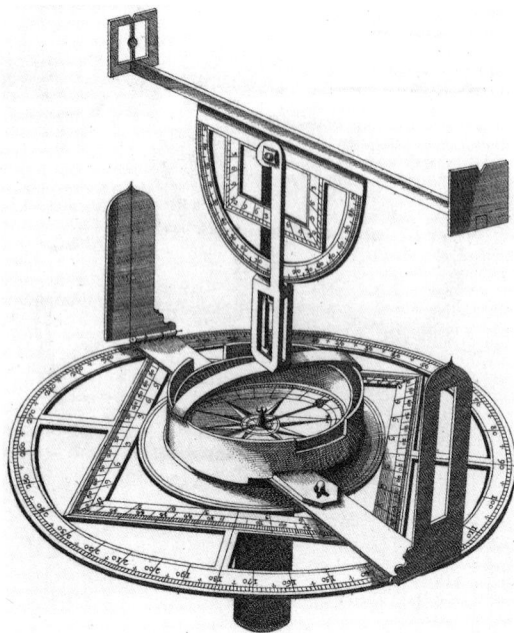

35 b Engraving from Dudley, *Dell'Arcano del Mare* (1647). BL Oxford.

This theodolite has been used to illustrate a number of works over the years. It appeared in an engraving by Lucini, who worked for Sir Robert Dudley. This was first published in *Dell'Arcano del Mare* (1647), Vol. 3, Book V, p. 18, prop. V, fig. 44. It is quite obvious that the engraver had Ryther's theodolite in front of him, since the detailing is exact (compare Figs **35**a and **35**b). The Italian's engraving appears in Gunther (1920–23, p. 367), where it is said to have come from the Blaeu *Atlas* of 1664. There then occurs an unfortunate leap into the conjectural:

> The next representation of an early form of theodolite is that given in Blaeu's atlas and presumably manufactured by him in Amsterdam. The construction of Blaeu's theodolite was practically the same as that of Digges, although it was about eighty years later. It included all the parts of Digges's theodolite with the addition of a magnetic compass and of improved sights.

Needless to say, a search into the atlases of Blaeu does not reveal this engraving.

have a notch at the top, and one, the back-sight, a pin-hole, with the foresight a bead with a hole.

The compass box (diameter 82, height 16) is made from a bent and soldered strip of brass attached to the base disc, which has a peg to push into the base plate of the instrument. There is a notch for a location lug, and a pair of cuts to take a horseshoe clip for fixing. The compass card is of paper, and hand-drawn; eight of the directions are coloured beige, blue, red (Plate 14d). The scale is in 120 units, beginning at North, with half-unit divisions alternately shaded. The numbering is in two rows, every 5 in the upper, and every 10 in the lower. The magnetic needle (70) has a spade end to the North, and an arrow to the South. The cover-glass is probably original as it contains air bubbles, one nearly 1 mm in diameter.

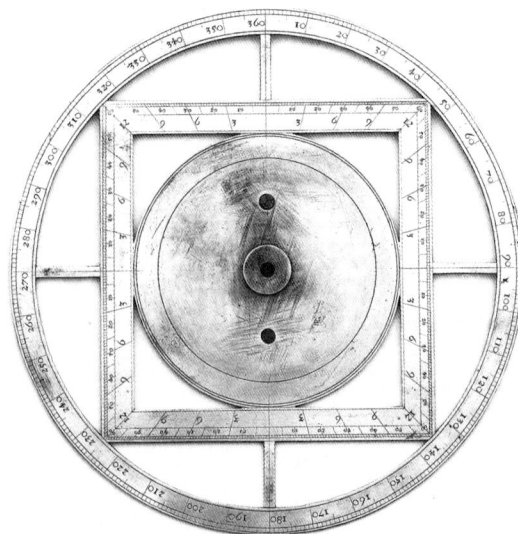

35 c. Azimuth circle.

Exactly the same picture appears in Kiely (1947, p. 193), with the same caption as Gunther's. Kiely writes: 'The first Continental instrument which shows a refinement of construction over the early English theodolite is the one diagramed [*sic*] in Blaeu's *Atlas* of 1664.' The illustration of the actual theodolite in J. Brown (1979) is probably the first time it was properly appreciated, especially in the context of the London trade.

Augustine Ryther was a skilled map engraver and instrument maker, whose maps for Saxton are dated 1576, and whose map for Hood is dated 1592. Only two of his instruments are known to exist. His burial is registered at St Andrew Undershaft on 30 August 1593. Ryther was head of the succession of mathematical instrument makers in the Grocers' Company in the City of London; he took as his only apprentice Charles Whitwell on 17 December 1582, who served nine years to 10 November 1590 (J. Brown, 1979, pp. 24, 58).

Provenance

One of the instruments taken to Florence by Sir Robert Dudley in 1606.

Literature

J. Brown (1979, p. 24); G. Turner (1991, p. 203).

Location

Museo di Storia della Scienza, Florence (240).

36 Two cards from the 1590 pack: Royal Arms; map of England.

36 Playing cards 1590

Signed: A/ RYT=\ H=

Inscribed and dated: W.B. *inuent* 1590

Augustine Ryther was the engraver who, having produced county maps for Saxton, engraved, some ten years later, a pack of playing cards for the designer, William Bowes (G. King 1996, pp. 60–1). The four suits each contain county maps of England and Wales (listed in Skelton 1970, pp. 16–18). There are eight extra cards, which comprise a title card with the royal arms, Queen Elizabeth on a throne, a map of England, a bird's-eye view of London (see Plate 1), and four cards with historical notes. The title card reads, in part of the cartouche: ENGLAN= FAMOVS PLAC=; below the arms: W.B. *inuent* 1590. The card with the map of England has Ryther's name on a banner between the legs of dividers above the scale. The letters and numbers are characteristic of Ryther's style of engraving.

The miniature county maps at the centre of the cards are copied, in so far as the size will allow, from Christopher Saxton's *Atlas of the Counties of England and Wales* (London, 1579), for which Ryther cut the plates for four counties in addition to the map of the country. A complete list of the engravers is in Skelton (1970, pp. 14–16). Although not an instrument, this pack of cards is included because the country map resembles those with the diptych dials, **56**, and **99**.

According to Hind (1939, p. 2) and Skelton (1970, p. 17), this is the earliest known set of geographical cards by 70 years, and the earliest with English county maps by 85 years. One bound set, in a private collection, has a title page, dated 1595, that provides a brief description of the pack, and the page is reproduced as a fold-out in Skelton (1970, p. 18). Much sundry information was added in this edition on separate sheets, such as a perpetual calendar, a table of the Sun's entry in the Signs of the Zodiac (both reminiscent of astronomical compendia), coats of arms, the trees of Vertue and Vice, etc. A fine set of 60 cards was acquired by the British Museum in 1938; in this pack the county boundaries are coloured. It was published by Hind (1939), but he did not discover Ryther's signature, and nor did Skelton (1970). Only one other pack is known (lacking the suit with the map of England), and this belongs to the Royal Geographical Society, London; see Heawood (1932, p. 13, plate 21).

Known copies

Private collection; Royal Geographical Society, London, Map Room; The British Museum, Print Room, London (1938, 7–9.57). The cards illustrated here are from the British Museum.

37 Sector 1597

Signed and dated: ✳ *Robertus* ✳ *Beckit* ✳ *fecit* · *1597* ✳

Brass and iron; radius of arms including points 329; radius of brass arms 314; radius of arc 155; width of arms 16, of arc 33; thickness of arms 3.2, of arc 1

This instrument, modelled on the figure and description in Thomas Hood, *The Making and vse of the Geometrical Instrument, called a Sector* (1598), is one of four described in the present work (see **46**, **69**, **70**). The arms are made of three strips of brass riveted together. They are connected through a compass joint; the arc (Hood's 'circumferentall limbe') is fixed into one arm and passes through the other. At the ends of the arms are iron points, and nearby are threaded holes into which sights could be screwed. The arm running over the arc has a small threaded hole, probably to lock the placement. The smooth hole in the compass joint will have been for mounting on a staff and to take a sight. None of these attachments is present.

37 Side 1.

Side 1

Both arms are engraved with lines of equal parts, from 0 at the joint to 110, with subdivision into half units; numbering is every ten. The arc has one scale of degrees, 0–110°, divided to a quarter degree, and labelled every ten.

Side 2

This side bears the signature and date. Both arms are labelled with their function. The arms have two scales, which Hood called 'internal' (inner sides of arms) and 'external' (outer sides). The inner scale, *Coards of a circle*, is non-linear, marked from 3 to 12, and represents the division of a circle into chords, the diameter being the opening of the tips of the arms. For instance, the distance between the 5s will give the chords for an inscribed pentagon. The outer, *Power of Lynes*, is a scale written as fractions from $\frac{1}{2}$ to $\frac{1}{10}$. By 'power' is meant 'square', as explained by Thomas Hood. For example, with the arms open, the line between the iron points is taken as 1, so the fractions give the area ratio; the line between the pair of positions shown by $\frac{1}{3}$ when squared gives an area a third of that obtained from the tips of the arms. The arc has three engraved scales, divided 1–180, 0–144, 0–108, each subdivided to half units. These are for scaling, and represent 18 inches between the tips divided into tenths, eighths, and sixths. The inside of the arc has three positions marked, 3, 4, 5, which indicate the distance between the tips for drawing polygons of 3, 4, or 5 sides.

This sector is similar to that attributed to James Kynvyn, **70**, and to the description published in 1598 by Thomas Hood. Little is known of the maker, Robert Beckit, except

on this sector is high, and comparable with that of Ryther or Whitwell.

Literature

G. Turner (1983, p. 103), plate 52.

Location

Museum of the History of Science, Oxford (80–8; 38251).

38 Alidade 1592

Attributed to Charles Whitwell; dated: 1592

Inscribed with a monogram: SP

Brass; plate length 240, width 43; compass box diameter 99

This alidade, for use with a plane table, has one edge bevelled, divided 0–100, and labelled in tens, with subdivisions to half a unit. Engraved by the maker on the underside is a explanation of the scale: *This scale is of 12 in a inche.* In line down the centre of the plate are five threaded holes, one of which retains an edge sight (height 46). This sight can be screwed into three of the holes, but of the other two one is too narrow and one too wide. The holes are variously identified by punched dots. At one end is the date, and the monogram. It is difficult to decide the sequence of the engraved lines, but the impression is that the S was cut first, and the P second. It is conceivable that the monogram stands for *Societas Pembrochiana*. The same

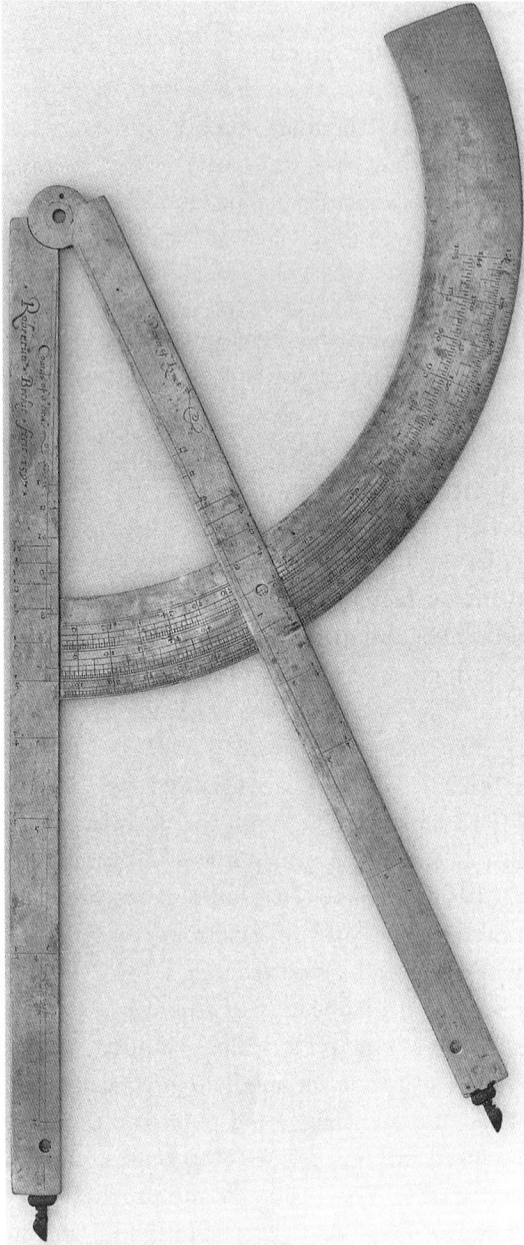

37 Side 2.

that he engraved five maps for the London edition (1598) of Jan Huygen van Linschoten, *Discourse of Voyages into ye Easte & West Indies* (see Chapter 3). The quality of the engraving

38

38 Monogram and date.

monogram is on the protractor **39**. The records of the Whipple Museum show the loan was made in the 1950s by Pembroke College. The College Archivist cannot find any record of these two instruments, and explains that the records are very sparse.

The base of the circular compass box is integral with the plate, and a band is attached to support a glazed lid (missing). A separate brass disc (diameter 95, thickness 1) carries at its edge a degree scale, 0–360°, numbered in tens, and subdivided to half a degree. Cross lines mark the cardinal points, but these are not labelled. At 11° and 191° are small lines to mark the magnetic declination. Still visible is the stump of the pivot for the magnetic needle. The underside of the plate and of the compass disc are as beaten and roughly filed.

The numerals and the lettering are typical of the hand of Charles Whitwell.

Provenance

Presumed to be on loan from Pembroke College, Cambridge.

Literature

O. Brown (1982), catalogue no. 23

Location

The Whipple Museum of the History of Science, Cambridge (Wh: 1006).

39 **Protractor** *c.* 1592

Attributed to Charles Whitwell

Inscribed with a monogram: SP

Brass; diameter 101, thickness *c.* 1.5

The upper surface is polished, and the bevelled edge is divided to 360° in single units. The scale is numbered in tens in both directions, with 360° at the North point. The following division is the 32 wind directions, or rhumbs, labelled with the four cardinal points only. Below the first division to the East of North is engraved the letter I, which indicates the declination of the magnetic needle at the end of the sixteenth century (Malin and Bullard 1981). The circular plate is solid except for a 45° sector cut out between West and South West. The underside is roughly filed. Adjacent to the open sector is the monogram, presumably that of the owner; this is discussed under **38**.

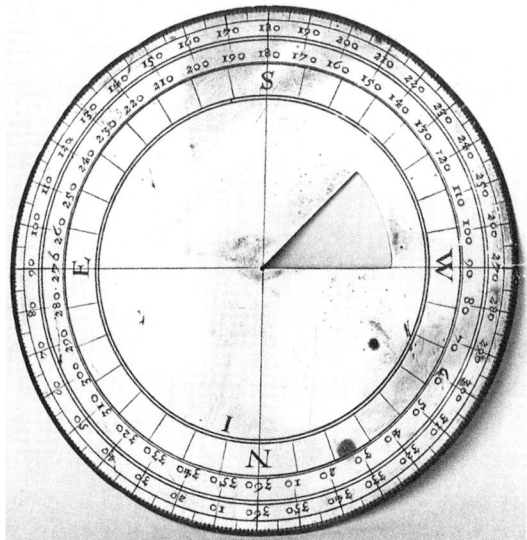

39

The monogram connects this instrument with the alidade, **38**, which is dated 1592. Presumably the protractor is also datable to 1592, in which case it is an early example.

The numerals and the four upper case letters are typical of the hand of Charles Whitwell.

Provenance

Presumed to be on loan from Pembroke College, Cambridge.

Literature

O. Brown (1982), catalogue no. 23.

Location

The Whipple Museum of the History of Science, Cambridge (1015).

40 **Pair of globes** *c.* 1592

Unsigned and undated; attributed to Charles Whitwell

Silver; diameter 63 (2¹/₂ English inches)

The single sphere is constructed out of two silver hemispheres that are joined at the equator by three bayonet catches. The outer surface is engraved with a map of the world. At least 900 holes of different sizes have also been drilled into the surface from the outside, representing the starry sky. Evidently this globe was intended to be used whole as a terrestrial sphere as well as in two parts in order to see, from the inside, the stars against a background light, exactly as seen from the Earth. The meridian, equator, and ecliptic are all divided in degrees, labelled every 10°. Lines of Right Ascension are every 15°; lines of latitude every 10°. The tropics and polar circles are engraved, but only the former labelled: *Tropicus Cancri, Tropicus Capricorni*. The only sea named is the northern Pacific: OCEANVS SERICVS, or China Sea.

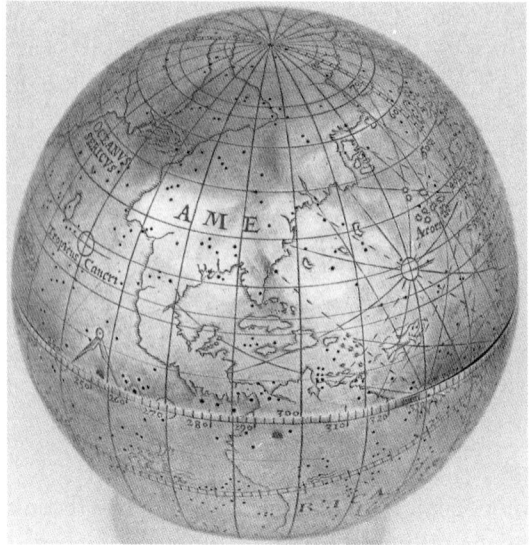

40 America on the silver globe.

The terrestrial and celestial maps of this 'pair of globes' have been copied from the well-known pair of globes made in Louvain by Gerard Mercator in 1541 and 1551, respectively. The overall outlines of land and sea depicted on the silver globe, as well as the sea monsters and most of the names engraved on it, are identical to those on Mercator's terrestrial globe. The inclusion of loxodromes is another feature connected with that globe. The identification of the mapping of the celestial globe is less straightforward, because only one constellation figure, Eridanus, is drawn on the inside of the sphere. It is obvious that the engraver had difficulty in using his burin. However, a number of peculiarities of Mercator's celestial globe, such as the shape and location of the stellar magnitude table, are found in exactly the same manner on the silver globe.

A special feature of the terrestrial globe by Mercator is that, in addition to the outlines of land and sea, some 50 stars are marked on it.

182

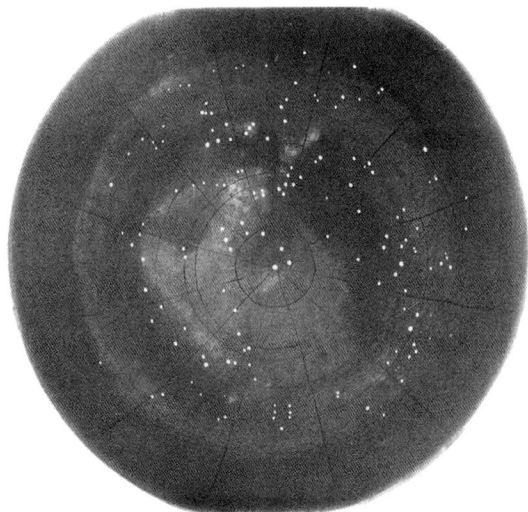

40 View from inside the globe, showing the U-shaped magnitude diagram at the bottom.

This idea was taken over by the maker of this silver globe, but instead of marking the stars on the surface, he drilled holes in the terrestrial sphere. He drilled far more, estimated at over 900, and probably the 1022 in Ptolemy's *Almagest*. By adding the catches so that the two hemispheres can be separated, an unusual concave celestial globe was created. The normal, convex celestial globe shows the stars from the viewpoint of an imaginary observer outside the sphere, hence not as they are seen by an observer on Earth. When the two hemispheres are viewed from the inside against an external light, one sees the stars, represented by the small holes, as they appear to an observer on Earth.

Dating and maker

The date *ante quem non* is 1551, when Mercator's celestial globe was made. The latest date is likely to be 1592, when Emery Molyneux (died 1598–99) introduced his new pair of 24-inch globes to the court of Queen Elizabeth. After this one may presume that the Mercator globes would no longer be acceptable as a model. For reasons considered in some detail (Dekker and Turner 1997), Charles Whitwell is chosen over his master, Augustine Ryther, as the engraver of the silver sphere. Whitwell took his freedom on 10 November 1590. The earliest instrument signed and dated by Whitwell is his pocket vertical sundial of 1593 made for Nathaniell Torporley; see **41**. Whitwell died in 1611. Apart from his mathematical instruments, Whitwell is known for a number of maps, three produced in 1595 or 1596. He illustrated books published in 1597 and 1598. Particularly important is the group of seven Whitwell instruments taken to Florence by Sir Robert Dudley in 1606. Of these, three are dated 1595. It is clear that by 1595 Whitwell, then in his middle twenties, was executing major commissions, some of which were grand, and others small and ingenious; all are described in this book.

The production of globes during the decades before and after 1600 is explained in Dekker and Turner (1997) and in Dekker (1999, pp. 101–3). In summary, the Molyneux globes of 1592 made the Mercator globes obsolete. They were so well advertised that a London engraver and instrument maker could not have remained unaware of them. This points to a date for the silver globe of before 1592, and so makes it the earliest still extant globe manufactured in England.

Provenance

A very small Dutch silver mark is between the legs of dividers at 257° longitude on the upper hemisphere, and on the lower at 295° longitude. The mark is a boar's head, introduced by the Koninklijk Besluit on 11 May 1831. In two sizes, the smaller was in use from the date of the Resolution to 1 November 1893 (Koonings 1964, p. 100). The Dutch customs stamped any silver object coming in or going out of the country.

Part of the Bonnefoy Collection, exhibited at the Musée des Arts Décoratifs, Palais du Louvre, Paris, March–April 1936. In May 1936 acquired by Sir James Caird, who presented it to the National Maritime Museum.

Literature

Dekker and Turner (1997); Dekker (1999, pp. 101–3).

Location

National Maritime Museum, Greenwich (GLB0025).

41 Altitude sundial and perpetual calendar 1593

Signed: ✴ *Charolus Whitwell Sculpsit* ✴; calendar begins at 1593

Also inscribed: ✴ *Nathaniell Torporley inventor* ✴

Copper, gilded; overall length 110; width 69.5; thickness 4

Side 1

The side with the signature has two vertical altitude dials arranged symmetrically about the vertical axis. The gnomons are set into pivoting arms that are adjusted to the season of the year by placing their pointers over the Zodiac scales at the outer rims. The hour circle is in two parts, the outer for morning, and the inner for afternoon. The central portion is labelled as:

A Table showeing the moueable feastes for euer

Calendrical details are arranged in six concentric circles, labelled:

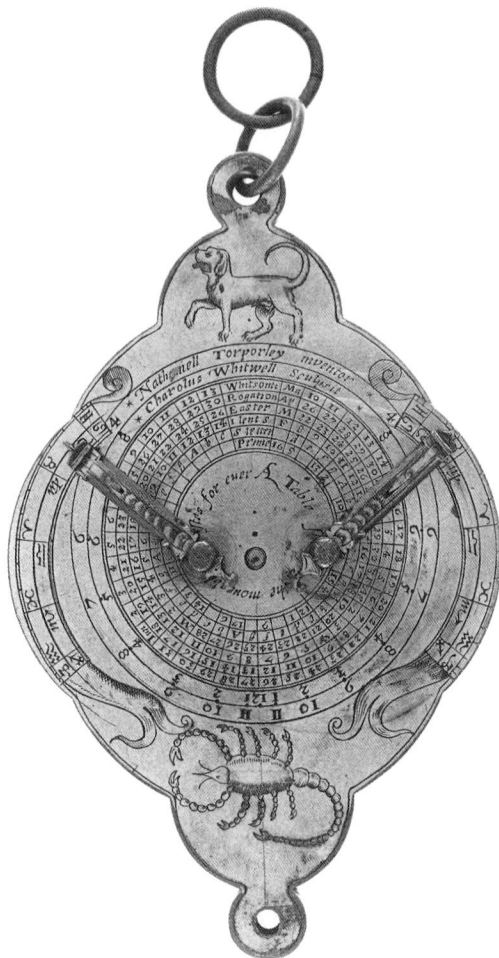

41 Side 1.

Whitsonti; Rogation; Easter; 1 lent S; S letter; Prime

This table gives the Old Style dates, etc. for Easter from the earliest to the latest date, that is from 22 March to 26 [*sic*] April. This last date is incorrect. The last possible date for Easter is 25 April. Whitwell missed out 23 in the Easter day sequence: the other dates are correct.

The gnomons are delicately made, with the slender pointers pulled into place by a tiny pin. The inner ends are moulded, one with a fleur-de-lis. The surfaces are engraved with a pattern of overlapping discs. The lobes contain above, *a talbot passant*, the badge of the Talbot earls of Shrewsbury, and below *a scorpion* (a birth Sign?).

Side 2

This side has a quadrant divided 0–90° by ones numbered in tens; a scale of the 24 hours in the day numbered to 12 and back; and a Zodiac scale disposed over two circles: from the equinoxes at the top (end with suspension

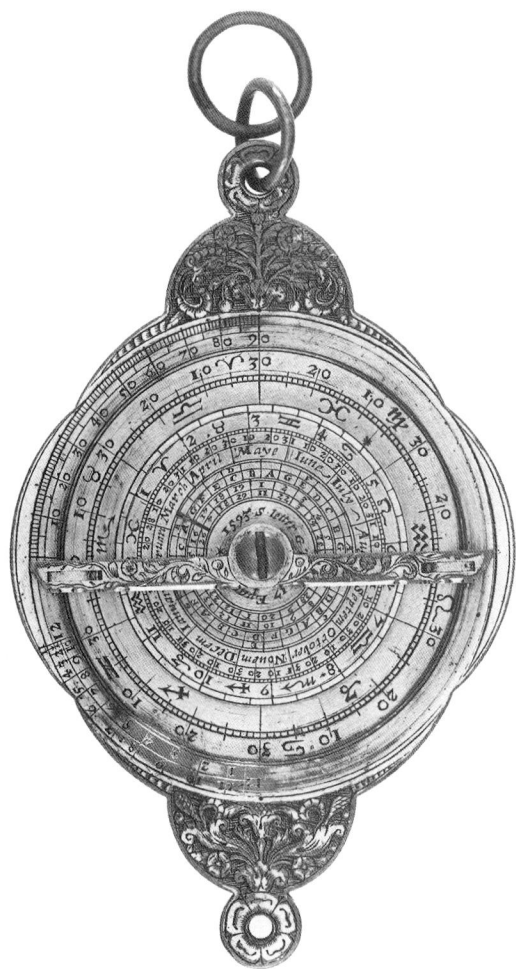

41 Side 2.

ring), the outer circle holds for the summer and the inner for the winter. The innermost seven concentric circles are engraved on a rotatable disc. The circles contain (from the outer): the Zodiac numbered from Aries 1; days of the months; the names of the months; Dominical (or Sunday) letter for leap years; the Dominical letters; Prime; Epact. At the centre are the indications for the beginning of the sequence in the year: · *1593· S· letter G Prime 17 Epact· 7* ✳. A rule with a pair of lift-up sights pivots at the centre of this face. The rule is decorated with a floral pattern. The curved extensions on this side are engraved with elaborate floral patterns. The rotatable disc can be removed, when one sees that neither its underside nor the space left are gilded. The underside of the rule is not gilded. The round-headed screw is hand-turned, and the depression from the tailstock is clear; the cut in the head is slanted.

Nathaniel Torporley (1564–1632) was an Oxford University mathematician (MA 1591) and cleric, who was patronized by the 9th Earl of Northumberland.

Provenance

From the Spitzer Collection, Paris; bought by Lewis Evans for £9 10s in April 1899 through Harding, King Street, London. Lewis Evans Collection.

Location

Museum of the History of Science, Oxford (LE 65; 52101).

42 Quadrant 1595

Signed and dated: ✳ *Charolus Whitwell fecit 1595* ✳

Brass; 175 square, thickness 1.5; alidade, radius 236

Side 1

This bears a horary quadrant with two separate parts. The smaller scale is for planetary or unequal hours, the larger for equal hours. The hour lines are straight, in the arrangement described by Johann Stöffler, *Elucidatio fabricae ususque astrolabii*, 2nd edition (1524), p. 66. A quadrant arc is below the larger scale, and is divided in half-degree intervals, labelled every 10°. Alongside each hour scale is a declination scale divided into the Signs of the Zodiac, each in 5°, marked by the Zodiac sigils. The equal hour scale runs from 4 to 12 then back to 8, with half-hour subdivisions. The hour lines are positioned for the latitude of $51\frac{1}{2}°$.

At the apex is riveted the alidade, which reaches to the shadow square junction, the scales of which are engraved on two edges of the plate. These scales are to 120 units, labelled in tens, with a separate labelling in units of ten to 12 at the 45° position. The rivet has a hole for a peg to hold a plumb-line (missing) to set the side of the plate to the vertical. Fixed to the alidade are two sighting vanes, with slit and pin-hole apertures.

Side 2

This is an instrument for establishing the port. An outer circle is divided into the Zodiac, each in units of 1° marked in tens, each Sign named in full. Next are the months in the year, named in full, divided in days, numbered in tens (28, or 31). The third circle numbers in roman the 24 hours, twice 12, subdivided in quarter hours. The fourth circle is the wind rose, with 16 of the points labelled with the initial letters of the direction. The central part is occupied by two volvelles. The first gives the age of the Moon, 0–30 in quarter-day intervals. As a single lunation occupies $29\frac{1}{2}$ days, the figure of 30 is unusual, especially at the date of 1595; however it was used by Humfrey Cole *c.* 1570 (**12**). Whitwell may have considered that it was easier to divide to 30 on an instrument intended for a rather crude measurement of the times of the high or low tides. The index attached to the volvelle at the New Moon position is for locating the Sun in the Zodiac. The second volvelle has an index to point to the Moon's age, and a lunar phase aperture. The surface is

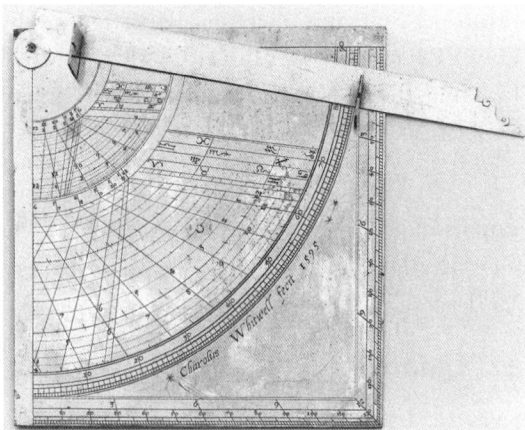

42 Side 1.

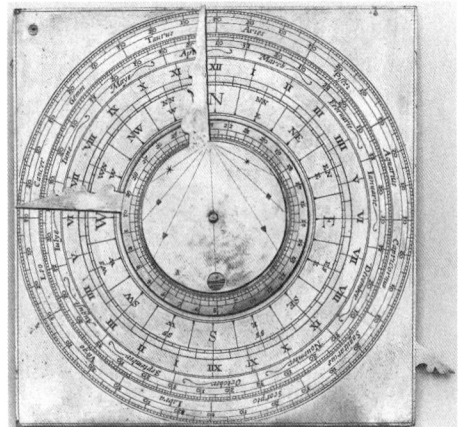

42 Side 2.

marked with the aspects: opposition, trine, quartile, sextile.

Whitewell generally uses the Latin form of his first name, Carolus. However, because of comparison with the English form, Charles, he also signs instruments Charolus, as here. On his map of Jerusalem he engraved Charolus, but on his map of Surrey he has Carolus with the *a* clearly inscribed over a previous *h*.

Provenance

One of the instruments taken to Florence in 1606 by Sir Robert Dudley.

Literature

G. Turner (1985), plate 99.

Location

Museo di Storia della Scienza, Florence (2519).

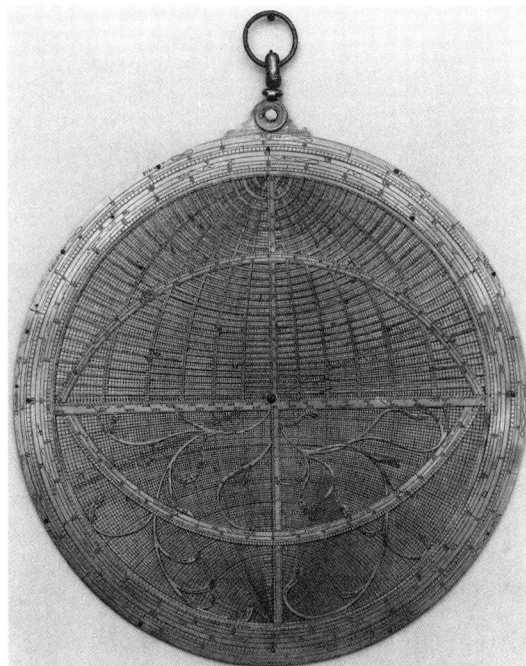

43 front

43 Astrolabe 1595

Signed and dated: ✳ *Charolus Whitwell fecit 1595* ✳

Brass; diameter 387; width of rim 27; thickness of rim 3

With this instrument, Charles Whitwell has attempted to make a novel astrolabe to the design of the mathematician, John Blagrave (*c.* 1558–1611), who described and illustrated his model in *The Mathematical Jewel* (1585). The Whitwell version is a single disc, with a universal projection engraved on one side, and originally with a plain back.

Throne

Integral with the disc is a small scrolled extension that holds a double swivel. The ring runs through a brass knob in the form of an acorn.

Limb

The annulus that forms the limb is secured to the disc with 12 hand-made round-headed screws. There are also four extra holes. The outer edge is divided into months, March at the top, and each month is divided in single days. The next scale is of hours twice XII in roman numerals, VI at the top. The Zodiac occupies the next band, labelled with both sigil and full name. The First Point of Aries is at 11 March, and of Libra $13\frac{1}{4}$ September. Finally comes a degree scale numbered in quadrants from the top 90°–0–90°–0–90°.

Mater

The whole of the front of the disc is engraved with the universal stereographic projection known as the *Saphea*, with the hours and the Zodiac marked. Coordinate lines are at single degree intervals, but there is no numbering.

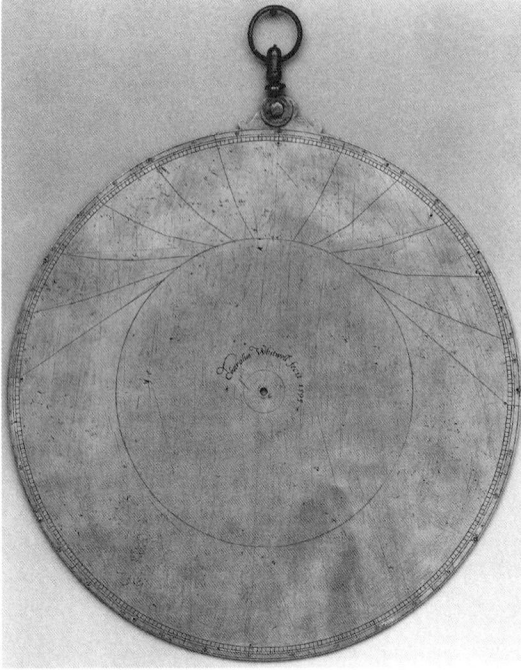

43 back

However, lines every 10° in altitude and every 15° in azimuth are emphasized by cross marks. The hour scale along the Tropic of Cancer contains an error: the 5 is in mirror image. Such a mistake is sometimes made by engravers used to producing maps, as is the case with Whitwell. The ecliptic is shown by a series of parallel lines every degree, with the lines of various lengths to show the 5°, 10°, and 30° intervals. This follows Blagrave's engraving.

Back

Round the edge is a degree scale of single degrees numbered in tens, divided in quadrants in the same way as on the front of the limb. Near the centre is Whitwell's signature and the date. A later attempt has been made to scribe a set of unequal hour lines. These are

very crude. Several casting fissures are visible on this surface.

Pin

The rete is held in place by a hand-made, round-headed screw of brass (length 10.3), with a four-winged nut. The depression made by the tailstock is clear, and the saw-cut in the head is angled.

Rete

The rete is in two sections. The lower section comprises the more conventional arrangement with extended star pointers. The names of the 29 stars discernible are either on pointers or placed on the various bars, and are listed below. The lower half is crossed by an arc engraved with the Zodiac from Aries to Virgo, named in full and with sigils. The equatorial bar is divided in single degrees, and numbered from 0 at the centre to 90° at the right and back to 180°, thence to 270° at the left and back to 360° at the centre. The meridian bar is divided in degrees from 0 at the centre to 90° both ways. The upper arm only is numbered in tens.

The upper half of the rete is fretted, with 2° gaps and bars 1° wide in latitude. The azimuth

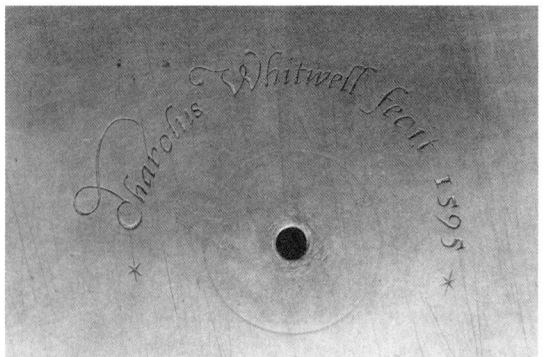

43 Signature

arcs occupy 2° in width and are spaced every 15°. The names of the 27 stars engraved on the fret are listed below. An arc engraved with the Zodiac from Libra to Pisces crosses the fret. At the right-hand end of the equatorial bar, adjacent to the rim, is a small hole for a plumb-line to read the scales on the meridian bar. At the lower end of the meridian bar is a small hole for a plumb-line to read the scales on the equatorial bar.

Commentary

It is clear that Whitwell had trouble making this instrument. Cutting the fretwork must have been exceedingly difficult; several bars were miss-cut and needed repair, and one small section was cut to single instead of double width. A further difficulty was in posi-tioning the stars on the fret. In many cases, Whitwell had to cut not only round the bars, but also round a pointer. This would have required a fine saw, a steady hand, and great patience. The extended star pointers on the lower half of the rete are remarkably sinuous and foliate, following Blagrave, and end in buds with a small star. These are most unprac-tically thin, and some have broken off and become lost. Blagrave's design was, in fact, unrealizable. To make it was extremely time consuming and therefore expensive; only Dudley would have had the interest and the money to commission it.

Blagrave numbered the stars on his rete, and in his book he presented a table in a double-page spread listing 76 stars, with name, magni-tude, longitude and latitude, Right Ascension,

Stars on the pointers

Ceti Venter	*Cin Orio* [3 pointers; additionally,
Andro Scapu:	on rim: *Cingu Orionis*]
[broken pointer]	*hu: dex: Ori*
3 *Cauda Ceti*	*Cano: in argo naui* 1
etis [broken pointer;	1 *Canis maior*
Cornu Arietis intended]	*Apollo* 2
4 *Ceti Iuba* [Algenubi]	*Hercules* 2
Caput Medulse 2	1 *Canis minor*
Extre: Eridani Acarn	*Asellus Boreus* [θ Boo]
[in this region a broken-off pointer	*Asellus Austrinu* [χ Boo]
leaves the letters *oe* on a strap]	*Lucida Hidra*
5 *Pleiadum* [7 stars, part on ecliptic,	*Cor Leonis*
part with pointers]	*Ceruix Leonis* 2
Oculus Tauri 1	*Hume: Vrsa ma:* 2
Haedorun [broken off; ζ Aur]	*Cauda Leo* 1
Sinister pes Orionis	*Corui Rostrum* 3
Hircus 1	*Corui ala dextra*
Hume: sinister Orio: 2	

Note: The stars are listed above as they occur on the rete. Most are in order of Right Ascension, but there is some disorder amongst the first four. Numbers before or after the star names are magnitudes, but not all stars are so marked, and no magnitudes are on the fret.

Stars on the fret

Cauda Vrsa Maio: [3 pointers]	*Caput Engouna* [α Her]
Preuindemiat [ε Vir]	*Caput Ophiuci*
Spica Virginis	*Caput Dra*
Arctu: Bootes	*Fidicula* [α Lyr]
Sinis: hu: Bootes	*Aquila*
Lances Borea: [β Lib]	*Cuspis Sagittari*
Chele Aust: [α Lib]	*Cauda Delphini*
Hastile Bootes [η Boo]	*Cauda Cigni*
Lucida Coro: Gnos: [α CrB]	*Dex hu Cephea*
Frons Scorpij [ω¹, ω² Sco]	*Cauda Capricor pre* [α¹ Cap]
Borea [β Sco]	*Se:* [α² Cap]
Aust [δ Sco]	*Fomahand*
Palma Ophiuchi	*Crus Pegasi*
Cor Scorpi	

and declination. Blagrave's heading reads: 'A Table of the most notable fixed starres taken out of *Stadius*, rectified by obseruation to the yeare of our Lord 1558.' The number of stars as given by Blagrave, 76, does not match the number on this instrument, 56. Whitwell did not use 13 of the stars (*Stella Polaris* is an obvious omission). Whitwell's *Pleiadum* has seven asterisms on the rete, but just the one name; Blagrave numbered only four stars in this constellation. Such duplication accounts for seven numbers, thus leaving 56 names on the rete, of which 29 are placed on the pointers, and 27 on the fret.

Gunther (1932, pp. 285–6) illustrated and briefly described an Islamic precursor (IC 140) to Blagrave's design with a fretted rete. It had already been published in the *Burlington Magazine* (October 1928), and was said to be dated AH 729, which runs from 5 November 1328 to 24 October 1329. The astrolabe was made by Ibn al-Sarrāj, and it is in the Benaki Museum, Athens (King 1987, IX; 1991, p. 3).

The Adler Planetarium & Astronomy Museum, Chicago, has an unsigned and undated astrolabe (IC 309) very similar in appearance to Whitwell's. On this the First Point of Aries is at about 20 March, which shows that it was not made for use in England, where the Gregorian calendar was not accepted until 1752. The Chicago astrolabe is not Whitwell's work, and is not English (see Chapter 5, and Figs 5.6–5.8).

Provenance

One of the instruments taken to Florence by Sir Robert Dudley in 1606.

Location

Museo di Storia della Scienza, Florence (1095) IC 482.

44 An addition to a 1399 quadrant attributed to Charles Whitwell *c.* 1595

There are two English, horary quadrants for equal hours, one dated 1398, and the other 1399 (Ward 1981, pp. 55–6, no. 146; Ackermann and Cherry 1999). The first is as originally engraved, while the second, from the same workshop, has additions: a solar

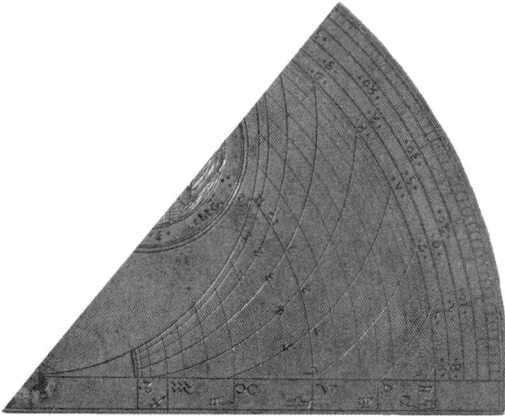

44 Segment of quadrant dated 1399 with additions by Whitwell.

45 Side 1, chronology.

declination scale at one side, and along the equinoctial line the numbering on the hour lines. The Zodiac sigils in the declination scale and the numerals allow the engraving to be attributed to Charles Whitwell (G. Turner 1985). The engraving on these parts is closely comparable to that on the horary quadrant **42**, which is signed and dated: *Charolus Whitwell fecit 1595*.

It is of interest to know that a serviceable, medieval quadrant was 'modernized' some 200 years after it was made, and that Whitwell was commissioned for the task.

Location

The British Museum, London (1860, 5–19.1).

45 Chronological and astrological tables 1596

Unsigned and undated: attributed to Charles Whitwell; chronology up to 1596

Gilt brass; diameter 67

A small disc, convex on the side bearing the chronological tables and concave on the side bearing the astrological table. At the top is a hole beside a notch. The engraved letters and numbers are tiny (average height 0.6 mm), and they correspond to the characteristics of Charles Whitwell. To produce the minuscule characters would have required the skill of such a master. The date rules out Ryther, and Kynvyn would not have had the deftness at this scale.

Chronology

The central table lists the monarchs of England from William the Conqueror to Queen Elizabeth, with a later addition to include James I. The heading reads: *A Chronicle of ye Kīg of Eng=*. To the left is a list of key events relating to England from the creation of the world; below and to the right is a list of key events during the sixteenth century.

The calendar is the Old Style, where New Year begins on 25 March. The left-hand column gives Anno Mundi, but it is 200 years too high, and not consistent. The increments do not correspond to those in the AD column. Anno Mundi, the date of the creation of the world, is claimed to be 7 October 3761 BC at 11 hours 11 minutes pm. (Richmond 1956, p. 77). J.J. Bond, a Keeper at the Public Record Office, wrote: 'there are as many as one hundred and forty different dates given for the Mundane era' (Bond 1875, p. 269). The year when a reign began is accurate, but the month and day is when the previous monarch died, yet not always. For a modern list, see Cheney (1961, pp. 12–31). The place of burial is accurate. The bowels of Henry I were buried at Rouen, the body at the church of the monastery he had founded at Reading (*DNB*).

William Bourne, *A Regiment for the Sea* (1574), includes 'A Table of the reigne of Kings since the Conquest' (Taylor 1963, p. 156). The layout differs from that engraved

on the disc, and some dates are not identical, so one cannot claim it was copied, though it might have been the inspiration. According to Capp (1979, p. 30) the list of kings began to be included in almanacs from 1571.

The engraving on the disc is in italic, but this has not been copied in the tables. However, italic has been used in the table at the bottom for data added after 1596 by the same hand. Black mastic filled the engraving on the original parts, but not in the later addition.

The key events in England are to the left, below, and to the right of the table of monarchs. The italic has not been reproduced.

Evil May Day = 1 May 1517, when the apprentices of London rose against the privileged foreigners, whose advantages in trade caused jealousy.

	Creacion of the world		5558	
	Generall floud of Noah		3902	
	First arriuall of Brute	vno	2703	
Since yᵉ	Conquest by Iulius Cæs	1596	1647	Yares
	Beginning of the Saxons	are	1140	
	Entrance of the Danes		c578	
	Coming of Wᵐ Conque		528	
	Taureñ, France and Floddē Field Hā		83	
	Euill Mayday, yᵉ tumult of prēt		78	
	The suppression of the Abbies		60	
	King Henry 8 went to Bullen		52	
Since	Muskelborow field tempʳ ·E·6·	vnto Anᵒ	49	Yares
	The côuxō of Sʳ Tho=wyat QM	1596 are	43	
	The loss of Callis by yᵉ Eng=to yᵉ Fr		39	
	The siege of Lieth, tempʳ QE		36	
	The buřing of Paules Steeple		35	
	Going to Newhaure by yᵉ Eng=		34	
	Buylding of yᵉ Royall Exchāg	vnto	30	
Since yᵉ	Rebelion of yᵉ 2 Earles, Noʳ & W	1596	27	Yares
	Duke of Norfolk was behead=	are	24	
	E·of Lecest= went first into H·		11	
	The Spa=fleet attey[?] towarde Engl dᵒ 1588			
	The Spa=ivd Callis & yᵉ Eḡ Cale 1596			

AᵒM	Their names	Their entry	Anᵒ dm̃	Their reig Y M D	Æ	Their place of bu=
5028	Willm Conq·	14·Oct	1066	20·11·22	74	Cane in Nor.
5049	Willm Rufus	9·Sep	1087	12·11·18	43	Winchester
5002	Henry 1	1·Aug	1100	35·4·11	67	Roã & Read
5097	King Stephẽ	2·Dec	1135	18·11·18	46	Feuersham
5116	Henry 2	25·Oct	1154	34·9·2	61	Fonteuerard
5151	Richard 1	7·Jul	1189	9·9·22	43	Fonteuerard
5161	King John ·	6·Ap	1199	17·7·0	51	Worcester
5178	Henry 3	19·Oct	1216	56·1·0	65	Westminster
5234	Edward 1	16·No	1272	34·8·9	69	Westminster
5269	Edward 2	7·Iul	1307	19·7·6	43	Gloucester
5286	Edward 3	25·Iañ	1326	50·5·7	70	Westminster
5337	Richard 2	21·Iun	1377	22·3·16	33	Westminster
5359	Henry 4	29·Sep	1399	13·6·4	46	Canterbury
5362	Henry 5	20·Maʳ	1412	9·5·24	36	Westminster
5371	Henry 6	31·Aug	1422	38·6·16	39	Wyndsor
5409	Edward 4	4·Mar	1461	22·1·9	50	Wyndsor
5431	Edward 5	9·Apr	1483	0·2·18	13	not knowne·
5431	Richard 3	12·Iun	1483	2·2·5	51	Leystʳ gray fri:
5433	Henry 7	22·Aug	1485	23·8·19	52	Westminster
5477	Henry 8	22·Apr	1509	37·10·1	56	Wyndsor
5515	Edward 6	18·Ian	1547	6·5·19	16	Westminster
5521	Q·Mary	6·Iul	1553	5·5·22	47	Westminster
5526	Q·Elizabeth	17·No	1558	*44.4.14*	70	*Westminster*
5566	*Iames · 1*	24 Maʳ	1602			

Newhaure = Le Havre, a village developed as a port from 1516, and in 1562 placed in the keeping of Queen Elizabeth by the leader of the Huguenots.

Cale = Cales, a spit of land from La Isla de Cadiz, where the combined English and Dutch fleets defeated the Spanish in 1596.

Astrology

At the top of the concave side is a Sun in Splendour, the rays pointing to the hours, labelled in roman from IIII to VIII, forming an equinoctial dial. Beyond the hour circle are four concentric circles with divisions to half and quarter hours, and even to five minutes!

45 Side 2, astrology.

		Sonday	Moday	Tuseday	Wednes	Thurs	Frida	Sater	
Sol	☉	1	12	9	0	10	0	11	♃
Ven=	♀	2	0	10	0	11	1	12	♂
Mer=	☿	3	0	11	1	12	2	0	☉
Luna	☽	4	1	12	2	0	3	0	♀
Satu=	♄	5	2	0	3	0	4	1	☿
Iupt=	♃	6	3	0	4	1	5	2	☉
Mars	☽	7	4	1	5	2	6	3	♄
Sol	☉	8	5	2	6	3	7	4	♃
Venus	♀	9	6	3	7	4	8	5	♂
Mer=	☿	10	7	4	8	5	9	6	☉
Luna	☽	11	8	5	9	6	10	7	♀
Satu	♄	12	9	6	10	7	11	8	☿
Iupit=	♃	0	10	7	11	8	12	9	☽
Mars	♂	0	11	8	12	9	0	10	♄
XI	X	IX	VIII	VII	VI	V	IV	III	II	I

At the mid-point of this circle is attached an index with the Sun's face around the pivot. There is a hole at the pivot point, presumably for a pin gnomon. The index is divided in hours 0–6 and subdivided as before to 5 minutes.

The engraved parts of the rim and side panels retain most of their black mastic, whereas the central table has lost nearly all the mastic, precluding a clear photograph and ease of reading. The engraving on the disc is in italic, but this has not been copied in the table below.

The rim and two side panels provide astrological information (see Capp 1979; Eade 1984):

Rim. *Capricorne & Aquaries be the Signes of Saturne· Sagiter & Pisces of Iupiter· Scorpius & Aries of Mars· Leo of Sol· Taurus & Libra of Venus· Gemini & Virgo of Mercuri and Cancer of Luna.*

Panel, left: *The best planetts are* ♃ ♀ *The indiffer* ☉ ☽ ♄ *The woorst* ♄ ♂

Panel, right: ☽ *goueṝs The head* ☿ *Lungs* ♀ *Kidnes* ☉ *ye hart* ♂ *ye gall* ♃ *Liuer* ♄ *Milte* [spleen]

The central table (see Eade 1984, pp. 95–7, for the explanation) is labelled:

Left: *Rulers of the day*

Right: *Gouerners of the night.*

Below is a motto: *Vanitas vanitatum, et omnia sub Sole vanitas—Ecclesiastes.*

Location

Musées Royaux d'Art et d'Histoire, Brussels (5886).

46 Sector 1597

Signed and dated: *C · Whitwell fecit 1597*

Brass and iron; radius of arms including iron points 164; width of arms 13, thickness 3; radius of arc 103, width of arc 21

The construction follows that of the Beckit (**37**) and Kynvyn (**69**) sectors, and is yet a third example of the Thomas Hood sector published in 1598. There are two brass arms made of three strips, with iron tips, connected through a compass joint. Attached to one arm and passing through the other is a wide arc of brass. Near the ends of the arms are holes into which sights could be screwed (now missing). The hole in the compass joint is for a back-sight and tripod support.

Side 1

Each arm is engraved with a line of equal parts, from 0 at the join to 110 just short of the end. The divisions are labelled every five units, which are subdivided to halves. The arc is divided 0–100 degrees in units. Each degree is divided to ten minutes by transversals passing across six bands.

Side 2

The signature is near the joint. The engraved scale is non-linear, as is the line of chords

46 Side 2.

46 Side 1.

running from three to ten. The arc has three scales extending over a quarter circle and marked out in half units 0–60, 0–80, 0–100. These scales are used to divide the line, defined by the opening of the tips of the arms, into scales of one inch divided into sixths, eighths, or tenths.

Charles Whitwell was the premier London mathematical instrument maker between 1590 and his death in 1611. He engraved the illustration of a sector and its accessories that was published by Thomas Hood in his *The Making and vse of the Geometricall Instrument, called a Sector* (1598). On the title page is printed: 'The Instrument is made by Charles Whitwell, dwelling without Temple Barre against S. Clements Church.'

Provenance

Presented by T. Hansen in 1936.

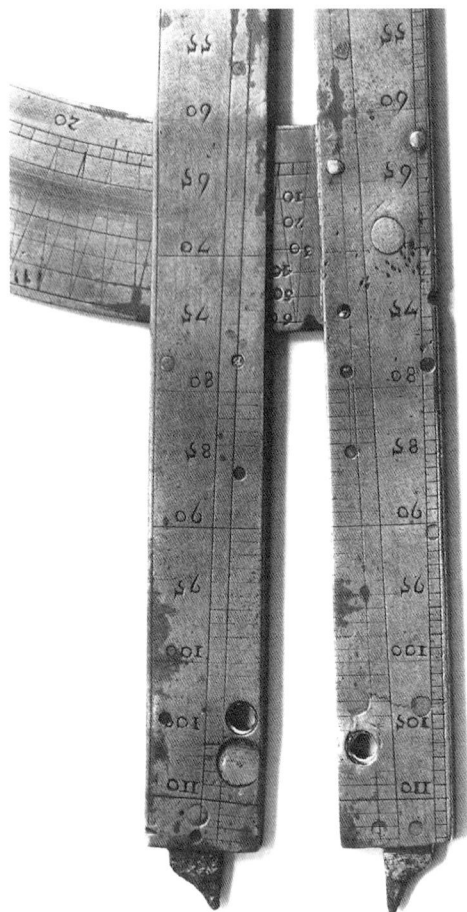

46 Bottom of side 1.

46 Top of side 2 showing Whitwell's signature.

Location

Rundetaarn, Copenhagen, Denmark (den historiske samling nr 224 [P1–3]).

47 Nautical hemisphere *c.* 1597

Signed: *C. Whitwell fecit*; undated

Brass; diameter of main plate 283 (excluding projections), thickness 2.6–4; diameter of secondary plate 208

This instrument follows exactly the design described and illustrated by William Barlow,

The Navigators Supply (1597). All the illustrations in this book were engraved by Charles Whitwell, whose advertisement is on the title page. For this Whitwell instrument see Plate 7.

The main plate is signed on the plain back. It is constructed from two sheets of brass extensively riveted: 38 in the edge, 18 in the centre, 14 in the projections. In the middle is a square hole (5 × 5). The uppermost of the four projections carries the shackle and ring to hold the instrument vertical. Riveted on the lower back are three bars that form a socket into which a compass box (missing) could be slotted. On the four projections are brass catches to hold the ends of two semicircular arcs simulating the meridian and the equator. The meridian is hinged 16 mm above the projection, and moves over the other semicircle,

47 Complete instrument.

47 Central plate.

which is not hinged. At the apex of the meridian there is a spring grip on one side. The meridian has to move to allow the second plate to be inserted.

The meridian arc is bevelled, and engraved on both sides, 90°–0–90°, by 1° intervals, numbered every ten. The equator arc, also bevelled, is graduated on one side in degrees, 0–90°–0, and also in the wind directions (rhumbs), numbered 0–8–0.

The plate has at its outer edge a degree scale numbered in quarters from the top 90°–0. Next is the Zodiac scale, beginning from the top with Cancer. There follows the calendar, the 24 hours, and the 32 winds (16 named by initials). All the letters are upper case, and both letters and numbers show great control. The central part of the plate is not engraved, but has some deliberate burin marks as try-outs for hardness. This region is hidden by the second plate.

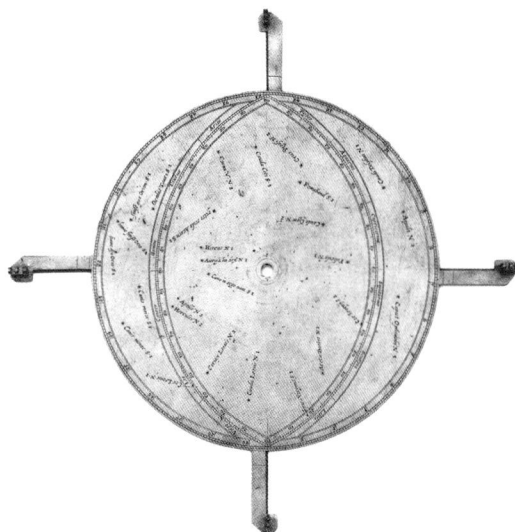

The rotating second plate has extension bars on four sides, bevelled for reading the scales on the outer plate. These support two semicircular arcs, both of which are hinged. The meridian arc has a cross-bar at the base, and a wide section at the top. It originally had a spring grip, now broken. The wide section has the solar declination scale divided into the Zodiac on both sides. The arc is divided in degrees on both sides, 90°–0–90°. The other arc is divided in hours, XII–VI–XII, and also in rhumbs, 0–8–0. At its apex is a thickened portion with a hole to lock by a pin into a similar hole in the equatorial arc on the outer plate.

There are two alidades, one (radius 102) runs over the inner plate; it has a foresight with a slit and three pinholes to cast sunlight onto the backsight, which is not pierced. The other alidade (radius 140) has two vanes, slit at the top and three pin-holes in each, the holes countersunk on the outer sides. This alidade reads the declination scale.

The inner plate is divided round its edge in single degrees. It is also divided into the days of the Moon's age, 29½. Across the middle of the plate are two symmetrical arcs divided into the Zodiac. The remaining surface is engraved with 27 star names in this form:

name of star/ North or South of ecliptic/ magnitude.

Stars on central plate

Cauda Ceti S.3	Cor Leonis. N. 1
Cornu [α Ari] .N.2	Ceruix Leonis N 2
extre erida Acarn S. 1	Cauda Leonis N 1
Oculus Tauri .S.1	Spica Virginis S 1
Hircus N 1	Arctu: Bootes N 1
sinister pes Orion .S.1	Cor Scorpij S 2
hume sinist Orio .S.2	Caput Ophiuchi N 3
Auriga hu dex N 2	Fidicula N. 1
hume dex Orio .S.1	Aquila N 2
Cano in argo naui .S.1	Cauda Delphini N 3
Canis maior .S.1	Cauda Cigni .N.2
Canis minor .S.1	Fomahand .S.1
Apollo N 2	Crus Pegasi N.2
Hercules N 2	

47 Back, showing attachment for compass, and signature.

The construction and mode of using this nautical hemisphere are described and illustrated by William Barlow, whose figure is reproduced on the title page of Robert Dudley's *Dell'Arcano del Mare* (1647), Vol. 3 (1647), Book V. His engraver, Antonio Francisco Lucini, obviously copied the illustration in Barlow rather than risk the difficult perspective presented by the real thing. Some writers on instruments have taken the Dudley picture as their example, forgetting that it was not original to him, and that Barlow was the author of this model. In fact, the origins are earlier. Michiel Coignet (1549–1623) showed such a hemisphere in his *Nieuwe Onderwijsinghe* (see Coignet 1580, 1581), but in this case the main plate was horizontal not vertical

(see also Meskens 1992; 1997, p. 147). This instrument in turn goes back to Martin Cortés, *Breue Compendio de la Sphera y de la Arte de Nauegar, con nueuos Instrumentos y Reglas* (1556). This was translated into English by Richard Eden as *The Art of Nauigation* (Cortés 1561); there were many subsequent editions. For a similar instrument by Cole, and the Cortés version, see **30**.

Provenance

One of the instruments taken to Florence by Sir Robert Dudley in 1606.

Literature

G. Turner (1991, p. 233).

Location

Museo di Storia della Scienza, Florence (1099, 1122).

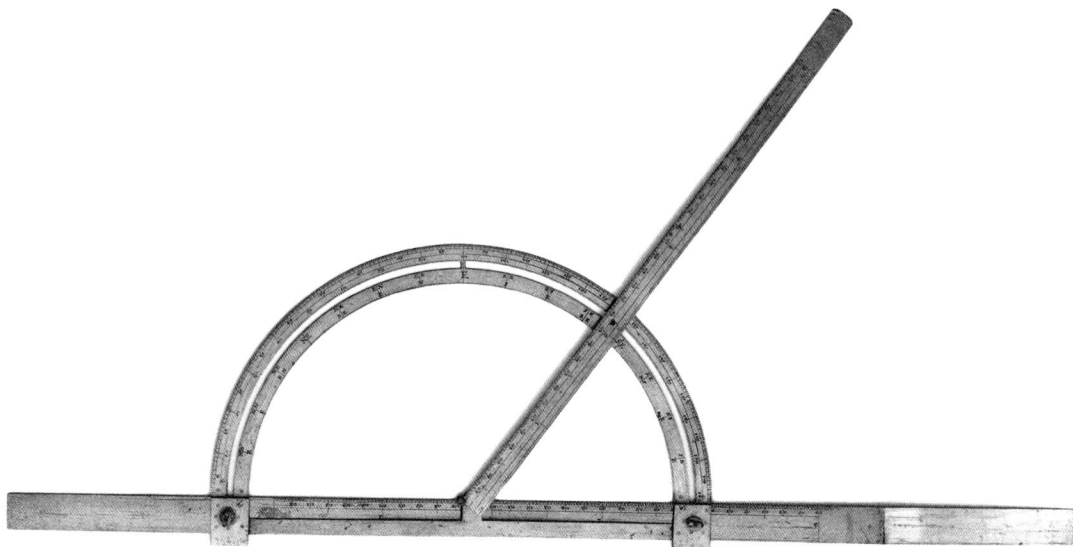

FIG. 48

48 Chart protractor *c.* 1600

Unsigned and undated; attributed to Charles
Whitwell

Brass; base bar length 506, width 17, thickness 0.7;
radius of semicircle 118

The base bar has two scales over the middle
portion: a scale of 0–60 repeated five times,
divided in units and numbered every ten;
below, a scale 1–5 that counts the groups of
60. A semicircle is attached to a pair of sleeves
that traverse the bar and can be clamped in a
chosen position by turn-screws. The arc is in
two sections separated by a narrow aperture
that locates a radial arm (length 288) pivoted at
the centre. The arm has scales identical to
those on the bar. The outer part of the semicircle is divided in $\frac{1}{2}°$ units from 0 to 180°,
labelled every 5°. The inner part is marked
out with the points of the compass from N
through E to S, each point lettered with the
initials of the direction.

This protractor is illustrated in Dudley,
Dell'Arcano del Mare (1647), Vol. 3, Book V,
p. 9, fig. 14. Its use for great circle navigation
is described on p. 9, where it is called the
spherical protractor.

The form of the numbers and letters
(especially E) and the manner of construction all point to Charles Whitwell as the
maker.

Provenance

One of the instruments taken to Florence by
Sir Robert Dudley in 1606.

Location

Museo di Storia della Scienza, Florence (1121).

49 Protractor *c.* 1600

Unsigned and undated; attributed to Charles
Whitwell

Brass; diameter 92; thickness *c.* 0.8

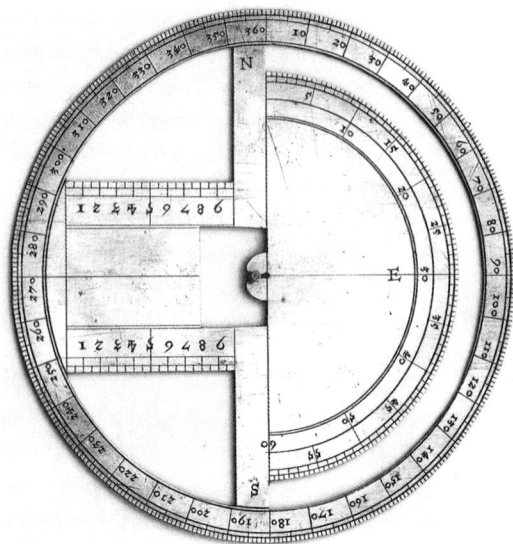

49 Side 1.

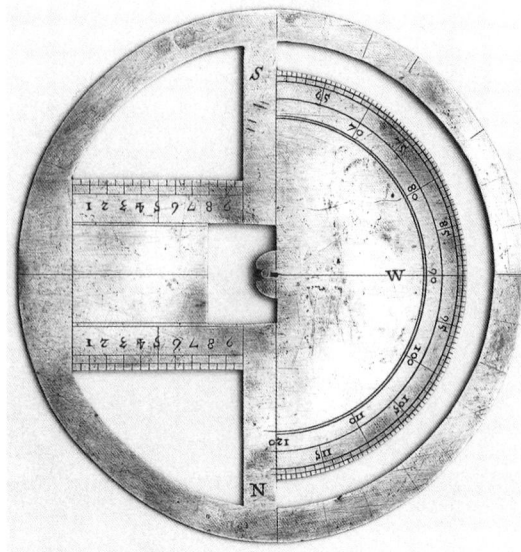

49 Side 2.

Side 1

The outer edge of the circle is divided in single degrees, 0–360°, labelled every ten. The diameter, 0–180°, has a bar marked N and S, with to the East a semicircle whose edge is divided with a scale, 0–60, in half unit divisions, and labelled every five units. To the West is a rectangle with top and bottom edges divided 0–10 in half units, labelled every unit.

Side 2

This side has no outer scale, but continues the semicircle scale, now to the West, from 60 at S to 120 at N. The rectangular scale is as before. At the centre is a pin-hole notch to locate a pencil or stylus. On the reverse the otherwise blank outer edge has the West half roughly marked with scratches corresponding to the points of the compass, i.e. 16. See also **50**.

The instrument comes from the same workshop as the chart protractors, **39** and **48**,

judging by its craftsmanship, which is that of Whitwell.

Provenance

One of the group of instruments taken to Florence by Sir Robert Dudley in 1606.

Location

Museo di Storia della Scienza, Florence (613).

50 Two watches before 1606

a) Pocket watch, unsigned and undated; inscribed: *Charles Whitwel*

b) Pocket watch, unsigned and undated; inscribed: *R. whitwell*

These watches have been included because of their association with an Elizabethan instrument maker. They are similar, and each is engraved with the name of its owner, but not in the hand of Charles Whitwell or any other identified craftsman. It appears that the same hand has engraved the names. It is

50 Wooden outer case, metal case, and movement.

reasonable to suppose that the Charles Whitwell of the watch is the maker of the instruments described in this book, and of several maps and book illustrations. He is linked with the watch trade through the watch dial, **55**. Charles was apprenticed in 1582, took his freedom in 1590, and died in 1611. His brother, Robert, died in 1606. J. Brown (1979, p. 24) quotes from the Court Minutes of the Grocers' Company (MS 11,588/2): 11 Aug. 1606 'Agreyd that Charles Whitwell grocer shall have the 50*l* for ii years wch his brother Robert Whitwell deceased latelie had. And William Whitwell and George Budd salters are alowed his sureties'.

Both watches are known through their appearance at auction, which has left photographs and concise descriptions. The first was seen by the present author, but not the second.

a) Christie's King Street, St James's, London, sale 14 July 1986, lot 114.

An English early gilt metal Puritan oval verge watch, the movement signed *Charles Whitwel*, with three-wheel train, fusee and chain, symmetrically pierced cock with stud and pin, steel ratchet and click, engraved border, the simple dial with single hand, the case with applied bands to band and plain covers, solid ring pendant; first quarter 17th Century, restorations; with rare boxwood outer case; crank key with later handle. 76 mm over pendant.

50 Engraved name on movement.

My measurements of the oval parts are (major axis, minor axis, thickness): the movement, 50 × 39 × 18; the plain metal case (excluding pendant), 57 × 46 × 25; length including pendant 73; outer boxwood case, 70 × 56 × 35. I am grateful to Mr Richard Garnier for photographs.

This watch is illustrated in Cedric Jagger, *The Artistry of the English Watch* (1988, pp. 29–30, plate 17), and the impression is given that Whitwell was the maker: 'The only other English watch from this exceedingly early period to pass [through auction] was one by Charles Whitwell, for whom speculative dates of 1593–1606 have been quoted. Very little is known about him.'

b) Christie's King Street, St James's, London, sale 21 November 1990, lot 157.

A James II [*sic*, read I] giltmetal Puritan oval early verge watch, the gilt movement signed *R. Whitwell* with baluster pillars, fusee and gut to large barrel, short train with plain steel balance, open bordered semi-symmetrically pierced pinned cock and foot, blued-steel and giltmetal ratchet to cock, foliate engraved border, the dial with similarly engraved centre, touch pieces to Roman chapter ring, single blued-steel arrowhead hand, the plain case with applied ringed band, swivel wind cover and shaped pendant. 75 mm over pendant. Probably Robert Whitwell, circa 1600.

The metal cases and the movements are the same, including the engraved floral motifs on the border to the movements. One may assume that both were made at the same time for the two brothers, and before Robert's death in 1606. I am grateful to Mr David Thompson for pointing out that the term 'Puritan watch' in the sale catalogues is incorrect, since such a term is inappropriate before 1630. On the contrary, the style of the watches is similar to a watch by Randolf Bull dated 1590 (BM 1874, 7–18.11). He also commented that the two watches are part of a small group of early English watches to pass through the auction houses in recent years.

51 **Lunar computer** *c.* 1600

Unsigned and undated; attributed to Charles Whitwell

Inscribed: *Sr Robert Duddeley was the inuentor of this Jnstrument*

Brass; overall diameter 726; diameter of plate 638, thickness *c.* 2

This is the largest and most complex of all the Elizabethan precision instruments, and one that does credit to the inventor and most particularly to the craftsman who engraved it. It was made between the alleged date of invention, 1596, and the time when Sir Robert Dudley left England in December 1605. The epoch for which it was made is engraved on it: 1601. At this period there were only two London craftsmen who could have produced this computer: James Kynvyn and Charles Whitwell. Both men were supplying Dudley, as is clear from signed instruments with the Dudley collection in Florence. This computer required the engraving of not only medium- and large-sized elements, but also exceedingly small ones. As has been made clear in the present study, Whitwell had great control over his burin, and with it could consistently cut minute numbers and letters. By examining these in detail, and comparing them with the rest of his corpus, Whitwell is revealed as the maker. The required mental and eye control, and the patience over a long period would have defeated most crafts-man. It seems likely that this computer took half a year to construct. The maker will have had to work under the guidance of Dudley, who intended this instrument to achieve a novel result: nothing less than the ability to calculate the place of the Moon over a period of 30 years. Dudley's intention was to refine the lunar distance method of finding the longitude. The inscription claiming Dudley as the inventor is

51 Side 1, close-up of rim, bottom right.

not in Whitwell's hand, and no identification has so far been made.

A version of this computer is illustrated in Dudley, *Dell'Arcano del Mare* (1646), Vol. 1, Book I, p. 38, fig. 21. It is described as an instrument for estimating the primary and secondary motions of the Moon. Book I is on the problem of finding the longitude: 'Nel primo de'quali si tratta della Longitudine praticabile in diversi modi, d'invenzione dell'Autore'.

A complete study of this instrument would require a considerable time, and result in a monograph. This is beyond the scope of the present work. What will now be provided is the labelling given to the various features, together with some photographs of details on the instrument.

Side 1

(On the rim)

The yearly motion of the ☽ middle place from the sunne by 360 for the merid; of London according to the tables of Renaldus which differeth 8′ 37″ from Maginus

The yearly motion of the ☽ Anomalae from the Apogeon of the ☽ great Epicicle by 360 and meridian of London

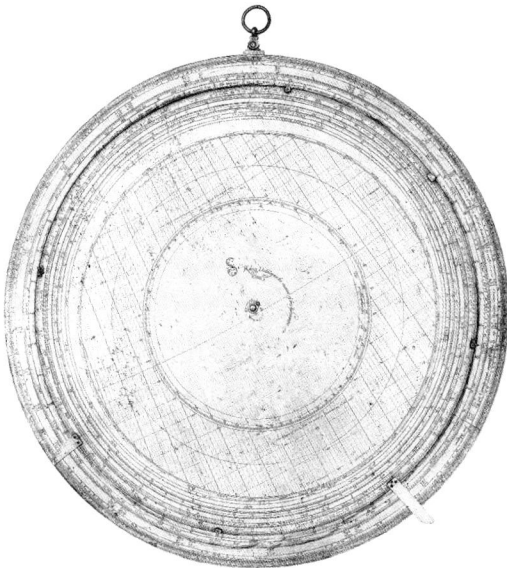

51 Side 1

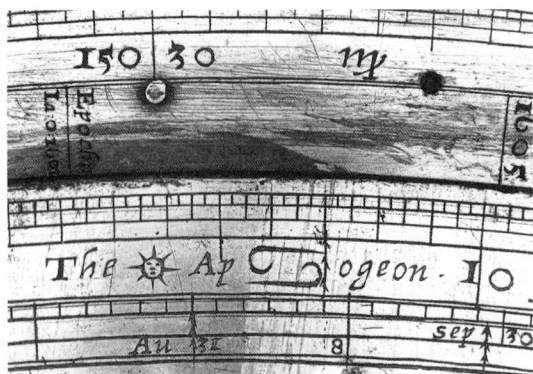

51 Side 1, close-up of rim, bottom right.

The yearly motion of the first starr of Aries beginninge 1600

The yearly motion of the ☉ *Apogeon beginninge 1600*
(followed by a short scale of 12 divisions, numbered 0–60)

Epocha / IA: ☉: 1601

The ☉ *declination*

The dayly motion of the ☾ *middle latitude*

The ☉ *middle place*

The ☾ *middle motion annd vnited to it the 24 howers*

Side 2
(inscription)
The first hower of Ianuary in all the ☾ *monthly motion beginneth with the first of Aries in this inst:*

The ☉ *Apogeon serving from the yeare 1600*

☉ *Peregeon*

A deuision for the true place of the ☉

for the dayly and 1 monthly motion of the ☾ *middle place*

To give the ☉ *middle place*

The dayly motion of the ☾ *Anomilae*

The 1 starr of ♈ *mag: 3*

The one monthly deuisions for the ☾ *Anomilae*

The 2 monthly deuision for the ☾ *Anomilae*
(first band)
The prosthaphaereses or aequations of the ☾
(repeated)

(second band)
The distance of the ☾ *from the earth in semi diameters of the earth*
(repeated)

(third band)
The prothaphaereses of the Sunne

(centre)
the inscription as above, and:
The cen: of ye ecliptic; The cen; of ye ☉ *excentric*

Rule

Now separated from the plate is a rule (length 360) with an integral circular protractor (diameter 91). At one end is a rivet for attachment to the camb at the centre of the plate; the other end has a portion broken off. The

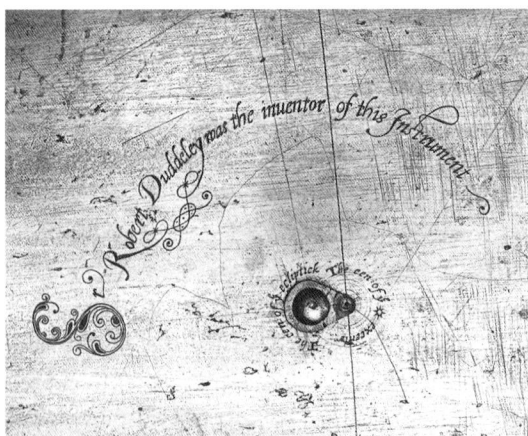

51 Inscription naming Dudley, not in Whitwell's hand.

back shows that the brass casting has a multiplicity of fissures, a result of poor annealing. The rule has a non-linear scale, 0–86, marked by engraved numbers every five. The protractor circle is engraved:

The ☾ little Epicikcle in Quadratura; and: In the midle Coniunction & in opposition.

None of this engraving is by Whitwell, and from the spelling it is probably the work of an Italian. The degrees are marked 0–360° after a double circuit, and the labelling, every 10°, is by punched numerals poorly executed. These are the same punches found on some other instruments of Dudley (e.g., the copper wind rose; see Chapter 5, Fig. 5.4).

Side 2

This side is occupied by a tide computer, now incomplete, that was intended to be more accurate than the common sort. The plate, which drops into the rim, is engraved, from the outer edge: a calendar scale; a non-linear scale labelled: *Caput Draconis, Cauda Draconis, Lunes Borea, Lunes Australea*; a degree scale divided to a quarter degree; a scale with 12 bands divided into the Zodiac with angled divisions (straight at the solstices). This scale is labelled: *The oblique Zodiack only vsed with the howers to giue the tyme of tydes.* The remainder of the plate is blank, and was originally occupied by two

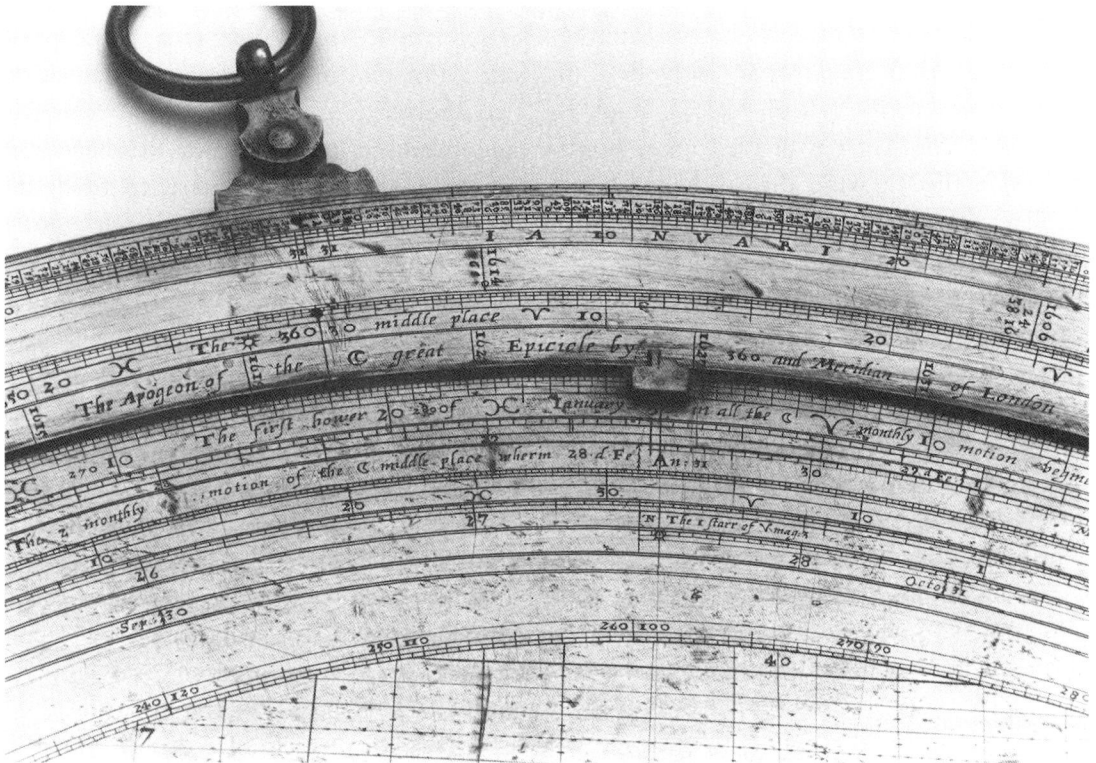

51 Side 1, close-up of top.

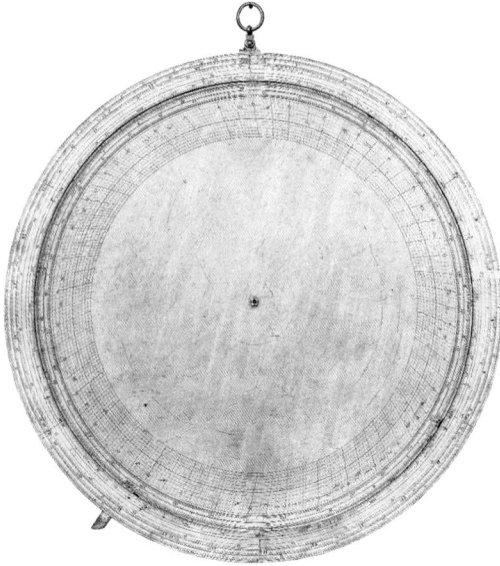

51

52 Nocturnal and rectifier of the Pole Star *c.* 1600

Unsigned and undated; attributed to Charles Whitwell

Brass; diameter 83; thickness of rim 4

An outer ring is marked with a 24-hour scale, twice 12 hours in roman numerals. On its opposite side are 24 small pegs with which to feel the hour positions during the night. Alongside the pegs is a scale of single degrees, not labelled. Within the outer ring fits a cast second ring with a flange. The side adjacent to the pegs is divided into the months and single days in the months. A small ring is fixed between the centre of the second ring and the position of March. At the middle of the instrument are two rotatable pointers. The longer is called by Wright (1610) 'the guard index' that pins into a hole in the day circle. The shorter is called by Wright 'the day index', and it reaches from the centre of the day circle to the circumference. Inside the small ring rotates a solid semicircle with a central notch. To the outer ring is fixed a strip of brass (length 70) that acts as a handle, which can fit into a slot in another instrument.

The side with the hour scale can be used as a nocturnal, and the other side as the rectifier of the Pole Star. It is this part that gives the measure of the difference between the altitude of the true North Pole and the altitude, as measured, of *Polaris* (α UMi). An adjustment, or rectification, is necessary because *Polaris* appears to rotate about the true Pole (represented by the centre of this instrument), so that when it is North or South of the Pole, about $3\frac{1}{2}°$ has to be subtracted or added to the altitude as measured by a quadrant or

volvelles. These missing parts can be identified from the paper volvelles in *Dell'Arcano del Mare* (1646), Vol. 1, Book I, figs 20–1. The larger is divided in 24 hours and with the 32 wind directions, and the smaller with the age of the Moon in days to 30, and an index at 30 (new moon). Finally, there is a rule that rotates about the centre.

Provenance

One of the instruments taken to Florence by Sir Robert Dudley in 1606.

Literature

This instrument has appeared in the literature as a mariner's astrolabe. For an explanation of this, and the reason why this computer was given the number IC 2055, see **75**.

Location

Museo di Storia della Scienza, Florence (1116; rule 1127).

forestaff; when East or West no correction need be made to the value obtained for the altitude (Waters 1958, pp. 45–6). The notch in the semicircle in the small ring is to be set to view the star *Kochab* (β UMi), the premier guard in the Little Bear. This star also serves as an indicator when the instrument is used as a nocturnal.

An instrument for correcting the variation between the true Pole and *Polaris* was described by Martin Cortés in his *Arte de*

52 Side 2.

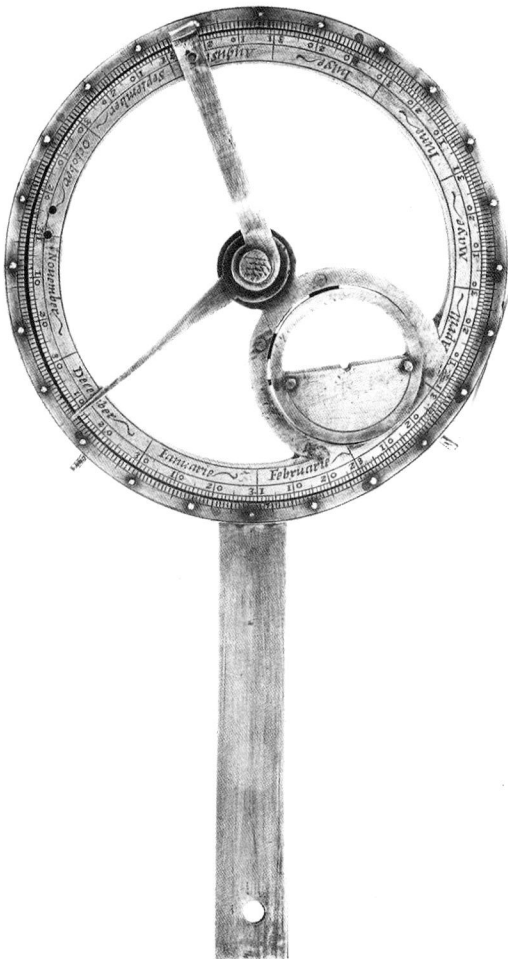

52 Side 1.

Navegar (1556). The first English edition, translated by Richard Eden, was published at London in 1561. There are many later editions; see Adams and Waters (1995, pp. 454–60). The variant as described here was published by Edward Wright, *Certaine Errors in Navigation*, 2nd edition (1610); it is not in the first edition of 1599. On the title page, and in a separate illustration, this ingenious device is shown attached to the seaman's bow, a form of forestaff. A similar bow is shown in Dudley, *Dell'Arcano del Mare* (1647), Vol. 3, Book V, p. 23, fig. 75. Here Dudley says that the bow is the invention of an English mariner.

The craftsmanship and letter forms are typical of Charles Whitwell.

Provenance

One of the instruments taken to Florence by Sir Robert Dudley in 1606.

Location

Museo di Storia della Scienza, Florence (2500).

53 Azimuth dial and planisphere *c*. 1600

Unsigned and undated; attributed to Charles Whitwell

Inscribed: *An Azumoth dyall for the latitude of 52· deg· inuented by Sr Ro: Duddeley Ano 1598* ✳

Brass; diameter 145, thickness 3

Side 1

On the dial side the rim is divided in single degrees, numbered every ten in quadrants, 90° at North. The plate is engraved with the upper hemisphere of the globe projected upon the plane of the horizon. The azimuth hour lines are cut to 20 minutes of time, the hours labelled in roman numerals. Each hour is marked by a straight line running through a wavy line; the subdivisions are by dotted line. The declination lines are at 1° intervals. Declination lines are continuous, with arrow heads every 5°. The equinoctial line is divided in single degrees alternately shaded. Above is engraved: *The Tropick of Capricorn*, and below: *The Tropick of Cancer*. The blank area at the South is occupied by the inscription, see above.

53 Front.

At the centre is a pin, secured by a wing-nut, about which rotates an over-long index arm. This was noted as a replacement in the Museum's Register when purchased: 'nut and index both modern'. To be complete, a means of showing the azimuth is required. The correct index is shown on an engraving of a dial of this type printed in Dudley, *Dell'Arcano del Mare* (1647), Vol. 3, Book V, p. 4, fig. 4. The use of such a dial is explained in Book V, p. 13, chapter XI, fig. 21.

Side 2

What may be regarded as the mater is engraved with the universal stereographic planisphere known as the *Saphea*. Lines are drawn every 2° in longitude and latitude. Lines at 10° of latitude are distinguished by arrows, and in longitude every hour (15°) is so marked. The hours are numbered in Arabic

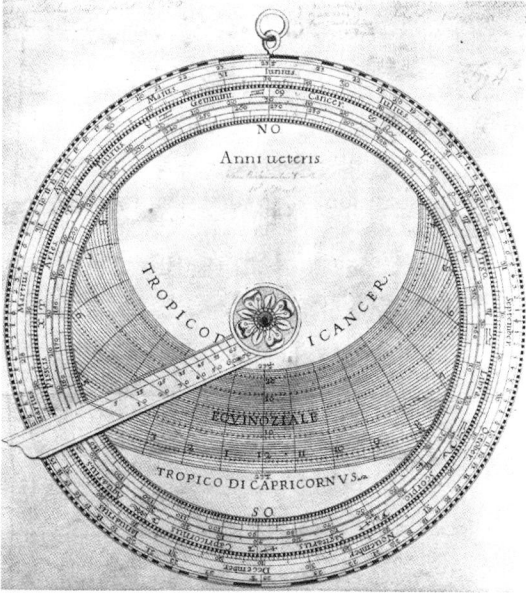

53 Engraving from Dudley, *Dell'Arcano del Mare* (1647).
BL Oxford.

numerals. The ecliptic is within a band, divided into the Zodiac, with the sigils. The raised rim is divided in degrees, numbered in tens both ways, quarterly 0–90°. Within the rim may be set a semicircular plate, finished and engraved solely on one side, with the same stereographic projection as before, but with lines of longitude at 2° only. These lines are emphasized every 10° by arrows. This plate can, therefore, be an alternative to the Right Ascension markings on the main plate. The shackle, with hanging ring, is centred on side 2, but not on side 1, where this is of no consequence.

It is generally accepted that this form of dial was invented by William Oughtred (1575–1660), a mathematician of King's College, Cambridge, where he studied from 1592 to 1603, when he took an ecclesiastical living near Guildford. That he may have achieved the

invention between 1597 and 1600 is revealed in the *Epistle dedicatore* written by William Forster, the translator of Oughtred's book from the Latin: *The Circles of Proportion and the Horizontall Instrvment, Both inuented and the vses of both Written in Latine by Mr. W.O.*, printed for Elias Allen, and sold at his shop opposite St Clements Church, in 1632. In his dedication, Forster said he often travelled to visit and learn from Oughtred, and in the summer of 1630 he was shown notes, written in Latin, on the circles and the horizontal instrument, which Oughtred told Forster had been projected about 30 years before (i.e., about 1600). Forster offered to translate Oughtred's work, which was done by May 1632. A priority dispute then arose between Oughtred and Richard Delamaine (d. 1645), this book having angered Delamaine. A second edition was published in the next year (Oughtred 1633), and to this was added what

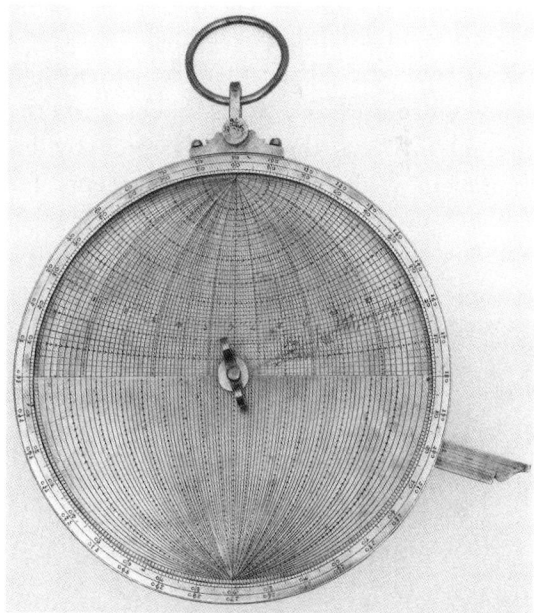

53 Mater.

Oughtred called *An Apologeticall Epistle*, where he denounced Delamaine. The matter of this dispute between Oughtred and Delamaine is treated at some length by A. Turner (1981).

What exactly the present dial has to do with Sir Robert Dudley is not at all clear. Perhaps Oughtred and Dudley, just seven months older, knew each other. Dudley published an engraving of such a dial, and there are resemblances between the two. It is noteworthy that both the engraving and the present dial do not have ecliptic arcs, whereas the earliest printed diagram (Gunter 1624, p. 65), and all known other examples of the dial, are provided with the ecliptic. Oughtred, in his *An Apologeticall Epistle*, says that in 1618 Gunter had seen a drawing of the Horizontal Instrument, which was then made into a copperplate that was used, years later, to produce the illustration in Gunter's book on the sector published in 1624.

Of the engravings in *Dell'Arcano*, several are copied from English Elizabethan texts, and many from the Elizabethan instruments in Dudley's possession. It is likely that the present dial was the model copied by his engraver. Whitwell made instruments for Dudley, who took a large group to Florence in 1606; this dial must have been one of them, so a terminal date is provided. The attribution to Whitwell rests on the calligraphy and on the layout of the coordinate net on both sides of the instrument, which resembles that on the signed and dated astrolabe **43**.

Provenance

Purchased by the British Museum from Henry Heilbronner in 1894.

Literature

A. Turner (1981, pp. 115–16, 122); Ward (1981, p. 120), no. 346.

Location

The British Museum, London (1894, 6–15.2).

54 Tide computer *c.* 1600

Unsigned and undated; attributed to Charles Whitwell

Book illustration

The only evidence for the existence of this instrument is plate LII after p. 200 in Waters (1958). The caption reads: 'Brass Tide Computer of the Latter Half of the Sixteenth Century'. Lt Cdr Waters told the author, in December 1983, that he was given the print in 1950 but did not know the source.

The design of this computer is exactly the same as that on the 1604 Whitwell compendium **60**, and on the 1602 Whitwell compendium, **59**, where the only slight variation is that the half hours are alternately shaded. The indices are thinner in the version illustrated in the book.

For a description of the construction and use of such a computer, see **59**.

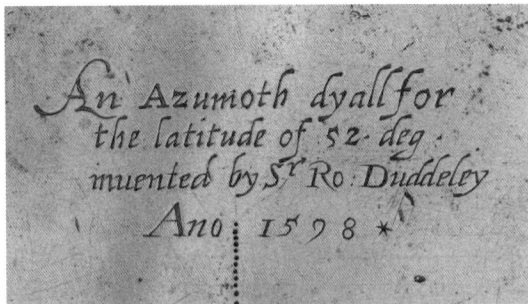

53 Inscription naming Dudley, not in Whitwell's hand.

EDWARD *the 3 did add this armes of France*
And Henry 5 made Fraunce to Brittaine yeelde:
But civill warres, did breede vs this mischaunce,
Small: gaynes remaynes, of that well fought: feeld

ENGLAN: FAMOVS PLAC:
In the 52 Shieres are marked
With the first letter of thir names
As A for exsample
Bishopricke & citie thus A
P Pallaices A cast: & c townes A
Ther-be chiefe places thus A

W. B: invent   1590

KINGE Henrie 7. from strife the state restorde,
To Henrie right, he left it in great welth,
He wise and boulde, good happ his fates afford,
But in his doughter most: Lord bless hir helth

THIS mayden queene, like Debora doth raign,
She by hir wisdom, and hir constant zeale:
In peace, and plentie, doth gods worde maintaine,
Would god J could hir vertues all reveale·

TWISE sixteene yeares 5 scepter in hir had,
No traitors could, nor forraie foes wrest out:
Great warres abrode, yet god defends hir land,
Lord let thy Angells, compasse hir aboute·

THVS much in Miles whole Englad it cotaies 340
Thus much in Miles will reatch aboute it rounde: 1390
Hir Length from Lizard point to Barwick strais, 338:
Fro: Douer Holyhead the breed: h is founde· 2 5 0

AMOGST good neigbors thus doth Englad stand,
Each shirr presents first letter of hir name·
For other worthie places in this land
My fistie two paticulers haue the same

FROM East to West Londo is most in Length,
The Bredth therof is from the North to South:
What place so fit for pleasure store & strength,
And for all trades, and trayninge vp of youth·

FOR fruitfull ground, for riuer, and good ayre,
For store of welth, of people, and of powrr,
Did Troye att any tyme, seeme halfe so faire,
As doth hir doughter Londo att this howr·

PLATE I   36 Playing cards by Ryther, 1590. BM London.

PLATE 2   **5** Gemini astrolabe, front, 1552. ORB Uccle.

PLATE 3   **5** Gemini astrolabe, back, 1552. ORB Uccle.

PLATE 4   **13** Cole azimuth sundial, *c.* 1574. SM London.

PLATE 5   **72** Nonius quadrant attributed to Kynvyn, *c.* 1600. IMSS Florence.

PLATE 6   **64** Case of drawing instruments, sector by Whitwell, *c.* 1600. SM London.

PLATE 7   **47** Nautical hemisphere by Whitwell, *c.* 1597. IMSS Florence.

PLATE 8   **26** Cole compendium, 1579. PC.

PLATE 9   **34** Ryther compendium, 1588. SM London.

PLATE 10   **59** Whitwell compendium, 1602. Christie's.

PLATE 11　**99** Diptych dial showing map of England, *c.* 1600. HM London.

PLATE 12　**56** Diptych dial showing map of England, probably by Whitwell, *c.* 1600. MHS Oxford.

PLATE 13　**71** Altitude semicircle for Kynvyn's theodolite, 1597. IMSS Florence.

(a)

(b)

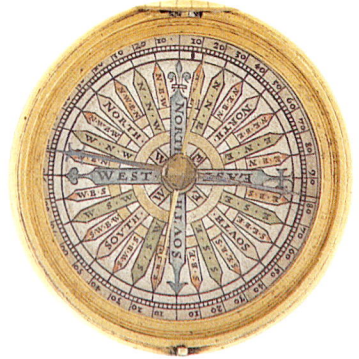

(c)

(d)

(e)

(f)

(g)

(h)

(i)

PLATE 14

(a) **25**, Cole, type A–1, 1575.

(b) **66**, Kynvyn, type A–1, 1593.

(c) **84**, Allen, type A–2, c. 1610.

(d) **35**, Ryther, type B–1, hand, 1590.

(e) **99**, RG, type B–1, c. 1600.

(f) **56**, type B–1, double, c. 1600.

(g) **58**, Whitwell, type B–2, 1600.

(h) **98**, TW, type B–3, 1602.

(i) **62**, Whitwell, type C, 1610.

55 Watch face *c.* 1600

Movement signed: *HRoberts*; undated; dial attributed to Charles Whitwell

Gilt brass and silver; diameter of face 65.5

An outer ring of twice 12 hours in roman numerals, is divided in quarter hours alternately shaded. Next are circles of days in the month, names of the months, degrees in the Zodiac and sigils. Two volvelles follow, the larger with a skeletal solar index, and the Moon's age to $29\frac{1}{2}$ days; the smaller with a skeletal lunar index and twice 12 hours in arabic numerals at the edge, a lunar phase aperture, and the rest of the surface covered with a fussy diagram of astrological dispositions (aspects). This volvelle is not by Whitwell, and is in an unidentified hand. At the centre is the blued-iron hour hand that points over the outermost hour ring.

The watch is regarded by Jagger (1988, p. 30) as:

55 Watch face.

one of the very finest early English watches, with a complicated dial, from the first years of the seventeenth century... . The splendid dial has both solar and lunar indications, including the age, phase and aspect of the latter; the chapter ring is divided into two periods of twelve hours ... and there are symbols for the zodiac.

I am grateful to Mr David Thompson for pointing out that the arms on the inside of the cover are those of the Gournay (Gurnay, Gurney) family of West Barsham, Norfolk. Henry Gournay had 16 children noted in his Will of 1614, proved in 1623.

Cedric Jagger was correct to give high praise to the maker of the dial: 'complex and superbly executed', because the craftsman was Charles Whitwell. For a different connection Whitwell had with the watch trade, see **50**.

The attribution to Whitwell rests on the characteristics of the letters, numbers, and sigils of the Zodiac, and comparisons with the signed vertical dial of 1593 (**41**) and the signed quadrant of 1595 (**42**) make this clear. The face is, in effect, an adaptation of parts found in many compendia, now moved by clockwork not by hand.

Provenance

Presented by Felix Slade in April 1856, having bought it at the Christie & Manson sale of 9 April 1856, lot 807, the property of the late Colonel Sibthorpe, M.P.

Literature

Jagger (1988, p. 31), plates 18 (dial), 19 (movement).

Location

The British Museum, London (1856, 4–29.1).

56 Diptych sundial *c.* 1600

Unsigned and undated; map attributed to Charles Whitwell

Ivory and silver; length 84, width 62, thickness 18

This dial is made in ivory, with hinge and catches in silver. The shape is octagonal, and closely resembles the slightly smaller dial, **99**. The top and bottom are plain, and have bevelled edges. At one end is a small ring for hanging onto a cord. Two catches engage studs in the side of the lid to close the instrument. Wires pierce the ivory to form a hinge to the top leaf, which is held vertically by the usual hook and peg. Inset in the inner side of the upper leaf is a map of England and Wales, and in the base is a magnetic compass. Both are surrounded by a simple decoration consisting of concentric circles. Similar decoration is on the dial **99**. Additional decoration is provided by a few indentations coloured in red, green, or yellow.

The base is provided with a turned well to hold the magnetic compass. The windrose is type B-1, and has a diameter of 41; it is not coloured. Remarkably, there is another wind rose underneath the first. By shining a light through the translucent ivory, the second card is made visible. Also type B-1, with a diameter of 41, it has 16 points like the first. The triangular points can be seen to be parti-coloured, so far as may be discerned, in black–white, red–yellow, and red–dark green. For a discussion of this and other windroses, see Chapter 4, and Plate 14f. The hidden card is positioned one point East of North ($11\frac{1}{4}°$), but the visible one shows zero declination. A position of one point East was accurate for 1580 and continued for another 20 years or more, but the declination reduced until the value was zero in *c.* 1650. The evidence given by the concealed card points to an Elizabethan date for the construction of the dial. The magnetic needle is a replacement, probably nineteenth century or later. The glass cover and gilt-brass retaining ring are presumably also replacements.

Around the compass well, and cut into the ivory, are the hour lines, divided in quarter hours, from V to VII, the loss of the normal hour at the ends of the day is occasioned by the restriction of space brought about by the presence of the compass. The string gnomon

56 The magnetic needle is a replacement.

(missing) is attached at the tip of the upper leaf to a metal loop adjacent to the South point of the compass. In the ivory at both ends of the string is punched the latitude: 52.

The map is the most striking feature of this diptych sundial (Plate 12). It is merely 56 mm in diameter, and is the only example known (G. Franks, private communication 1998). The details contained on it could only have been provided by an exceptionally fine engraver, such as Augustine Ryther, Charles Whitwell, or Elias Allen. The map displays the counties of England and Wales with their boundaries outlined in colour: red, blue, green, yellow. The counties are named in full or with abbreviations. Around the edge is the motto:

OF THESE THINGES FOLLOWINGE, THIS FAMOVSE
ISL IS FVEL, MOVNTAYNES FOVNTAYNES BRIDGES
CHVRCHES WOMEN & WOLL +

It is quite apparent that the original model for the map is the map of England and Wales cut by Augustine Ryther, dated 1579, in Saxton's *Atlas of the Counties of England and Wales* (London, 1579). This was the first printed atlas of the country, and the first uniform national atlas for any country. This was also the source for the tiny map in the set of playing cards by Ryther, dated 1590; see **36**. The only other similar dial belongs to the Horniman Museum, London (**99**). Ivory compass dials by Whitwell (**57**) and by Allen (**82**), suggest that all four were made in the first decade of the seventeenth century. With Ryther dead by August 1593, it seems that only Whitwell, who served his apprenticeship with Ryther, could have produced this map. Whitwell is known for engraving maps of Jerusalem, France, Surrey, and Kent. Elias Allen is not known for engraving maps. The calligraphy also supports Whitwell. Amongst the upper case letters the *S* tips over towards the right, and the bottom of the *E* is extended. In lower case letters, the *p* has a gap at the top between the risers, and the *r* has its upper curl distant from the riser. These, and other characteristics, are to be found on the products of Charles Whitwell.

The production of diptych dials with magnetic compasses is a trade particularly associated with Nuremberg during the sixteenth and seventeenth centuries. They were made in ivory, boxwood, or pearwood in the early seventeenth century (Gouk 1988). Most extant Nuremberg ivory dials are from the seventeenth century, and nearly all are rectangular, with a few oval in form. The Harvard University collection of ivory diptych dials has three octagonal in form, one *c.* 1620, and two *c.* 1650 (Lloyd 1992). Harvard's collection of 82 dials has nothing from England, and the Whipple Museum, Cambridge, has 50 diptych dials, but none are English (Bryden 1988). Before this acquisition, there were no English diptych dials in the large collection of over 120 German and French ivory diptych dials held by the Museum of the History of Science, Oxford.

Provenance

Christie's South Kensington, sale 30 May 1996, lot 176.

Location

Museum of the History of Science, Oxford (96–24; 47973).

57 Compass sundial *c.* 1600

Signed: ✶ *C · Whitwell* ✶ ; undated

Brass, gilded, and ivory; diameter 67, thickness 26

The ivory lid is fitted with brass discs on both sides, and the lower part of the instrument has a magnetic compass sunk into it; the bottom is plain ivory.

Set into the lid is a tide computer, with a 24-hour scale subdivided in half hours, marked in roman numerals twice XII. Inside the hour scale is the sequence of the 32 wind directions, alternately labelled with the initials of the directions. A volvelle has an index to point to the hour, and a lunar age scale, 0–29½. The second volvelle is missing; for a complete example, see **59**.

The disc inside the lid begins at its edge with a calendar scale divided into days and numbered in tens (28 or 31). The months are named in full. Next is a Zodiac scale, divided in degrees, numbered in tens, and named in

full. The First Point of Aries is at 10 March. The first (solar) volvelle has its edge divided into the age of the Moon, 0–29½ days, with an index to point to either the date or the position in the Zodiac. The second (lunar) volvelle has an index to point to the Moon's age, and a phase aperture. The centre is occupied with a diagram of the aspects: opposition, trine, quartile, sextile.

The chapter ring is set over the compass in the lower section. It is divided into quarter hours, alternately shaded, with the hours marked in roman numerals from IIII to VIII. The folding gnomon is held in two sockets, one placed by 12 o'clock, the other within the tracery over the south part of the dial. The gnomon is an open triangle with the short side composed of two S-shaped bars. This dial is designed for a latitude of about 52°.

The compass rose is printed from a copperplate and hand coloured (diameter 56). A degree scale (0–90° in quarters) is round the edge. The design is type A-2. The needle is straight, plain one end and with an arrow at the other. The cover is of mica.

Provenance
Purchased by the British Museum in 1854.

Literature
Ward (1981, pp. 23–4), no. 17.

Location
The British Museum, London (1854, 1–3.1).

57

58 Compendium 1600

Signed and dated: ✶ *Carolus Whitwell fecit* 1600 ✶

Brass, gilded; diameter 83, thickness when closed 30

Comprising from the top:

a) nocturnal with calendar scale and table of Prime Numbers and Epacts

b) the names and latitudes of 28 towns

c) equinoctial sundial

d) magnetic compass

e) tide computer

a) The lid is constructed to be the nocturnal. At the edge is a circle engraved with the names of the months: *Ianuarie, February, March, Aprill, Maye, Iune, Iuly, August, September, Octobe, Nouember, December*. The months are divided in 2-day intervals, and numbered in tens with 31/28 as appropriate. Next come two volvelles, the outer with a 24-hour circle, marked out as twice XII, the hours sub-divided in half hours by shading. At each hour is a protruding tooth, and an index is at 12 o'clock. The next volvelle is for sighting the Pole and aligning the Guards of

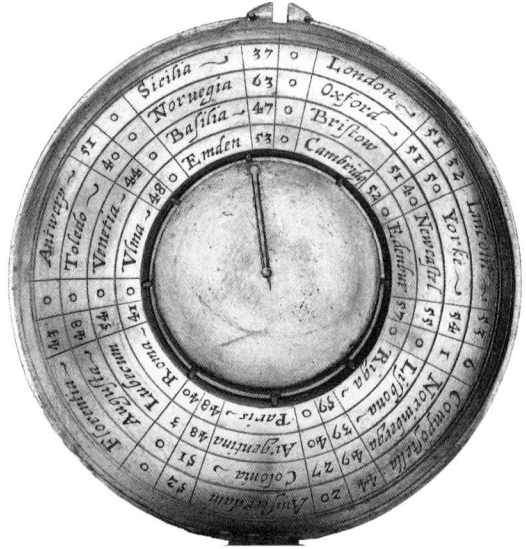

58 b.

the Lesser Bear (UMi) by means of a slit. An index at the edge is in line with the slit. The slit has been made by drilling two holes and connecting them by a saw cut. At the edge of this volvelle is a circular table in two bands, labelled *Prime* and *Epact*. The year, 1600, is engraved at the head of this table, and the year when this Julian Epact begins with: 5/25, 6/6, 7/17, and so on for the sequence of 19 Prime Numbers.

b) Inside the lid the sighting volvelle is held in place by eight clips. It is surrounded by a circular table of seven columns and four rows listing the latitude in degrees and minutes of 28 towns or regions.

c) The equinoctial dial is erected when the lid is fully opened. The gnomon with latitude quadrant is hinged to turn into the vertical position. The quadrant is in 2° divisions, numbered every 10°. The plummet is, as usual, missing. The chapter ring is divided in

58 a.

58

London	51 32	Lincoln	53 6	Compostella	44 20
Oxford	51 50	Yorke	54 1	Norinberga	49 27
Bristow	51 40	Newcastel	55 0	Lisbona	39 40
Cambridg	52 0	Edenbur	57 0	Riga	59 0
Amsterdam	52 0	Florentia	43 0	Antwerp	51 0
Colonia	51 0	Augusta	48 0	Toledo	40 0
Argentina	48 3	Lubecum	54 0	Venetia	44 0
Paris	48 40	Roma	41 0	Vlma	48 0
Sicilia	37 0				
Noruegia	63 0				
Basilia	47 0				
Emden	53 0				

24 hours with half-hour subdivisions, the hours marked in roman numerals.

d) This compendium is larger than most, so the compass card has also been made larger with a diameter of 69 mm. It is type B-2, where the 32 directions are drawn with lines alternating with extended triangular pointers that are particoloured (white–white, yellow–white, red–yellow). The directions are not named (a fleur-de-lis shows North); the degree circle is marked to 360° in 1° intervals alternately shaded (Plate 14g). The copperplate to print this card will have been cut by Whitwell judging by his characteristic numerals. The magnetic needle has an arrow at one end and a cross at the other. The brass boss is a six-sided pyramid with ears across the top. The cover is of glass with air bubble inclusions.

e) The face of the compendium serves as a tide computer, and has more engraved on it than the compendium made two years later (**59**). This may be because the calendar and Zodiac scales were not necessary. The calendar is the same as that on the top face (a), and

is followed by the Zodiac, with the names of the Signs in words. There follow a 24-hour scale, in half-hour divisions alternately shaded, hours numbered in roman, and the 32 wind directions, half named by their initials. Two volvelles complete this side. One with a long, solar index gives the Moon's age numbered

58 d.

from 1 to 29½ days; the final volvelle, with the lunar index, has a phase aspectarium. Also engraved are the planetary aspects: trine, quartile, and sextile.

Note. There is a separate brass disc that is loose in the compendium, with a Rojas projection and rule. It is inexpertly made, and is not part of the original Whitwell instrument.

Provenance

Bought through Maggs Bros, Ltd, in May 1936 for £65, and presented to the Museum by Sir James Caird.

Location

National Maritime Museum, Greenwich (AST0468; old no. D.35/36–194C).

59 **Compendium** 1602

Signed and dated: *Carolus Whitwell fecit / 1602*

Brass, gilded; diameter 62, thickness when closed 31

Comprising from the top:

a) nocturnal with calendar scale

b) table connecting Prime Numbers with Epacts, dated 1602

c) equinoctial sundial, with signature

d) magnetic compass

e) tide computer

a) The lid forms the nocturnal. On the outer side a circle is engraved with the names of the months: *Ianuarie, Februari, March, Apr=* (catch is centred at 25 April), *Maye, Iune, Iuly, August, September, Octobe=* (the hinge is centred at 26 October), *Nouemb=, December*. The months are divided in 2-day intervals, and numbered in tens with 31/28 as appropriate.

59 a.

There follow two volvelles, the outer with a 24-hour circle, marked out as twice XII, the hours subdivided in half hours by shading. At each hour is a protruding tooth, and an index is at 12 o'clock. The next volvelle is for sighting the Pole and aligning the Lesser Bear (*Ursa minor*) by means of a slit. An index at the edge is in line with the slit. The slit has been made by drilling two holes and connecting them by a saw cut. Inside the lid the sighting volvelle is held in place by five lugs with brass pegs.

b) Around the volvelle are two circular bands, the outer labelled *Prime*, the inner *Epact*, beginning: 7/17, 8/28, 9/9. This is the Julian Epact, beginning at 1602. Between the Epact circle and the volvelle is the date: 1602.

59 b.

reading winter hours, the ring is divided from VI to XII to VI. Hinged to the hour ring at its lowest point is a latitude quadrant with polar gnomon, the scale divided in single degrees and marked in tens. The plumb-line for setting the latitude is missing.

d) The compass box, which holds a printed paper windrose with coloured segments (diameter 56), showing the directions of the 32 winds, has the edge divided into four quadrants of 90° divided in 2° intervals, numbered every ten. The card has a fleur-de-lis at North, an arrow at South, and a cross at East. The compass card is type A-2; the colours used are beige, green, orange; circles blue and orange (inner). The card is now detached from its

c) Opening the lid erects a semicircular arc that supports the hour ring of an equinoctial dial, calibrated in 24 hours in half-hour divisions, and numbered in roman I to XII twice (see Plate 10). On the southern side, for

59 c.

59 e.

218

base, and the original glass cover is lost, replaced by plastic. The compass needle has an arrow at one end and a cross at the other, with a brass centre boss, formed into a six-sided pyramid, with two ears across the top.

e) The underside of the base is a tide calculator, with solar and lunar volvelles. The outer ring is calibrated in twice 12 hours, with half-hour divisions alternately shaded, and numbered in roman. The next ring has 32 points of the compass marked, with 16 points lettered. There follows the solar volvelle with an index to set the volvelle to the 'establishment' of a desired port, and on its edge are the $29\frac{1}{2}$ days of the Lunar Cycle. The innermost volvelle (lunar) has its index set to the age of the Moon; it has a lunar aspectarium to represent the phases of the Moon. The central portion of this volvelle is also marked with lines indicating the planetary aspects: trine, quartile, and sextile. For a similar tide calculator, see the Cole nocturnal **12**. The suspension ring passes through a lug on the side of the ribbed casing.

In fine condition, this instrument is virtually the same as Whitwell's compendium of 1604, described under **60**.

Provenance

Christie's South Kensington, sale 3 March 1994, lot 263.

Location

Private collection.

60 Compendium 1604

Signed and dated: *Charles Whitwell / 1604*

Brass, gilded; diameter 67, thickness when closed 30

Comprising from the top:

a) nocturnal with calendar scale

b) table connecting Prime Numbers with Epacts, dated 1604

c) equinoctial sundial, with signature

d) magnetic compass

e) tide computer

This instrument so closely resembles **59,** described above, that it is only necessary here to point out the minor differences.

a) The calendar is divided into single days. The hinge is centred at 20 October and the catch at 25 April, thus marking the meridian line of the nocturnal. The volvelle is held in place by six lugs.

b) The *Prime/Epact* table begins: 9/9, 10/20, 11/1. This is the Julian Epact beginning at 1604.

c) The latitude arc with gnomon is a modern replacement. It is punched with the monogram HW (Harriet Wynter), and engraved with the year 1975.

d) The compass card (diameter 60) is the ornate type C (see Plate 14i). The printed paper, with hand-colouring in blue, red, and green, has a sequence of triangular pointers to form the pattern of 32 wind directions, surrounded by a degree scale labelled in four quarters, 0–90°. The same engraving is found with Whitwell's compendium of 1610, **62**, on his chart published in Edward Wright, *Certaine Errors in Navigation* (1610), and with the Elias Allen compendium of 1617, **87**.

e) The tide calculator is the same as that on **59**, but on the hour scale the half hours are not alternately shaded.

Provenance

C. Holden-White Gift in 1935 to the Fitzwilliam Museum, Cambridge. Now on permanent loan to the Whipple Museum.

Literature

Bryden (1988), no. 303.

Location

The Whipple Museum of the History of Science, Cambridge (Wh: 1733).

61 **Compendium** 1606

Signed and dated: *Carolus Whitwell fecit* / 1606

Brass, gilded; diameter 63, thickness when closed 24

Comprising from the top:

a) aperture for a nocturnal, with calendar scale

b) table connecting Prime Numbers with Epacts

c) equinoctial sundial

d) magnetic compass

e) tide computer

This instrument resembles so closely that described under **59** that it is only necessary here to point out the minor differences.

a) The calendar is divided into single days. The hinge is centred at 25 October and the catch at 22/23 April, marking the meridian line of the nocturnal. Both volvelles are missing, leaving a hole.

b) The *Prime/Epact* table begins: 11/1, 12/12, 13/23. This is the Julian Epact beginning at 1606.

d) The compass card is type A-2, diameter 56 with degree scale, as on **59**. The points are hand-coloured in white, brown, apple green, grey, the rings in blood-red and bright blue.

61

Every point is named, and the declination one point East is marked. The compass needle is a replacement.

e) The tide calculator is the same as that on **60**.

Provenance

Purchased on 12 December 1903 for £20 by Lewis Evans. The Lewis Evans Collection was acquired by the University of Oxford in 1924.

Location

Museum of the History of Science, Oxford (LE 220; 44225)

62 **Compendium** 1610

Signed and dated: ✳ C ✳ W ✳ / 1610

Brass, gilded; diameter 68, thickness when closed 28

Comprising from the top:

a) nocturnal with calendar scale

b) table connecting Prime Numbers with Epacts

c) equinoctial sundial

d) magnetic compass

e) tide computer

This instrument closely resembles that described under **59**. It is necessary only to point out differences.

a) The calendar is divided in single days, and the catch is centred at 25 April, the hinge at 27 October. The nocturnal volvelle is held in place by five (originally seven) lugs with brass pegs.

b) The *Prime/Epact* table begins: 15/15, 16/26, 17/7. This is the Julian Epact beginning at 1610.

d) The compass card is type C, an ornate pattern. It is now detached from its base, and the original glass cover is present (the glass contains small air-bubbles and veins). The compass needle has an arrow at the South and an elongated fleur-de-lis at the North; there is a brass centre boss, formed into an eight-sided pyramid. To improve the balance, a brass needle is fitted at right angles to the magnetic needle. Its points bear an arrow at the West and a cross at the East. As with **60**, the outer circle is divided in single degrees, numbered in tens in four quarters. The card was printed from a copperplate, and the numbers are in Whitwell's hand. For more information on this type of rose, see Chapter 4 and Fig. 4.4.

62 Compass rose type C.

Literature

Wynter and Turner (1975, p. 135), fig. 156.

Location

Private collection.

63 Sector *c.* 1610

Signed on an inside edge with initials: *C · W*; undated

Brass; radius 205; width of each arm 25

The arms are made of five strips of brass riveted together. In one arm a rectangular blade (149 × 25) is inserted, which can pivot outwards to be gripped in a slot in the second arm. This form of construction is used on all

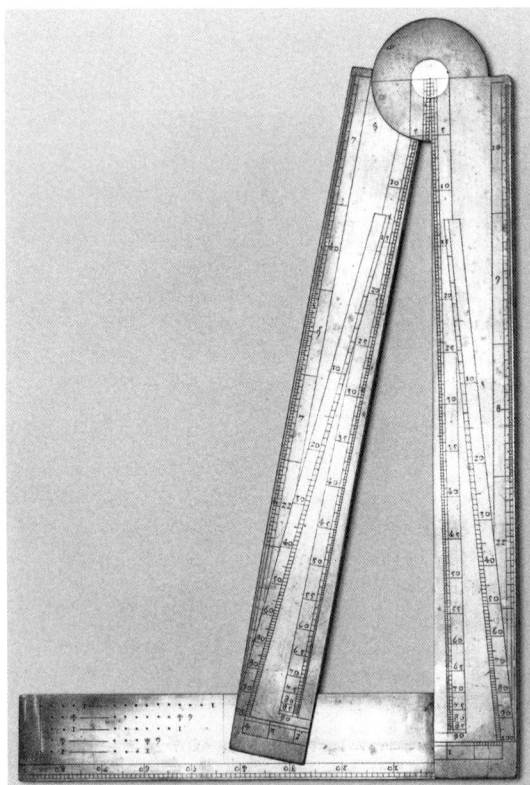

63 Side 1.

sectors made during the subsequent centuries. None of the scales is identified by a letter or name, as became the custom.

Side 1

This side has the blade fitted in the right-hand arm. The edges are engraved with a plotting scale, inches and $1\frac{1}{2}$ inches divided into parts. Each is labelled inconsistently, with the primary division or the number of the subdivisions.

Inches: *22* (11 × 2), *7* (7 × 4), *9* (9 × 4), *40* (10 × 4); $1\frac{1}{2}$ inches: *7* (7 × 10), *10* (10 × 10), *9* (9 × 10), *8* (8 × 10), *22* (11 × 2).

The other scales are: line of solids (volume), line of sines. Across the closed ends is a scale 0–4, not subdivided, with a further section in tenths; this is a decimal scale rule.

The blade has a non-linear scale, 0–90, at the edge, and five short scales with numbers and dots.

Side 2

The edges are engraved with two non-linear scales, one 0–8 in quarters, the other 0–25 in tenths. On one arm, near the joint, is a short scale, 0–20, one section of which is in tenths.

The other scales are: the line of superficies or planes, the line of equal parts. Across the closed ends is a scale, 0–4, divided in quarters. The blade has a linear scale, 0–20, undivided, with a further section divided in 40 parts; another measuring scale. Five short scales have numbers and dots.

On both sides of the compass joint is engraved the letter *S* and also at the end of one arm; they indicate the position to turn the arm to form a right angle, or square. Side 2 has the additional letter *l* on the joint and arm.

The initials show that the maker was Charles Whitwell, and the numerals are typical of those in his hand.

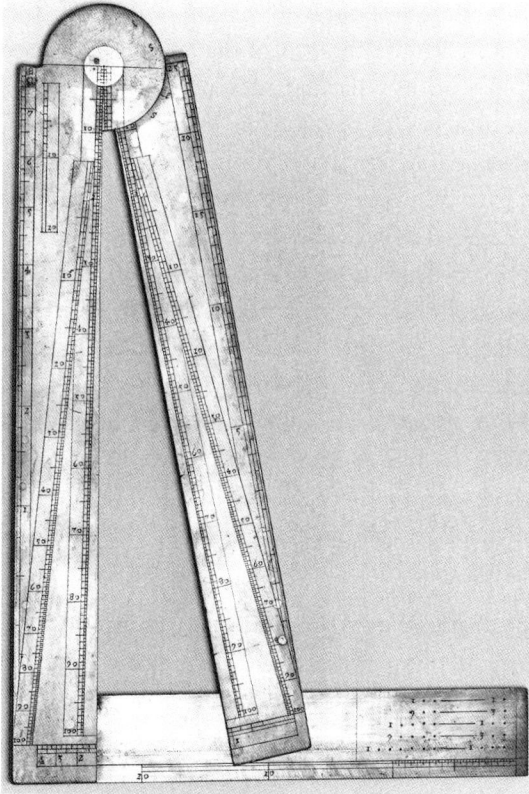

63 Side 2.

blade has its free end cut in a double ogee curve. The engraved scales are virtually the same, but with a few minor variations, which are described below.

Side 1
The plotting scale is in $1\frac{1}{2}$ inches only, variously subdivided:
22 (11 × 2), *6* (6 × 10), *7* (7 × 10), *10* (10 × 10), *9* (9 × 10), *8* (8 × 10).

Side 2
Both the line of planes and the line of equal parts end at 50 and not 100. The scale at the edge of the blade has 0–30 not subdivided, with a further section divided in tenths; this is a decimal scale rule.

This instrument was acquired by the Science Museum in 1914, when it was in a specially shaped leather sling-case with a partition expressly to hold an 8-inch sector. The associated items are: a pair of steel-pointed compasses with an ink groove in one point, and a substitute leg to hold a pencil (probably early seventeenth century); a three-legged dividers; small dividers; a lining pen (all three not part of the original set). See Plate 6.

Location
The Science Museum, London (1914–263).

This sector is identical with another in the Science Museum, **64**. These two instruments are an early form of Edmund Gunter's sector, made before Whitwell's death in September 1611. See Chapter 4.

Location
The Science Museum, London (1984–1296).

64 Sector, with case *c.* 1610

Signed on an inside edge with initials: *C · W*; undated

Brass; radius 205, width of each arm 25

The construction of this sector is the same as **63**. At either end of both arms are holes. The

65 Compendium *c.* 1590

Unsigned and undated; attributed to James Kynvyn

Brass; diameter 62, overall thickness 21

Comprising from the top:

a) nocturnal

b) latitudes of 31 world locations

c) equinoctial sundial

65

calendar is engraved at the edge of this side, with the months of the year divided into 2-day units, marked in tens except when a month has 28 or 31 days. In the case of July, there is an extra half unit, representing the 31st day; however, the total for the month is engraved as 30. The divisions are not regular, the 30 days in September are 19 per cent shorter than 30 days measured in October. March is another 'short' month. The months are named in italic letters that are fat and clear. Four months have been abbreviated. When used as a nocturnal, the meridian line will have cut 10 October and 8 April. Although inexact, this shows that the Julian calendar was being used.

b) Inside the lid are four rings of place names arranged in eight sections. Beginning from the hinge, these sections read:

Commentary

Kynvyn commonly places a dot over a one (*1*), making it appear as an *i*. Lisbon has been entered twice. For a comparison with Kynvyn's 1593 compendium, **66**, and a commentary on the names of the towns, see Chapter 4, Table 4.7.

c) Hinged between the top and bottom parts of the compendium is an equinoctial sundial, the chapter ring supported by a semicircular

d) magnetic compass

e) establishment of the port

a) The top has lost the volvelle for the 24 hours, and the index arm. The large central hole is equipped with a bush to take the fitting from the volvelle; a central hole will have remained for sighting the Pole Star. The

Ierusalem 36· 40	*London 51· 30*	*Briscil 44· 5*	*Patauia 44· 28*
Cesarea 31· 40	*Edenburge 58·*	*Venice 45· 18*	*Perusia 42· 30*
Florenca 43· 40	*Babilon 35*	*Naples 40· 36*	*Orleance 47· 13*
Malta 34·	*Rome· 41· 40*	*Paris · 48· 30*	*Lions 45· 10*
Brasilia 47· 41	*Norinbarg 46· 24*	*Cōstantinople 43·*	*Alexandria 31·*
Burdensi 45· 30	*Burgis 42· 48*	*Antioche 37· 20*	*Corinthe 35· 55*
Vienna 48· 22	*Lisbon 39· 38*	*Granata 37·*	*Niniui 41· 40*
Ments 50· 8	*Lisbō 39· 38*	*Athens 37·*	*Damascus 33*

bracket. The latitude quadrant, with the gnomon, is hinged to the ring to allow it to lie flat when the instrument is closed for the pocket. The quadrant is positioned in a slit to the desired latitude, thus placing the gnomon in the polar axis. The hour scale is divided into half-hour intervals, and is marked in roman numerals to twice 12 hours. A double scroll decoration occupies the northern part of the chapter ring.

d) The magnetic compass has a needle with an arrow head and a cross tail. The hand-drawn card has 16 points, parti-coloured, with additional lines between, so marking out the 32 points of the compass. The primary 16 are labelled with the initials of the winds, e.g., *nne, ne, ene.* The outer rim of the compass box

65 e.

is engraved with a stylized acanthus-leaf decoration, and a rope-twist.

e) The base of the compendium is for the establishment of the port. The highest tides occur when the Moon is new or full, and the timing of such tides depends on the location. As a convenient reminder, seamen noted the compass bearing of the Moon at the time of high water. The directions are marked out into 12 segments surrounding a central square. These compartments are in pairs of opposite directions (e.g., NE and SW) because the bearing of a new moon is diametrically opposite that of a full moon (Waters 1958, pp. 31–3). The arrangement and choice of ports is exactly the same as that on **66**, which is signed by Kynvyn and dated 1593.

Both this and Kynvyn's 1593 compendium have much in common, in particular the diagram for the establishment of the port. The casing, however, is not as good quality as it is on the later compendium. The compass card is very simple and marked out by hand, whereas the card on **66** is a printed version used by other makers. The selection of towns with latitudes is eight fewer, but the spelling and values of those engraved are nearly all the same. The execution of the engraving is rather less confident, so it is not unreasonable to accept an earlier date of around 1590.

Location

Historisches Museum, Basel, Switzerland (1982.538).

66 Compendium 1593

Signed and dated: *Iames Kynuyn fecit 1593*

Brass, gilded; diameter 59, overall thickness 27

Comprising from the top:

a) nocturnal

b) latitudes of 39 world locations (equinoctial dial lost)

c) magnetic compass

d) establishment of the port

e) perpetual calendar and table of fixed feast days

f) tide computer

g) motto engraved on the side

a) The nocturnal on the upper disc is no longer complete. The outer circle has the names of the months, followed by a circle divided into the days of the month in units of 1 day. May has 31 divisions, but is labelled as 30 days. The volvelle is marked out in two 12-hour periods with roman numerals. All the hours have protruding angular teeth, one 12 o'clock position having an extra long tooth. These enable the hours to be counted in the dark by feeling with the fingers, the customary method when using a nocturnal. The central part is decorated with floral motifs, while the central hole, for observing the Pole Star, is now blocked. The index is a poor replacement. The meridian line cuts the calendar at 20 April and 22 October, which shows that the Julian calendar is being used, and the star by which the time was found is *Kochab* (β UMi).

b) On the reverse of the disc are engraved, in four circular rows, the names and latitudes of 39 locations throughout the world, including Cuba and China. The choice is curious; exotic places and cathedral towns are particularly evident. Such a compendium would be a bespoke item, and the choice was no doubt

66 b.

that of the owner, whose arms are engraved at the centre of the disc. The shield has 16 quarterings for Robert Devereux (1556–1601), 2nd Earl of Essex. It is placed on a Garter surmounted by a coronet with three large and two small fleurons. His motto is engraved on two curved panels flanking the Garter: ★ *IN* ★ *VIDIA* ★ · *VIRTUTIS* · *COMES* · Devereux graduated MA at Trinity College, Cambridge, in 1581, and in 1598 was Chancellor of the university. For a time he was a favourite of Queen Elizabeth, and received the Garter in 1588. The arms are fully described in Bruce (1866, pp. 352–3).

The contents are considered more fully in Chapter 4, and compared with **65** in Table 4.7.

Between this plate and the compass, there was an equinoctial sundial, as on **65**. The remains of its hinge can be seen.

c) Around the edge of the compass box is a brass degree scale divided in 1° intervals from

66

London 51. 33	Perusia 42. 30	Lisbon 39. 38	Malta 34
Briscils 44. 5	Brasilia 47. 41	Athens 37. 15	Compostella 42. 15
Patauia 44. 28	Burgis 42. 48	Niniui 41. 40	Carthage 38
Burdensi 45. 30	Antioch 37. 20	Babilou 35	Heercules pillers 36. 15
Norinbarg 40. 24 ★	Corinth 35. 55	Roome 44. 40 ★	Tours 47. 30
Constantinople 43	Cesaria 31. 40	Paris 48. 30	Antwerp 51. 28
Alexandria 31	Florence 45. 40 ★	Lions 45. 10	Quinsey 37. 40
Ierusalem 36. 40	Napels 40. 36	Ments 50. 8	Cuba 23½
Edenburge 57	Orleance 47	Braga 43	
Venice 45. 18	Viena 48. 20	Granata 37	
		Daascus 33	

★ Nuremberg (*Norinbarg*) should read 49° 24′, not 40° 24′. Florence should read 43° 40′, as on **65**. Rome should read 41° 40′, as on **65** (42° would have been more exact); Jerusalem is just 5° too high on both dials.

Quinsey is the Chinese city now named Hangzhou.

0° to 360°, and marked every 10°. At the North is a fleur-de-lis. The compass card (diameter 55) is printed on paper, and hand-coloured. It is not now in the correct orientation. This card is type A-1, coloured in blue, gold, blue-black, and grey (see Plate 14b). This card is similar to those used by Humfrey Cole. The needle has a cross at one end and a point at the other. The glass cover is chipped.

d) Underneath the compass is the establishment of the port. The highest tides occur when the Moon is new or full, and the timing of such tides depends on the location. As a convenient reminder, seamen noted the compass bearing of the Moon at the time of high water for a given port. The directions are engraved in 12 segments surrounding a central square. These compartments are in pairs of opposite directions (e.g., SE and NW) because the bearing of a new moon is diametrically opposite that of a full moon (Waters 1958, pp. 31–3). The arrangement and choice of ports is the same as that on **65**, which is unsigned.

e) The outer three circles contain the feast days of the Church, and the inner three comprise the perpetual calendar. At the centre is engraved: '*This Tabell begin: eth at 1593 and so for euer*'. The circles are named: *Easter da*; *Prime*; *Epact*; *dominic*; *leape*. (Prime is another name for the Golden Number.) Above the first circle are the initials *a* or *m*, indicating April or March, the month in which Easter falls, to govern the dates below. The first date for Easter is 15 April; the first Prime Number is 17; the first Epact 7; the first Dominical Letter G. The figures are correct for the year 1593.

Several dates for Easter Day are incorrect: For 1596, *m* is over 11 instead of the following date, 27. For 1610, Kynvyn's 1 April should read 8 April. For 1614, 17 April should read 24 April. For 1617, 13 April should read 20 April. For 1627, *m* is omitted over 25. The Church calendar is arranged in three circles divided into quarters, with 3 months in each section. The information is similar to that published by Digges, or engraved by Cole on

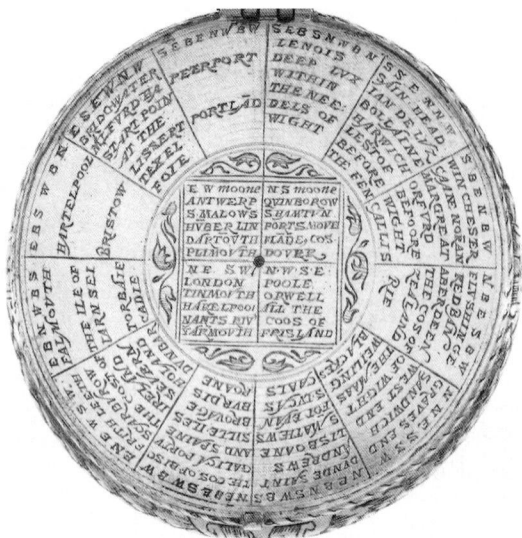

66 d.

66 e.

his compendia (see Tables 4.2, 4.3). However, a dozen dates are omitted, doubtless through lack of space.

f) The circles of information on this tide computer begin at the outer edge with the Zodiac, named in words and sigils. There follows a degree scale in units of 1° labelled in tens, for use with the Zodiac. Next are the months, named and with the days in each month divided in the next circle. Then come the 32 wind directions, each point marked with the initials for the winds. The innermost circle is divided into twice 12 hours, with roman numerals. Rotating round the centre pin is the solar volvelle with a pointer at the position of the new moon. The edge of the volvelle is marked out from 1 to $29\frac{1}{2}$ to give

I Circum	6 Epiphani	11 ☉ ♒	25 Con paul
F 2 purifi	9 ☉ ♓	14 Valentin	24 mathi
M 11 sun in ♈	25 Anunsiatiō		
A 11 ☉ in ♉	23 George	25 Marck euange	
M 1 Philip and Iacob	12 sun in ☉		
I 11 barā	12 ☉ ♋	24 Iō bap	29 pet pa
I 6 dog be	13 ☉ ♌	22 Mari mag	25 Iames ap
A 14 ☉ ♍	17 dog ēd	24 barth	29 Ion be.
S 14 ☉ ♎	21 Matew	29 micaell	
O 14 ☉ ♏	18 Luck euan	28 Simon and Iud	
N 1 all saints	13 ☉ ♐	30 Andrew apo	
D 12 ☉ ♑	21 thō	25 natiui	26 ste

228

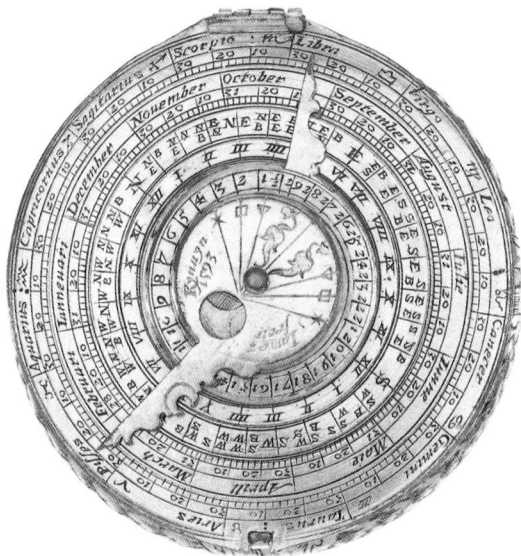

66 f, with signature and date.

the age of the Moon. Finally, a second volvelle, with a pointer engraved with a crescent, has a lunar phase aspectarium. At full moon one sees a complete face. It is this part of the instrument on which the signature and date are engraved. This face is completed with the astrological aspects: trine, quartile, sextile. This form of computer became quite popular, and an early version is on Humfrey Cole's compendium of 1569 (**9**), and also on the reverse of a Cole nocturnal of *c.* 1570 (**12**).

g) The suspension bracket is missing. The side of the casing is decorated with rope twists, and a stylized acanthus-leaf motif. Between the bands of decoration is engraved the motto: *HE THAT TO HIS NOBLE LINNAGE ADDETH VERTV AND GOOD CONDISIONS IS TO BE PRAYSED: THEY THAT BE PERFECTLI WISE DESPISE WORLDLI HONOR WHER RICHES ARE HONORED GOOD MEN ARE DESPISED*

Provenance

The original owner was, presumably, the Earl of Essex. Bruce (1866, p. 353) says there was a piece of paper inside the cover that read: 'formerly belonged to the Prince of Waldeck'. The instrument belonged to Edward Dalton, LL.D, FSA, when he permitted John Bruce to describe it for *Archaeologia*; the paper was read on 4 May 1865. The owner then presented it to the British Museum. Through a misreading of Bruce, Ward (1981) wrote that this dial was formerly the property of the Prince of Wales, later King Edward VII.

Literature

Bruce (1866); Ward (1981, p. 126), no. 361.

Location

The British Museum, London (1866, 2–21.1).

67 Mariner's platting board
1595

Signed and dated: *Iacobus · Kynvyn · fecit · 1595 ·*

Brass; 463 × 350

The double-sided plate is held to one side of a hinged brass frame by 30 rivets. There are five hinges of the simple pin type. The frame is engraved with scales only on the outer half that surrounds the front side of the plate (with signature). The blank half of the frame opens out to hold oiled paper or similar material. Five hooks keep the frame closed over the plate.

Two-thirds of the front face of the plate is occupied by a square with a pivot hole for a rule (missing) at one corner. The base line is divided into a linear scale of 500 units, alternately shaded, numbered every ten. At right angles to this scale are engraved parallel lines whose mutual spacing is non-linear. There are degree lines engraved from 0 to 85°, labelled every 5° on five lines that cut across the degree lines. The degree lines are distinguished in Kynvyn's style. The lines at 5° intervals are dotted; the intermediate degree lines are solid; between the degrees are two further solid lines denoting 20′ intervals, the central pair distinguished by a wavy line running between them. The square is bounded at the left and at the top by two scales on the frame. The inner is a geometrical square divided by half units, alternately shaded, labelled in five-unit intervals. The outer is a zig-zag degree scale, subdivided in 10′ intervals alternately shaded. Every 5° interval is numbered, and the degrees are indicated by three dots.

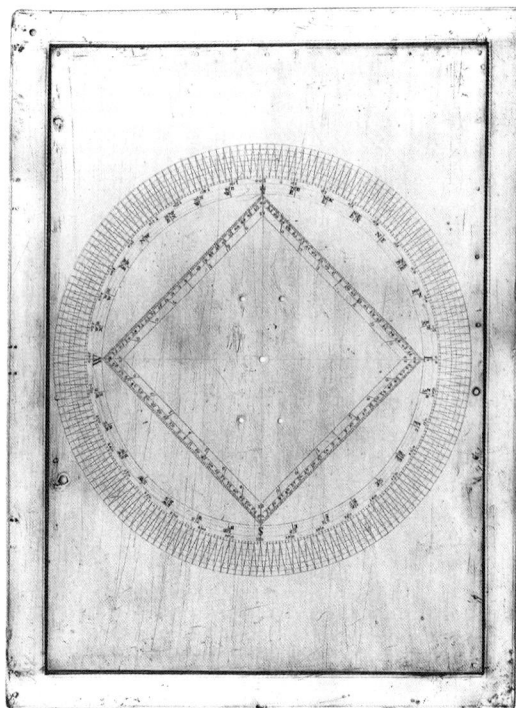

67 Back.

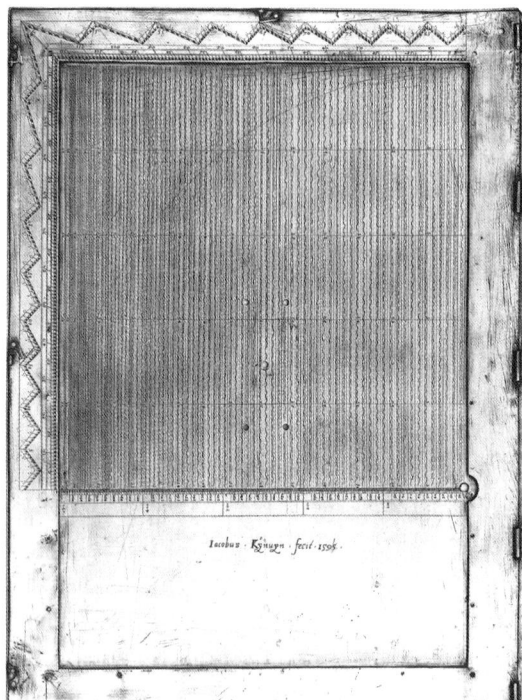

67 Front.

On the reverse of the plate is a wind rose comprising a set of circles with an outer diameter of 288. On the inside are 32 wind directions marked by initials, except for the North, which has a fleur-de-lis. Outside the winds is a degree scale, 0–360° from North, numbered every ten. Surrounding this scale are six further circles containing transversals that subdivide each degree into ten minutes of arc. Inside the wind rose is a geometrical square with the corners at the cardinal points. The divisions are 0–60 numbered in fives, with another labelling in threes to 12. At the centre of the plate is a circular hole with four more holes placed in pairs above and below. These may have held a magnetic compass.

The front side derives from Robert Dudley's scheme for perfecting navigation as recounted in *Dell'Arcano del Mare* (1647), Book V, which is subtitled: 'Nel quale si tratta della nauigazione scientifica, e perfetta, cioè Spirale, ò di gran Circoli'. This subject is explained in detail, in the manner of a teaching text. The platting board, dated 1595, was made immediately after Dudley's voyage to the West Indies, from early November 1594 to the end of May 1595 (Warner 1899). He travelled with the scholarly ship-master, Abram Kendall (died 1596), and it was said that he taught Dudley 'enough navigation for an Admiral' (Taylor 1954, p. 180). It seems Dudley put some of his new ideas into the commission for the platting board. An illustration of a version of this board is in *Dell' Arcano del Mare*, Vol. 3, Book V, fig. 11. This version is also found as a later addition to one of the semicircles of **75**. For early great circle sailing by the English, see John Davis (1595).

Provenance

One of the instruments taken to Florence by Sir Robert Dudley in 1606.

Location

Museo di Storia della Scienza, Florence (663).

68 Sector 1595

Signed and dated: *Iacobus Kynuyn fecit 1595* [*London*]

Brass; radius of arms 635; width of arm 29; thickness 6.5

The arms are composed of two plates, with a gap between to contain the short arms that form the geometrical square. The plates are riveted together, with about 30 rivets each side.

Side 1

This side bears the signature, where the word '*London*' appears to have been added later by a different craftsman, probably in Florence. Both arms are engraved with the line of equal parts divided to half units (100 units = 82 mm), 0 at the joint to 770 at the ends, labelled every ten. The compass joint is engraved on both sides with a 180° protractor scale, divided to 1° and labelled every ten. A clamping screw keeps the

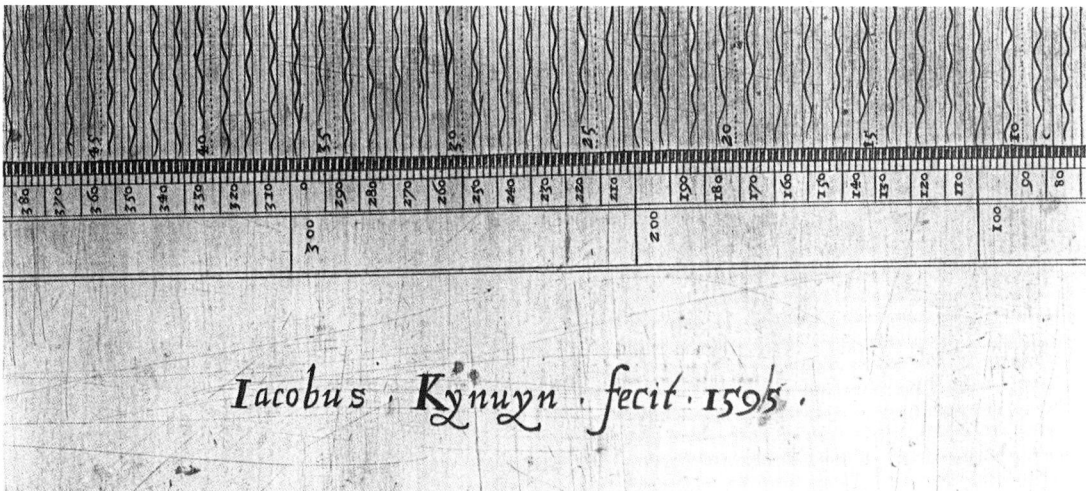

67 Signature.

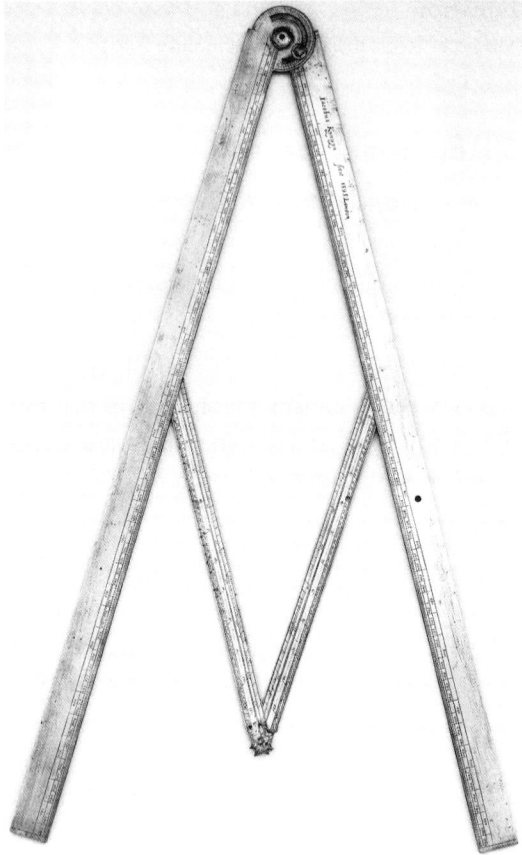

68 Side 1.

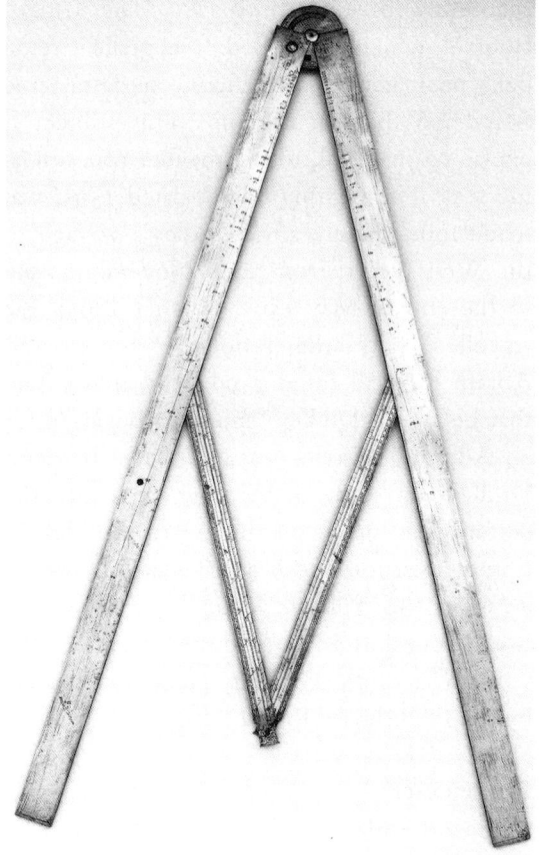

68 Side 2.

chosen angle firm. With the long arms set to 90°, the geometrical square is formed by the two short arms (each: length 28, width 17, thickness 2) contained within the others. The join at 45° is held by a bayonet catch. The square is divided on both sides, the outer edge 0–300 by ones; the middle scale to 12; the innermost a degree scale to 45° by ones. These scales are to be read with a plumb-bob (missing) pegged into the central hole of the joint. The right-hand arm has a hole at 480 that runs through the short arm, which stows away in a gap. A peg or screw in this hole would prevent the arms from falling open unintentionally.

Side 2

On this side the arms are engraved with a non-linear scale that represents the chords of a circle, marked out in a manner similar to Thomas Hood's description, published three years later in his *The Making and use of the Geometricall Instrument, called a Sector* (1598). There are two sets of calibrations. One series is distinguished by plain dots (•), and begins 31 mm from the joint at 20 running to 2 at 317 mm from the joint. The other series is distinguished by a dot in a circle (⊙), and begins 100 mm from the joint running to 3 at 546 mm from the joint. Where the scale points coincide,

68 Close-up of centre of side 2.

for example 6 and 2, the position is indicated by a dot inside a star.

Three other early English sectors of the Hood type are described elsewhere: Beckit **37**, Whitwell **46**, and Kynvyn **69**.

Provenance

One of the instruments taken to Florence by Sir Robert Dudley in 1606.

Literature

Schneider (1970, p. 48).

Location

Museo di Storia della Scienza, Florence (2516).

69 Sector *c.* 1597

Unsigned and undated; attributed to James Kynvyn, *c.* 1597

Brass and iron; radius of arms to iron points 268; radius of brass arms 255; radius of arc 112; width of arms 16, of arc 23; thickness of arms 4.5, of arc 2

The arms are made of three strips of brass riveted together, the centre strip narrower than the other two. A compass joint connects them at one end; the arc (Hood's 'circumferentall limbe') is fixed into one arm and passes through the other. The ends are iron points. There is a non-threaded hole in the joint, intended to accept a peg for affixing to a tripod. At the ends of the arms are threaded holes into which sights could be screwed. The movable arm has a rectangular hole. None of the attachments is present.

Side 1

Both arms are engraved with lines of equal parts, from 0 at the joint to 110, with subdivision into half units; numbering is every ten. The arc has two scales, one of degrees, 0–150, the other of polygons, 3–10. Each degree is subdivided into six parts by transversals. This side follows exactly the illustra-

233

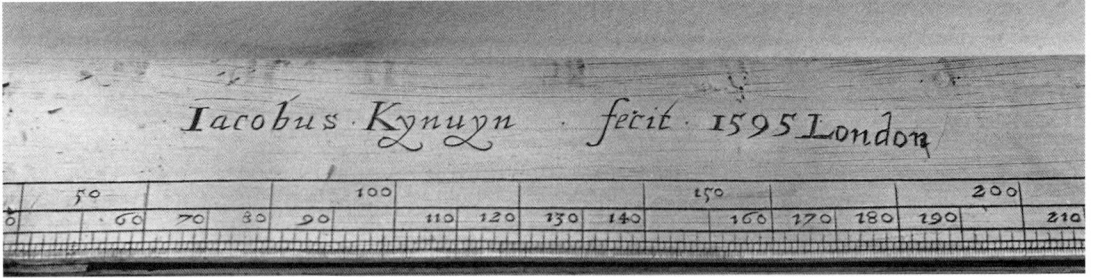

68 Signature; note clumsy late addition of *London*.

tion in Thomas Hood, *The Making and use of the Geometricall Instrument, called a Sector* (1598); the illustration was engraved by Charles Whitwell (see Fig. 4.5).

Side 2

The arms have two scales, which Hood called 'internal' (next to the inner side of the arms) and 'external' (next to the outer sides). The inner scale is non-linear, marked from 3–10, and represents the chords of a circle. The outer is a scale of fractions from $\frac{1}{2}$ to $\frac{1}{8}$. Thus by setting the points for a square of side 1, the fraction gives the side of a square that has one-half, one-third, etc., of the area of a square of side 1. The arc has four concentric scales, three of which are divided: 0–102, 0–136, 0–170, all three in half divisions. At the quarter-circle mark, the values are: 90, 120, 150. These scales divide inches defined by the arms into sixths, eighths, or tenths. The scale at the inner edge of the arc is non-linear, and runs from 0 to 32 in quarter units. Hood explains the form and use of these scales in chapter 1 of his book.

At the free end of the arc is a gunner's table, which is not completed with figures. Gunnery tables are reviewed in Chapter 4.

The rows read:
The names of peeces; *Height of bullot*; *Weight of shot*; *Quantety of pouder*.

The columns read:
Doubble Canon / Dub Can of Franc / Demi Canon / De. Cano of Fran / Culuerin / De. Culue / Sacar / Minion / Faucon / Fauconet

Two other English sectors of Hood's design, and of similar size, are described elsewhere, both being dated 1597. One is by Robert Beckit (**37**), the other by Charles Whitwell (**46**). An earlier (1595) large sector is signed and dated by Kynvyn (**68**). The present instrument can be identified as made by Kynvyn by the characteristics of his engraving hand. The novelty around the time of Hood's publication can be expected to have attracted all the best craftsmen of London, and presumably this piece was made at about the same time.

Literature

Bennett and Johnston (1996, p. 23), figs, 13, 14.

Location

Museum of the History of Science, Oxford (49–36; 44505).

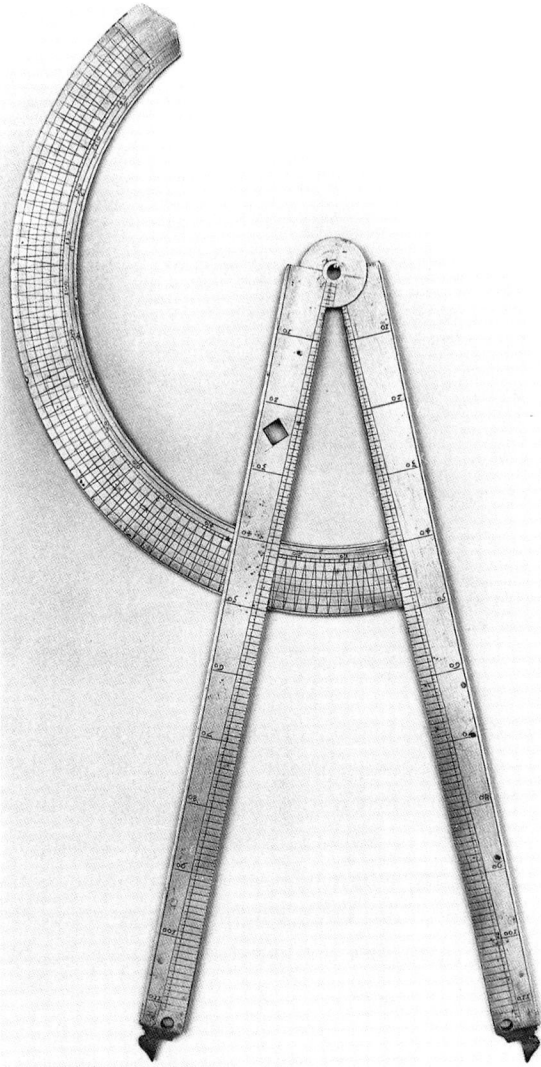

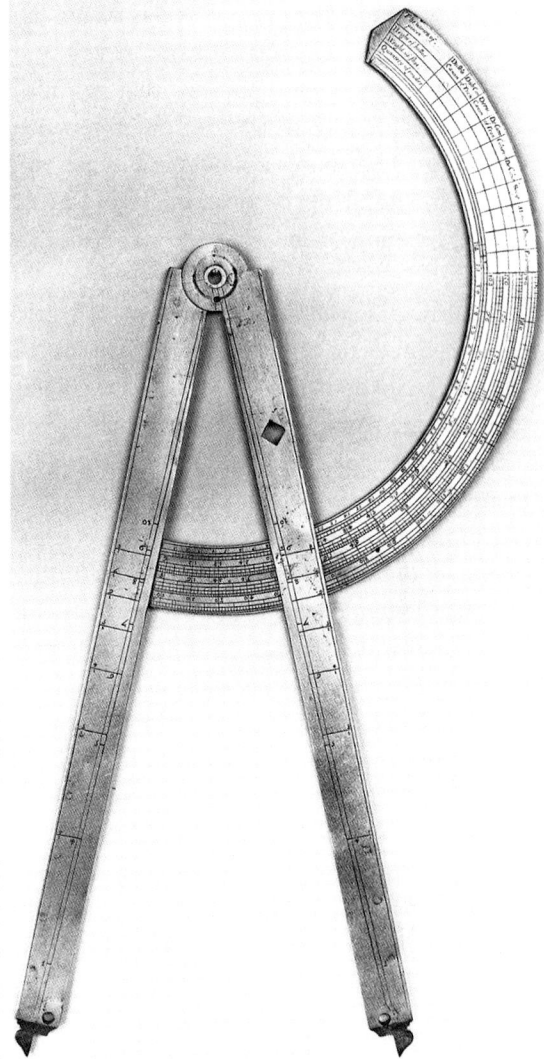

69 Side 1.

69 Side 2.

70 Simple theodolite 1597

Signed: *James Kenvyn fecit* 1597

Brass; diameter 308; thickness 1.9–3.2

The circular plate is polished and engraved on the upper surface and is very roughly filed on the under side. A bracket is integral with the plate; this probably held a magnetic compass.

On the outer rim is a degree scale, divided to $\frac{1}{4}°$ degree and labelled every 10°; 360° is at the top of the vertical axis below the bracket. A diameter divides the engraving into two parts. The upper left contains a shadow square, or geometrical quadrant, both edges divided in units, 0–60, labelled in fives, and also 0–12 labelled every three. The upper right quarter is

235

a sine table with degrees both sides to 90° by ones. The sines are engraved on an adjacent scale to the left of the table divided in units, 0–200. This is equivalent to 0–1.00 on modern sine tables. The spaces round these tables are filled with floral decoration and grotesques, and the lettering of the signature has elaborate flourishes.

The lower semicircle is filled with a grid 0–40, divided along the axes in half units, alternately shaded. The body of the grid is partitioned in squares of five units, each square further partitioned by dots into units. The squares have their coordinates marked every five units. Rotating about the centre is an alidade, counterchanged, with on one arm the sine scale extended from 200 to 284, and on the other arm the same 0–40 scale that defines the grid. Attached to the alidade by copper-coloured wing-nuts is a support for a sighting bar. At the middle is an L-section catch to slide into the base of a support for a sight. The rim, half the alidade, and the lower half of the plate, correspond to the surveying instrument invented by Peter Apian, which is fully described and illustrated in Röttel (1995, pp. 243–6).

The underside of the plate has a seven-sided boss that connects the plate to a supporting cylinder for fixing to a post. This is a piece of bent and soldered brass, length 59, diameter 32, thickness 1.3. It bears a diamond-shaped escutcheon for a locking key.

Kynvyn has used the underside to try the hardness of the brass with his burin. Several curved and straight cuts can be seen, also his name, *Jacobu*, *Iames*, and that of a brother, *Job* (Fig. 2.9).

In the illuminating paper by Gerhard Betsch, 'Instrumente aus Peter Apians Nachlass'

70 Azimuth circle.

(Röttel 1995, pp. 239–46), the part played by Georg Galgemair in furthering the knowledge of Apian's inventions is made clear. Philipp Apian (1531–89) was a professor at Tübingen University where Galgemair matriculated in 1583 to study under Philipp, son of Peter. It was Galgemair who edited some papers left by

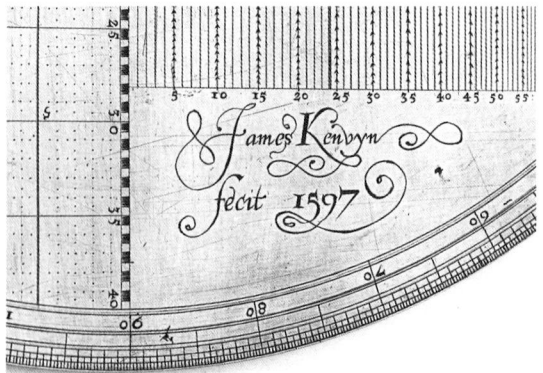

70 Signature.

Peter Apian (1495–1552) that were published in 1616 and 1619. Of present concern is *Inventum P[etri] Apiani ...* (Augsburg, 1616), where the base plate is closely similar to this Kynvyn instrument of 1597. This 'invention' is, in effect, a simple theodolite modified for use by surveyors without the need for calculation. The same principle of direct reading ('without Arithmeticall calculation') is put forward by Aaron Rathborne in his *The Surveyor* published in 1616, where he promotes his Peractor (see **76**). It is unclear at present just what linkages existed between Peter Apian, Galgemair, Sir Robert Dudley, who commissioned this Kynvyn instrument, and Rathborne.

It is likely that this instrument was made as an altazimuth theodolite that incorporated the altitude semicircle, **71**. Two other English altazimuth theodolites are described elsewhere, **31** (by Cole, 1586), and **35** (by Ryther, 1593).

Provenance

One of the instruments taken to Florence by Sir Robert Dudley in 1606.

Location

Museo di Storia della Scienza, Florence (3174).

71 Altitude semicircle *c.* 1597

Unsigned and undated; attributed to James Kynvyn, *c.* 1597

Brass; radius 78; thickness 1.6–1.9

The straight edge of the semicircle has riveted to it two strips of brass constituting a bar (length 156, width 5.6, depth 6.1) over which

71 Side 1, Tudor Royal Arms.

slide sight vanes that push on the bar, which has grooves on the underside. Only one vane remains; it has a single pin-hole. There is a hole at the centre of the bar for a supporting structure.

Side 1

This side is prominent because it displays the Tudor Royal Arms of England, supported by the English lion and the Welsh dragon (Plate 13). The arms are framed by a shadow square, with sides named in capitals with flourishes: ✴ *VMBRA* ✴ *VERSA* ✴ / ✴ *VMBRA* ✴ *RECTA* ✴.

The sides are divided 0–36 in single units, alternately shaded, and numbered every three; the scales are also marked 0–12. The edge of the curve is divided in degrees, alternately shaded, marked every ten, 90°–0–90° and 0–180°. The centre of the diameter has a hole to take a plummet.

Side 2

This is engraved with four altitude sundials, the central two inverted with respect to the outer two. They are drawn for the marked latitudes: *51 30; 50; 53; 55;* these fit London, Plymouth, Nottingham, Newcastle. Each dial

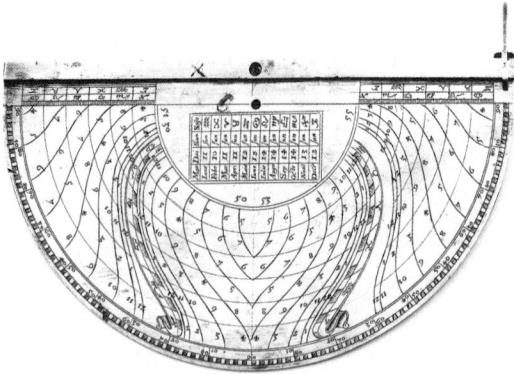

71 Side 2, four altitude dials.

consider these two items as forming another example of an English altazimuth theodolite, to compare with those by Cole and Ryther described elsewhere (**31**, **35**).

Provenance

One of the instruments taken to Florence by Sir Robert Dudley in 1606.

Location

Museo di Storia della Scienza, Florence (2528).

has its solar declination scale. Along the diameter are the two for the outer pair, divided in 2° intervals marked every 30° with the sigils of the Zodiac. The inner pair have curved scales, each Sign divided to 10°, alternately shaded. At the top centre is a table giving for each month the entry of the Sun into the Zodiac; it is headed: *Mon / Dai / Sun / Sige*. The dates are the same as those on some of Cole's instruments, for example **13**, **17**, **21**. The edge of the curve is divided in degrees, alternately shaded. The scale is labelled in tens, from the bottom centre 0–90° both ways to be used with the central pair of dials, and from the ends of the diameter 0–90° for the two outer dials. Measuring the Sun's altitude on the other side of the semicircle, the angle can then be found for the appropriate dial and the time read from the hour lines for a given time of year.

This altitude semicircle is an appropriate item to be set above the simple theodolite that is signed and dated by Kynvyn, **70**. No support structure remains, but the semicircle is obviously engraved by Kynvyn through the characteristics of his numerals, Zodiac sigils, and lettering. There are reasons, therefore, to

72 Pantometer or nonius quadrant *c.* 1597

Unsigned and undated; attributed to James Kynvyn

Brass; base circle diameter 305, thickness 2; quadrant radius 300, thickness 1.5

The parts of this instrument have been separated for a long period, and received separate inventory numbers. The base circle has been broken, probably deliberately, along the North–South diameter, leaving the West side and the degree scale, 180°–360°, numbered in tens. The division is to 15 minutes of arc. An inner scale is the wind rose, with a quarter of the points lettered.

Attached to the underside is a cylinder (length 54, diameter 36) with a square boss on the side with a threaded hole for a locking screw. This is to support the instrument on a pole.

The quadrant is divided on the front with 45 concentric arcs, in the nonius manner. At the edge is a degree scale in units of 1° from 0 to 90°. There are no numbers or letters. Each arc is marked at 5° intervals, ending at the innermost arc at the 45° point. Originally, an alidade with two sight vanes would have been attached at the apex of the quadrant.

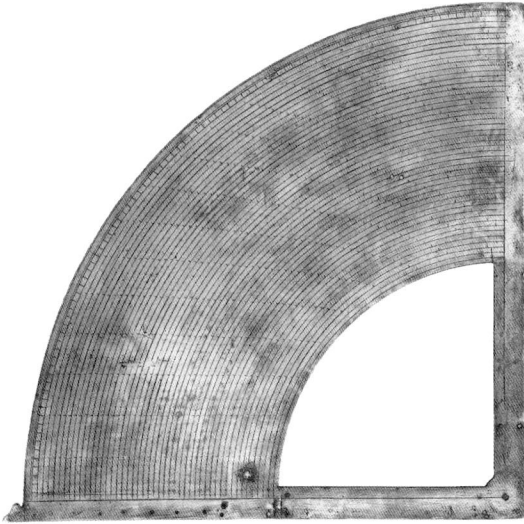

The reverse side is roughly finished, and fitted with two sockets for a horizontal bar on which would have been two sighting bars. Near the top is a peg to hang a plummet, and part-way down is a guide, over the fiducial line that indicated the vertical. The base of the quadrant is fixed to a bar that has a pair of dovetailed sockets for attaching a compass. Only the brass compass box remains. The front of the complete instrument is shown in Plate 5.

This type of angle-measuring instrument is the invention of the Portuguese, Pedro Nunes (1502–77), whose *De Crepusculis* was published at Lisbon in 1542. The design of this particular instrument is clearly described and illustrated

72 Nonius quadrant.

72 Azimuth semicircle.

72 Parts assembled.

93

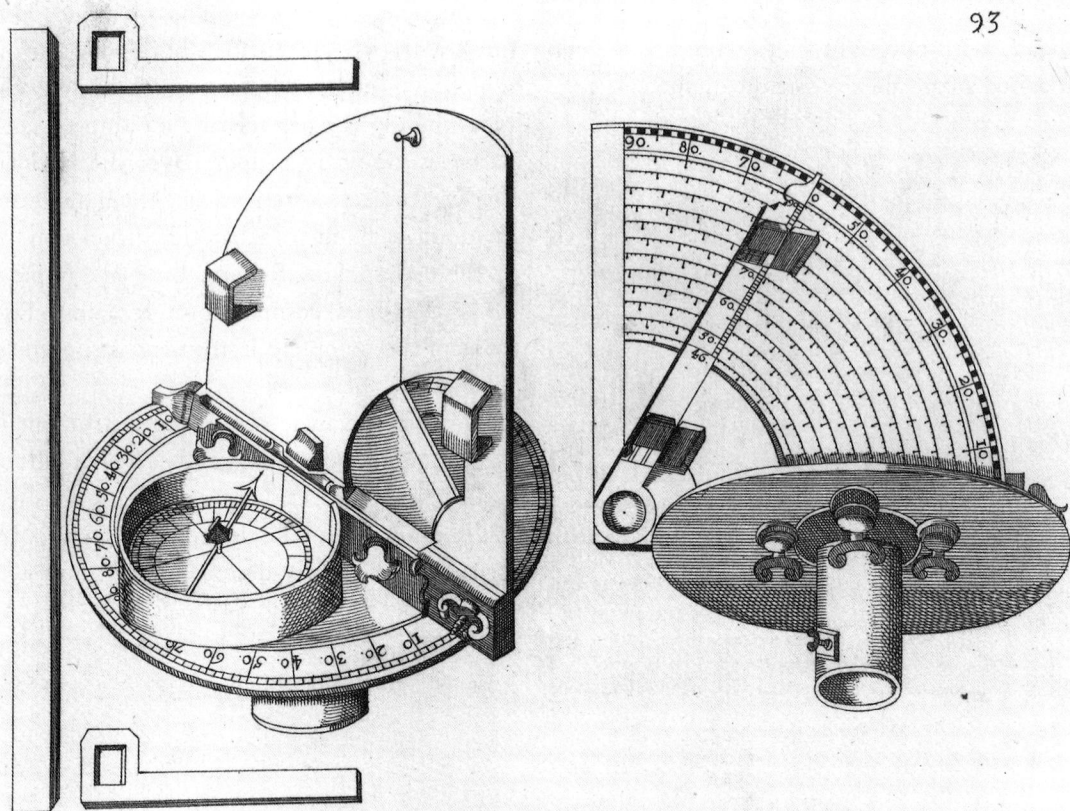

72 Engraving from Dudley, *Dell'Arcano del Mare* (1647). BL Oxford.

in chapter 4 of William Barlow, *The Navigators Svpply* (1597). The engravings for all of Barlow's illustrations were made by Charles Whitwell, whose advertisement for the items he can supply from his address 'over against Essex howse' is printed on the title page. Nevertheless, this instrument was made by James Kynvyn, judging by the characteristics of the letters and numbers on the base circle. Barlow calls the instrument a 'Pantometer', and says it is 'the proportionall Quadrant of *Nonius*', 'for the exact finding of Altitudes or Heights'. The illustration by Whitwell was copied exactly by Robert Dudley's engraver, and it appears in *Dell' Arcano del Mare* (1647), Book V, p. 18, prop. VI, fig. 45, but there is no explanation on its use (Fig. **72** c). It is this later, Italian, engraving that has been used by modern writers to show what the nonius was like, and there is no reference to the Barlow original. A nonius quadrant was displayed in Tycho Brahe, *Astronomiæ instauratæ mechanica* (1598), but to a somewhat different design; see Ræder *et al.* (1946).

An account of the nonius has been published by Captain dos Reis (1997), with a colour picture of the Florence instrument, and

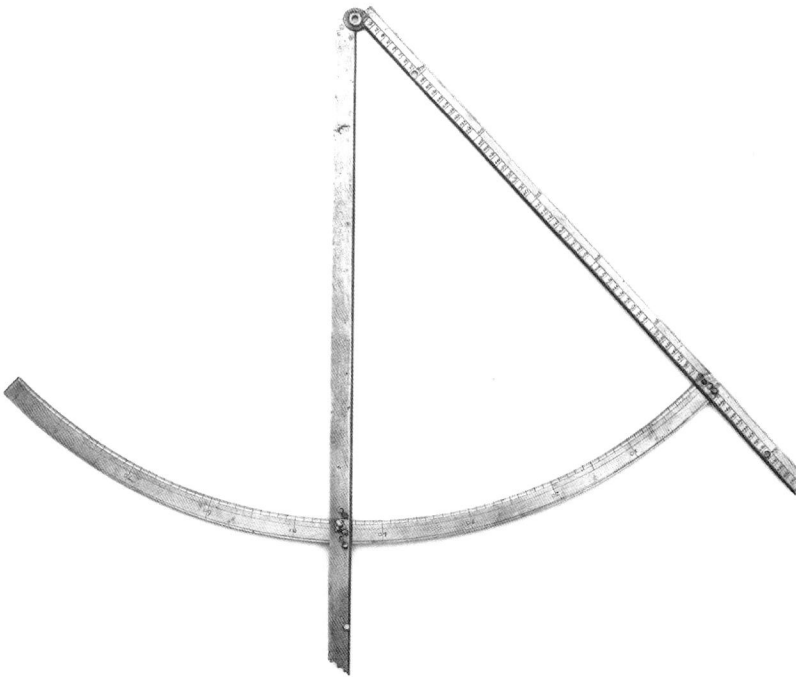

73 With close-up of section of scale engraved by Kynvyn.

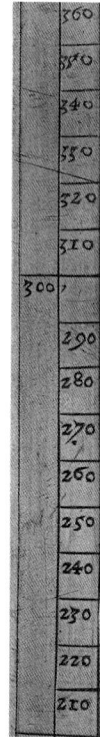

the plate from Robert Dudley's book. The working of the nonius, with a brief history, is provided, and the author points out that it was not a successful design, and that the Florence instrument is the only example known. The nonius was superseded by the diagonal scale. The two forms are compared in a drawing by George Waymouth in his manuscript, 'The Jewell of Artes' (1604). See Waters (1958, pp. 304–5), plate 72.

Provenance

One of the instruments taken to Florence by Sir Robert Dudley in 1606.

Literature

Estácio dos Reis (1997, pp. 119–29).

Location

Museo di Storia della Scienza, Florence (circle 3362; quadrant 242).

73 **Wing-dividers** *c.* 1595

Unsigned and undated; attributed to James Kynvyn and Charles Whitwell

Brass; length of arms 414; width 13; arc, radius 335, width 15

This instrument has been altered and adapted. The jointed arms have been cut down in length, and the cut ends finished with scalloped decoration. This cuts into the original numbering of one arm at the 750 division.

The unit divisions are on the bevelled edge, and numbered on the top of the arm in tens. The original length of the arm was probably 1000. A mistake has been made at 300, where a 3 has been abandoned, and at 350 there is a muddle. The other arm is blank, as is the whole of the underside of the instrument. The two arms are joined at a simple hinge, a bracket is riveted to one arm, and the join completed by two hand-made cheese-head nuts. Two holes in the numbered arm may be for sights, perhaps on the original version of the instrument. The arc is riveted to one arm, at 600, and a guide with a butterfly locking screw holds the arc in position against the blank arm. The arc is divided in degrees, 0–87° (the end is cut off) by ones, labelled in tens. The scale is read from the inside of the arc. The numbers on the arms are attributable to James Kynvyn, but the numbers on the arc are not by Kynvyn. They correspond well to the style of engraving used by Charles Whitwell.

Provenance

The parts are definitely English, and the engraving is attributable to Kynvyn (arms) and Whitwell (arc), so this compass is part of the group of instruments taken to Florence by Sir Robert Dudley in 1606. See **74** for a similar, but larger instrument.

Location

Museo di Storia della Scienza, Florence (599).

74 **Wing-dividers** *c.* 1595

Unsigned and undated; attributed to James Kynvyn and Charles Whitwell

Brass; length of arms 650; width 10; thickness 3–4; arc, radius 460, width 27

This instrument appears to be another adaptation, as is the similar instrument **73**. Short lengths of brass are riveted to the inner end of each arm, and these are joined at a circular swivel joint. The outer ends appear to have been cut short, and are at an angle. The arc is riveted to one arm, and held to the other by a guide and locking screw. The divisions are 0–90°, in units of a third of a degree, and the numbering is in 5° steps. The scale is read from the outside of the arc. Each arm has at its end a rectangular (65 × 25) sight screwed in place, with a wide aperture and a thin vertical bar. Half-way along each arm are attached brass rectangles (30 × 18) riveted to the underside of the arms; these could be used for alignment on the central bars of the sights. Between these rectangles and the junction with the arc, on both arms, is a non-linear scale, 0–90°, the units of 1° alternately shaded. Labelled on both arms: *The degrees for*

74

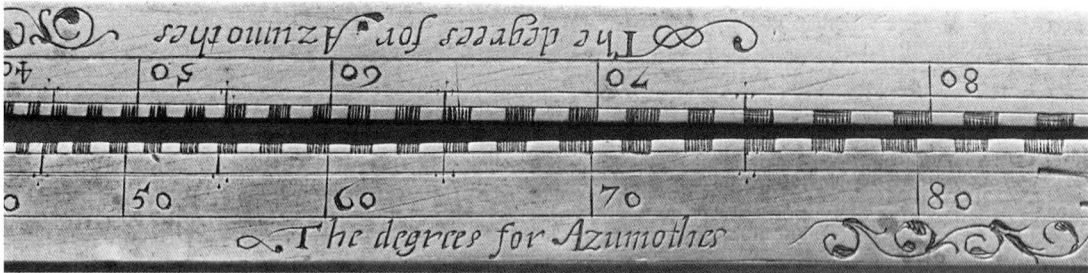

74 Close-up of section of scale.

Azumothes. These words are followed by arabesque decoration. The reverse side of the instrument is blank. There are signs of a previous use.

This and **73** are described here as wing-dividers because of their shape, and their possible use as adapted. What they were originally is a matter for conjecture. Each may have been part of an astronomical sextant, or even a triquetrum. An original version of this instrument, when complete, is illustrated twice in Dudley, *Dell'Arcano*. Firstly in Vol. 1, Book II, fig. 9, there described as the *Balestriglia* of the author; secondly in Vol. 3, Book V, p. 22, fig. 66.

The numerals and letters on the arms serve to attribute it to Kynvyn as the engraver, while the numbers on the arc are in the style of Charles Whitwell.

Provenance

The parts are definitely English, and the engraving is attributable to Kynvyn (arms) and Whitwell (arc), so this instrument is part of the group of instruments taken by Sir Robert Dudley to Florence in 1606. This appears to be confirmed by the engravings in his book. See **73** for a similar instrument of Dudley's.

Location

Museo di Storia della Scienza, Florence (600).

75 Astronomical computer
c. 1600

Unsigned and undated; attributed to James Kynvyn

Brass; diameter 670; thickness of rim 8–9

The large and heavy disc is held in a brass rim, slightly wider one side than the other. It is made up from riveted sections with a channel in the outer edge. The wider side (19–21) is unmarked; the narrower (*c.* 18) is divided in thirds of a degree, numbered every 10°. The whole scale has been revised, with punched numbers and punched letters for the winds. These punches occur elsewhere on this instrument's attachments, and on the copper wind rose (see Part I, Fig. 5.4). The explanation TO FINDE THE DECLINATION is punched, but there are a few traces of the original engraved numbers. The disc is dropped into the rim and held in place by brass pegs. There are 16 holes for these in the rim, but only eight pegs remain. The plate can be reversed within the rim, so there was no need to engrave both sides of the rim.

Side 1

There are circles around the edge of the plate containing engraved information. From the outside:

1) Roman numerals for the days of the Moon's age, ending: XXIX & $\frac{1}{2}$. Each Moon increment is divided into 48 units, alternately shaded. This represents 48 minutes, the time the Moon ages per day, thus the time between successive high or low tides.

2) The Zodiac in words and sigils; for each: the sigil / ☉ / name / ☉ / sigil.

3) Four concentric bands forming a diagonal degree scale. The diagonals become straight on the solstitial line, 90°–270°. The lines are given emphasis: 30° wavy, 20° and 10° by arrow heads, 5° by barbs. The four bands are separated centrally by a double degree scale, alternately shaded above and below, and to the sides.

4) The Zodiac shares the diagonal scale of degrees with the subsequent scale of 0–360°, giving the hour angle from the First Point of Aries. The scale is divided in $\frac{1}{2}$°, alternately shaded, and numbered every 10°.

The central portion of the plate is in quarters. One quarter is exactly as on the simple theodolite signed by Kynvyn, dated 1597 (**70**), with a square net 40 × 40 units. Another section has parallel lines, numbered in tens to 60, described as: *The measuringe of any parallel by the minutes of the equater.* The remaining two sections are the same, and have non-linear scales 0–75. The upper section is inscribed *parrallels*, the lower *meridians*. On the central line are the words *The vse of right lined triangles.*

Within the concentric markings, the whole of the central part is additionally engraved with 31 star names and positions. It is also traversed by 16 radial lines showing the winds, which are not identified. Those lines not marked for some other purpose have a wavy line running along them. The remaining 16 intermediate points are indicated by small triangles.

N 2 ☉ *Crus Pegasi*	S 2 ☉ *Lucida Hidri*	
S 1 ☉ *Foma hand*	N 2 ☉ *Pectus Cancri*	
N 3 ☉ *Dexter humerus*	N 1 ☉ *Canis Minor*	
N 2 ☉ *Cauda Cygni*	S 1 ☉ *Canis Maior*	
N 3 ☉ *Cauda Delphini*	☉ S 1 *Canobus in Argo naui*	
N 4 ☉ *Cuspis Sagitte*	N 2 ☉ *Hircus dexter*	
N 3 ☉ *Caput Engounasi*	N 1 ☉ *Orionis dexter*	
S 2 ☉ *Cor Scorpij*	N 1 ☉ *Hircus sinister*	
S 3 ☉ *Frontis Scorpij Borealior*	N 2 ☉ *Orionis sinister*	
N 2 ☉ *Lucida Coronæ*	N 1 ☉ *Oculus Tauri*	
N 1 ☉ *Arcturus Booetes*	N 2 ☉ *Dextrum latus Persei*	
S 1 ☉ *Spica Virginis*	N 2 ☉ *Caput Medulse*	
S 3 ☉ *Ala dextra Corui*	N 4 ☉ *Ceti Venter*	
S 3 ☉ *Rostrum Corui*	N 2 ☉ *Cornu arietis*	
N 1 ☉ *Leonis Cauda*	N 3 ☉ *Andromedæ*	
N 1 ☉ *Cor Leonis*		

Side 2

This side has two universal projections, presumably for comparison purposes. The plate is partitioned along the meridian line, with on the right the *Saphea*, and on the left a Rojas projection. On both sides the equator is divided in degree intervals alternately shaded. The azimuths and almucantars are to 1°. On the *Saphea*, the Zodiac is surrounded by a scribed band, which is 12° in width. The Zodiac band is engraved with sigils on both projections, as are the hours, in roman numerals, along the Tropics.

75 Side 1.

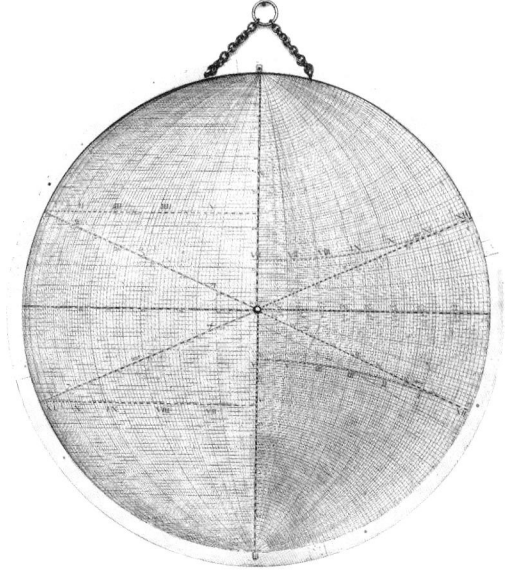

75 Side 2.

Fret

A long narrow fret (634 × 72, thickness 1.5) may be attached over this double projection, the parallel-sided part over the Rojas projection, and the curved over the *Saphea*. The edges are engraved with degrees numbered in tens. The Zodiac is along the centre line. The underside is rough finished, and has several soldered breaks in the thin brass. The fret is held in place by a round-headed copper bolt (length 12, diameter of shaft 6.8), and a four-winged brass nut. The head of the bolt has no saw-cut.

Detachable plates

Two semicircular plates are provided as accessories. One has a straight edge 634 in length, a radius of 317, and a thickness between 1.5 and 2.5. The whole of one side is a stereographic

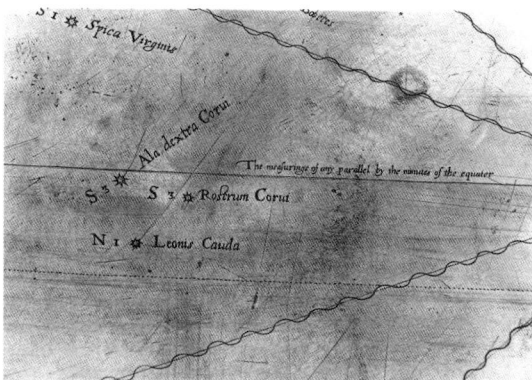

75 Close-up of side 1.

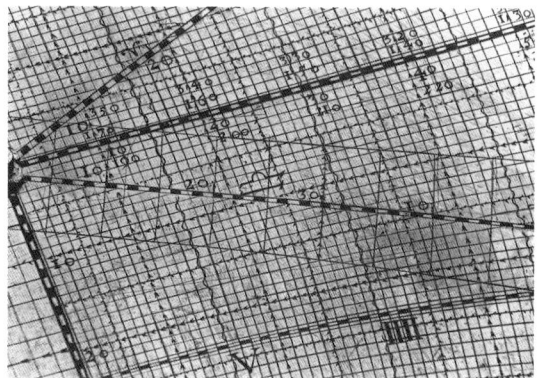

75 Close-up of side 2.

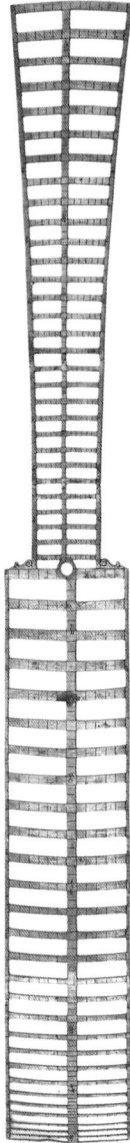

75 The fret.

projection, identical to the *Saphea* on the main plate; on the reverse is a polar projection. Azimuths and almucantars are to 1°, and hours are marked in roman and arabic numerals.

Some of the lines of latitude are numbered by punches, which will be later additions. This plate is to fit over side 2 (with the double projection).

The second semicircular plate has pins at the ends of the straight edge (length 630) to fit into sockets attached to the outer rim. The curved edge is bevelled on both sides, on one of which there is a degree scale engraved with numbers every 10°. This scale is original; later markings are on both sides. On the side with the degree scale is a rectangle, with a poor attempt at scribing the positions of eight wind directions from North to East. The points are in the form of triangles shaded on the right side. All letters and numbers are punched. This figure is the same as the engraving in Dudley's *Dell'Arcano del Mare* (1647), Book V, p. 8, fig. 11; see also **67**. The other side has the table of proportions of Matthew Baker (*c.* 1530–1613) in the form of a right-angled triangle (base 43, height 275, hypotenuse 480). All numbers are punched, and the impressions are obvious on the reverse. This was originally a blank semicircle, with only a scale on one edge. The blank areas of precious brass plate were put to use in Florence.

This table of proportions occurs, with little change, in the ship-building part of Dudley, *Dell'Arcano del Mare* (1647), Vol. 2, Book IV, p. 10, fig. 5. Matthew Baker was the Queen's chief shipwright, who had European fame. He was responsible for designing over 12 royal ships, and stressed the need for mathematics and geometry in shipbuilding (*Armada* 1988, pp. 152–3, 163). There seems to have been an attempt to persuade Baker to come to Florence, judging by a letter from Sig. Lotti writing from Deptford on 23 May 1607 (Leader 1895, p. 187).

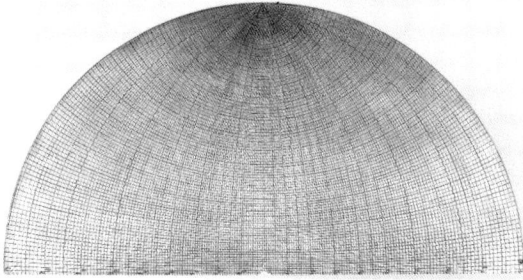

75 Stereographic projection.

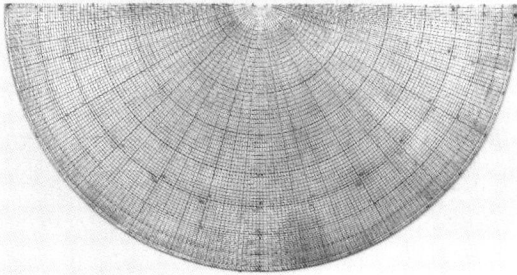

75 Polar projection.

75 Later rectangle.

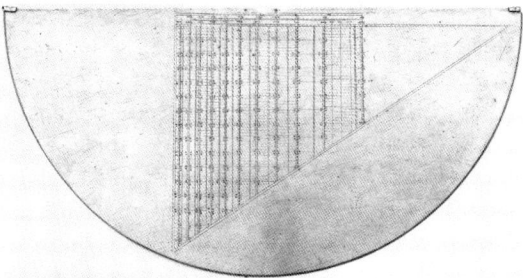

75 Later engraving of Matthew Baker's table of proportions.

Provenance

One of the instruments taken to Florence by Sir Robert Dudley in 1606.

Literature

This instrument has occurred in the literature as a mariner's astrolabe. The museum used to name both this and **51** as 'astrolabio nautico'. Consequently, both are included in Waters (1966). Price (1973), in the revised version of his 'International Checklist of Astrolabes', gives this instrument the number IC 2056. It is illustrated in the survey of mariner's astrolabes by Alan Stimson (1988, p. 95), where it is correctly named.

Location

Museo di Storia della Scienza, Florence (1123, 1124).

76 Surveyor's peractor
1/4 17th Century

Unsigned and undated; tentatively attributed to James Kynvyn

Brass; radius of quadrant 157; length of sighting bar 265

The instrument is the altitude measuring part that fits over an azimuth, or simple, theodolite. One face is engraved with a geometrical quadrant, and the other is blank, scraped smooth. The stand has a rectangular base (now bent) that is intended to span a large compass box. The quadrant is held at its apex in a double-sided vertical support; one side provides the fiducial edge, while the other side holds the plumb-line. The plumb-bob is viewed through a circular hole. The upper side of the quadrant carries the sighting bar, with sight vanes at each end; these are peg and pin-hole.

The quadrant has a rectangular net, the edges labelled every five units from 0–50. The

76

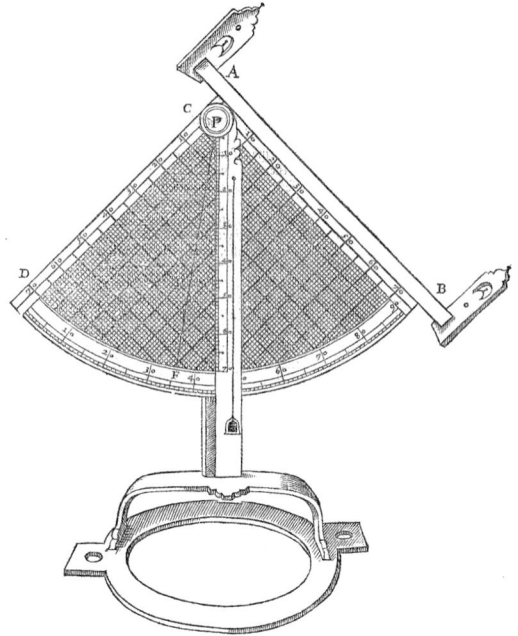

76 Illustration in Rathbone, *The Surveyor* (1616).

76 Close-up of scale.

support with the bevelled reading edge is graduated from the apex 0–50 in the same manner. The arc is divided in degrees, subdivided in 30′, and labelled in tens to 90°. The peractor employs the principle of similar triangles to link the incline with the horizontal distance and height. The incline is measured on the vertical support, the horizontal distance is read off the left side of the quadrant, and the height is read on the top of the quadrant (under the sighting bar) by following the lines on the square grid. The engraved scale to 50 can be converted to any length by simple multiplication, e.g., 100 yards = 50 × 2; 500 yards = 50 × 10.

The instrument closely resembles the illustration in Aaron Rathbone, *The Surveyor in Foure Bookes* (1616). In the preface to his Third Book he wrote:

This Book tendeth chiefly to matter of survey, wherein is first described and declared the severall Instruments, fit for that purpose (with their use in practice) as the Theodolite, the Playne Table, and Circumferentor, whereunto I have added an absolute Instrument, which I call the Peractor, together with the making and use of the Decimal Chayne, used only by myself.

Rathborne's intention was to speed surveying by avoiding unnecessary calculation, so by using this quadrant measurements along inclines can be converted to horizontal for mapping, 'reducing Hypothenusall lines to Horizontall'.

There are only numbers on this instrument on which to base an attribution. Some of these are very close to Kynvyn's style; others, notably *o, 6, 9*, are too well rounded to be typical. Kynvyn died in 1615, so he could have made this piece as a prototype to Rathborne's order. Alternatively, this peractor could be the work of a maker not yet known to us. It is not the work of Elias Allen, who made the peractor on a complete theodolite, see **88**.

Location

Private collection.

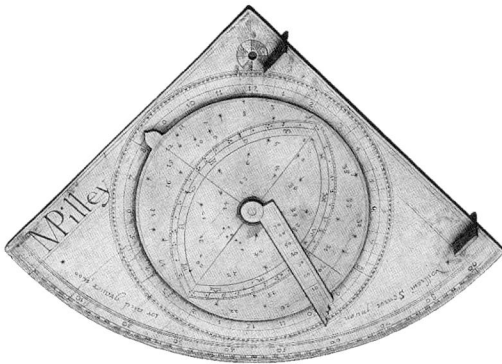

77 Side 1.

77 Horary quadrant 1600

Signed and dated: ✶ *William Senior jnven = tor and grauer 1600* ✶

Inscribed: MPilley

Brass; radius 120

This is the only instrument known with the signature of William Senior. He is recorded by Bendall (1997, ii, p. 458) as a surveyor, who flourished 1600–41, and who was employed by the 1st and 2nd Earls of Devonshire, and 1st Earl of Newcastle, among others. Bendall (p. 409) also records Michael Pilley (*c.* 1745–*c.*1795) of Lincoln, another surveyor. He is also noted by Wallis and Wallis (1993, p. 108).

Side 1

The side with the signature carries a pair of sight vanes with two pin-holes each, and the curved side is divided in degrees, alternately shaded, labelled in tens, 0–90°. At the centre rotates a volvelle that passes over a scale of 24 hours (twice 12 labelled in arabic numerals), divided in units of 3 minutes of time, alternately shaded. The volvelle has an index to point to the time, and engraved across the face are two arcs representing the ecliptic, which is divided into the Zodiac, labelled by sigils. The scale is divided in 2° intervals, the beginnings of each ten marked by one, two, or three dots. The rule is divided in 2° units of latitude, labelled in tens, 0–70°. The entire face of the volvelle is marked with asterisms, numbered to correspond to a table that would give the names; highest number 36. Several of the asterisms have a dotted trail to the appropriate number; a short line extends from the asterism to show the exact place of the star. An exception is the Great Bear, which is indicated by its seven prominent stars, only *Alkaid* (η UMa) being numbered, 21, via a long string of dots.

Side 2

This side is engraved with an upper right quarter of a universal, Rojas projection of the

77 Side 2.

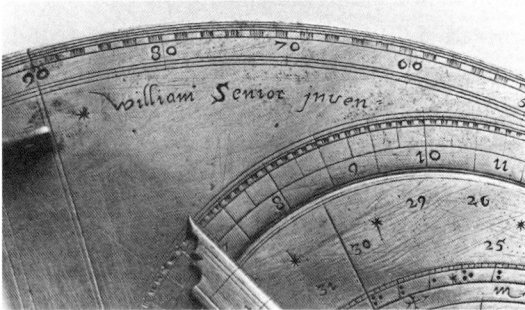

77 Signature.

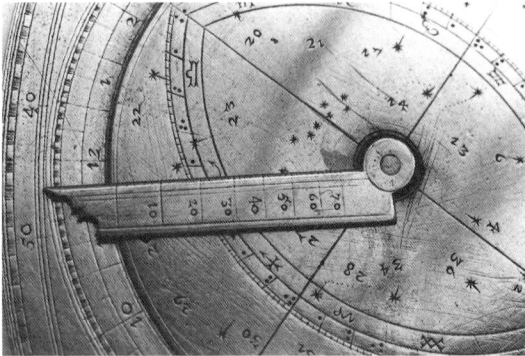

77 *Ursa maior* is near the centre at 11 o'clock.

sphere, the latitude scale being divided in degrees, alternately shaded, labelled in tens, 0–90°. The equatorial line is divided in hours, subdivided in quarter hours, from 6 at the

apex to 11 at the bottom edge, and along the 40° line from 6 to 1. There is a hole for a rule, now missing.

Location

Optical Museum, Ernst Abbe Foundation, Jena, Germany (03/0731).

78 Horizontal sundial 1609

Signed and dated: *Isaack Symmes Gouldsmith &c Clockmaker at London / Anno 1609*

Bronze; 310 square; thickness 2–4

The rectangular plate is fimbriated at the sides, and the hours are placed along the sides, the inner boundary being a double and two single lines. At the four corners are later fixing holes; the corners have been clipped so losing the original holes. The hours are numbered IIII–VIII, and are divided to quarter-hour intervals on the inner bands. The engraved hour lines are suitable for the latitude of London, 51° 30′; the angle of the gnomon is the same. The point of the gnomon is at the centre of a double circle, which contains the initials: W E (West, East). The point also touches an ellipse with a double-line edge, which encloses a coat

78

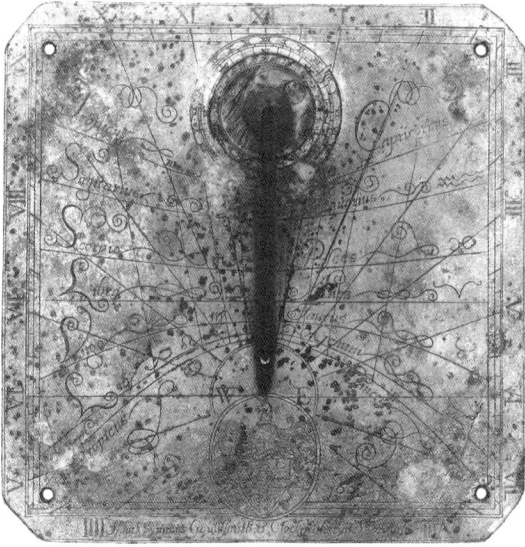

78 The plate.

underneath. The northern side is decorated by a series of deeply cut curves, similar to the gnomon on another dial by Simmes (**80**).

At the centre of the North side of the dial, below 12 o'clock, is a lunar volvelle and aspectarium. The edge of the volvelle is divided to $29\frac{1}{2}$ days, with an index that passes over a circle engraved with twice 12 hours. The age of the Moon and its phase is to be set by another volvelle, now missing; the device is complete on **79**.

Isaac Simmes (*c.* 1580–1622) obtained his freedom in the Goldsmiths' Company in January 1604; see D. Thompson (1987). His quality as a watchmaker is known through a fine example in the British Museum.

Provenance

Presumably made for Sir William Leigh of Stoneley, Warwickshire. Purchased from a dealer by the Science Museum in 1976.

Location

The Science Museum, London (1976–414).

of arms: *[Gules] a cross engrailed in first quarter a lozenge [Argent], with a mullet for cadency on the cross.* The close helm is surmounted by the *crest on a torse [Or] and [Gules] a unicorn's head erased [Or].* The tinctures are not on the dial; these have been added from a full description of the arms, which belong to the Leigh family of Stoneley, Warwickshire. The mullet at the top of the cross is a mark of cadency reserved for a third son, in this case Sir William Leigh (d. 1632), the third son of Sir Thomas Leigh (1504?–71), who was Lord Mayor of London and knighted in 1558.

Another set of lines shows the declination of the Sun throughout the year, from the Tropic of Capricorn to the Tropic of Cancer. Each line has the Zodiac marked in name and sigil. The declination at any particular date is indicated by the position of the shadow of a spur set in the edge of the gnomon. The brass gnomon (vertical height 177) has lugs to attach it to the plate, which is roughly finished

79 **Horizontal sundial** *c.* 1610

Signed: *Isaack Simmes Gouldsmyth and Clockmaker*; undated; inscribed 1735 over an original date

Bronze; 182 × 180; thickness 3–4.5

This dial is of the same design as **78**, but about half the size. The rectangular plate has the hours placed along the sides, bounded by single and a double lines. At the four corners are fixing holes. The hours are numbered IIII–VIII, and are divided to quarter-hour intervals. The engraved hour lines are for the latitude of about 53°, and the angle of the gnomon is the same. The point of the gnomon is at the centre

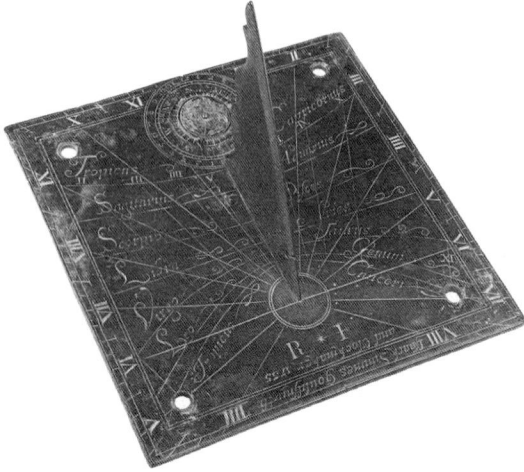

79

of a double circle, and below this are the initials: R ✳ I, which are not original lettering.

The Sun's declination through the year is provided by a set of lines from the Tropic of Capricorn to the Tropic of Cancer, and the marker is the shadow cast by a notch and small hole cut in the gnomon.

At the centre of the North side of the dial, obscuring the 12 o'clock position, is a solar volvelle that passes over a circle engraved with twice 12 hours in arabic numerals. Also engraved, below the hour scale, are the 32 wind directions. The edge of the volvelle is divided to $29\frac{1}{2}$ days, with an index (partly broken) for setting against the lower circle. The age of the Moon is set by the index of the top volvelle, and its phase viewed by a circular hole. A planetary aspect diagram fills the remaining space, showing trine, quartile, and sextile. With this device it is possible to convert readings by moonlight into solar time.

The brass gnomon (vertical height 104) has two lugs to attach it to the plate, which is

roughly finished underneath. The northern side is decorated by two ogee curves, similar to **91**.

Provenance

The dial was placed in two gardens belonging to Lewis Evans (1853–1930), at Sidmouth and Dawlish. The Lewis Evans Collection came to the University of Oxford in 1924.

Location

Museum of the History of Science, Oxford (LE 233; 50900).

80   Horizontal sundial 1610

Signed and dated: ✳ *Isaack Simmes* ✳ / ✳ 1610 ✳

Bronze; 161 × 162; thickness 4–5

The rectangular plate is fimbriated at the sides, and the hour circle is contained between bands; the outer and inner edges each have two concentric circles. At the four corners are the fixing holes, still containing hand-made screws. The hours are numbered IIII–VIII, and are divided to quarter-hour intervals on the inner band, and to 5 minutes on the outer. The angle of the gnomon is about 52°, and the hour angles engraved on the plate show that they were calculated for a latitude of about 53°. The

80

gnomon has a thickness of 3, with a bevelled edge, and the meridian hour line makes no allowance for its thickness. The point of the gnomon is at the centre of a double circle.

The gnomon (vertical height 86) has lugs that fit into a pair of slots in the plate, which is rough, as cast, underneath. The northern side of the gnomon is decorated by a series of deeply cut curves, partly ogee, similar to **78**.

Location

Private collection.

81 **Horizontal sundial** 1606

Signed and dated: *Elias Allin fecit 1606*

Brass; 127 square

Now octagonal, this dial may have been square originally. The four fixing holes are close to the hour ring. The surface is finely pitted.

The hours are engraved in roman numerals, IIII–VIII, between a double circle to the outside and three circles below. These circles have the divisions of an hour into eighths and quarters, and further radial lines extend for the half hours. The point of the gnomon is at the centre of a double circle (the present gnomon does not reach the centre). The hour lines engraved on the plate are for a latitude of about 53°. Also engraved in the space at the southern side are the initials: I ★ B.

The signature is remarkable because the surname is spelt with an *i*. This spelling is not found on any other extant instrument by this maker. The date 1606 is in the course of Allen's apprenticeship to Charles Whitwell, therefore it is an early product. There is nothing in the forms of the letters and numbers to suggest that they are not engraved by Allen. At this period

81 Dial plate.

variations in the spelling of a name are common; compare with Cole and Kynvyn. This dial is laid out in a similar way to several other horizontal dials of the period.

Provenance

Bought by the National Maritime Museum in 1980.

Location

National Maritime Museum, Greenwich (AST0240).

82 **Compass sundial** *c.* 1610

Signed: *Elias Allen fecit*; undated

Silver and ivory; diameter 72, thickness 22

The compass and dial are fitted into a turned well in a round block of ivory. The screw-on ivory lid has a low relief turned button at the centre and a roll at the edge.

The chapter ring (diameter 56) and gnomon are in silver. The hours are divided in quarters, and are numbered in roman numerals,

82

IIII–VIII. The folding gnomon is held in two sockets placed at 12 o'clock and within the tracery over the South part of the dial. The tracery is foliate and zoomorphic, and may be compared with the tracery on **99**, which is also zoomorphic, and with **57**. The gnomon has a bevelled shadow-casting edge, ogee curves on the northern side, and a onion-dome aperture at the bottom. It is kept vertical by turning a castellated nut that moves a stub to give resistance to movement of the gnomon. This sundial is designed for a latitude of about 52°.

The compass rose (diameter 56) is type A-2, printed from a copperplate, and is hand coloured in blue, red, apple green, and parchment. It has a degree scale divided in quarters, 0–90°. There is no cover glass because of the gnomon tensioning adjustment. The magnetic needle is blued iron (length 40), with a crescent at the South and a drop at the North.

The brass pivot head is castellated to match the screw head above it. I am grateful to Mr S. Waterman for pointing out the gnomon adjustment.

Location
Private collection.

83 Compendium *c.* 1610–*c.* 1615

Signed: *Elias Allen Fecit*; undated

Brass, gilded; diameter 80, thickness closed 30

This is the first of a group of five Elias Allen compendia; the fifth is dated 1617. After considering variations in the values of the latitudes, the instruments have been ordered as follows. There is a marked improvement in accuracy of the values given. Most significant is the fact that, on the 1617 instrument, the figures given for Scottish towns are almost equivalent to modern ones, while values on earlier compendia are inaccurate, For example, Glasgow has changed from 57° to 55° 20', modern 55° 52'. Therefore it has been assumed that the four undated compendia were produced between *c.* 1610 and *c.* 1615.

Comprising from the top:

a) nocturnal

b) table of latitudes of 30 towns

c) equinoctial sundial

d) magnetic compass

e) tide computer

a) The nocturnal is of the pictorial type, which is simpler to construct. At the edge is a 24-hour circle, divided in quarter hours, and labelled in roman numerals. There follows a large volvelle, with its rim divided in months, indicated by their initial letter, and subdivided in days, numbered in tens (28, 31 as appropriate). From the centre, radial lines are cut to the beginning of each month. Disposed over the central part are pictorial representations of six constellations, with a simplified star pattern on each: Cassiopeia, Perseus, Ursa Maior, Ursa Minor, Draco, Cepheus. The time is found by matching the orientation of a readily observable constellation on the instrument with that seen in the sky, and then reading off the time from the date, in the manner of the modern Philips' planisphere. This form of nocturnal was figured by Allen in Edmund Gunter, *The Description and vse of the Sector* (1623), Sig. A4, and pp. 60–1, as 'a type for the vse of Sea-men'. To use it: 'For looke vp to the pole, and see what starres are neare the meridian, then place the rundle to the like situation, so the day of the moneth will shew the houre of the night.' See also **86**.

b) Inside the lid is a circular table of six columns and five rows listing the latitudes in degrees and minutes of 30 towns. Because it is larger than many compendia, the engraver has had space to engrave clearly.

c) When fully opened, the chapter ring in its cradle is erected, and the appropriate latitude set by a quadrant that turns to the vertical. The quadrant arc is divided in degrees, 0–90°, numbered every 5°. The gnomon is attached to the quadrant, and erects with it. The upper, summer, side of the ring is numbered in hours, IV–VIII, and the lower, winter, VI–VI. The inside of the ring has the divisions to half an hour.

d) The compass has lost its glass or mica cover, and the magnetic needle is a replacement. The card is type B-2 (Plate 14g.), and has a diameter of 69, including the circle of 360°. This is divided to 1°, alternately shaded, and labelled in tens. The 32 wind directions are subdivided in halves, alternately shaded. The pointers for the 16 main directions are triangles coloured in blue, red, or gold.

e) The base is a tide computer. The form and sequence of information is the same as on the Charles Whitwell 1600 compendium, **59**. Minor differences on the present instrument are the divisions of the calendar and the Zodiac into single units; the use of Zodiac sigils instead of the names; the age of the

Londinium	51 30	*Edenburgum*	56 0	*Burdegala*	45 30
Cantuaria	51 0	*Lutecia*	48 50	*Marsilia*	43 0
Oxonia	51 45	*Caletum*	50 0	*Madritum*	41 0
Eboracum	54 0	*Bruxela*	51 20	*Seuila*	37 0
Dublinium	53 30	*Haga*	51 25	*Lisbona*	39 40
Copenhaga	57 20	*Florentia*	43 5	*Heirosalima*	31 40
Heidelberga	49 35	*Neapolis*	41 0	*Constantinopō*	43 5
Brunsuaga	52 30	*Moscoua*	55 30	*Vennecia*	45 15
Praga	50 5	*Cracouia*	50 15	*Vienna*	47 45
Borges	61 0	*Geneua*	45 30	*Roma*	42 5

Borges is a cathedral city and port near Helsinki, Finland.

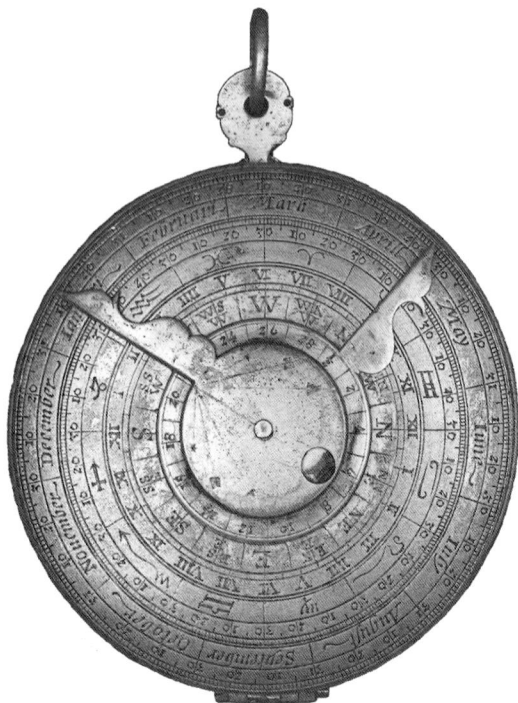

83 e.

84 e.

Moon labelled every 2 days; and May with no final *e*. See also **87**.

Provenance
Christie's South Kensington, sale 14 March 1985, lot 224.

Location
Private collection.

84 Compendium *c.* 1610–*c.* 1615

Signed: *Elias Allen fecit*; undated; also inscribed: W B

Brass, gilded; diameter 63, thickness closed 23; leather case diameter 72, depth 27

Comprising from the top:
a) nocturnal
b) table of latitudes of 24 towns
c) equinoctial sundial
d) magnetic compass
e) tide computer
f) leather case

a) The nocturnal is of the slit type, which is constructed as described for the Whitwell 1600 compendium, **58**. The calendrical table is not present on this Allen instrument.

b) Inside the lid the sighting volvelle is held in place by four clips. Either side of the slit are the letters: W B (probably the owner). There is a circular table of six columns and four rows listing the latitudes in degrees and minutes of 24 towns.

84 Case.

of North, correct for about 1620 at London. The pointers for the 32 wind directions are coloured in blue, pink, green, orange, both rings in yellow (see Plate 14c). The magnetic needle has a double crescent at one end, and an arrow head at the other. A brass rod shaped like a compass needle is set at right angles to steady the movement of the magnetic needle. The brass boss is an eight-sided pyramid. The original glass cover is present, and one air bubble is 5 mm in length.

e) A tide computer on the base completes the instrument. The form and sequence of information is the same as on the Allen compendium, **86**, and the Charles Whitwell 1600 compendium, **58**.

c) The chapter ring in its bracket is erected when the case is opened through two prongs set in the hinge. The latitude is set and the gnomon positioned in the customary manner. The upper, summer, side of the ring is numbered in twice 12 hours; the lower, winter, VI–VI; both sides with half-hour divisions. There is no provision for a plummet. The signature is on the winter side.

d) The compass card is type A-2, and has a diameter of 57, including the circle of degrees, 0–90°–0 twice. A rough ink line is at 6° East

f) Unusually, the case has remained with its instrument, which is extremely well preserved. Made of pasteboard, it is covered on the inside with claret-coloured velvet, and on the outside with brown leather, gold tooled. The edges of top and bottom are ringed by a running fox motif, and the central area by a diaper pattern of small rectangles. The rims are floriated. A simple brass hinge holds the two parts, closure completed by two hooks.

Location
Private collection.

Cantuaria	51 0	Edenburgum	56 0	Burdegala	45 30
Londinium	51 32	Dublinium	53 30	Marsilia	43 0
Oxonia	51 45	Caletum	50 40	Madritum	41 0
Eboracum	54 0	Lutecia	48 50	Seuila	37 0
Lisbona	39 40	Copenhaga	57 20	Heirosalima	31 40
Florentia	43 5	Heidelberga	49 35	Constantinop	43 5
Neapolis	41 0	Brunsuaga	52 30	Veanetia	45 15
Roma	42 5	Praga	50 5	Geneua	45 30

85 Compendium *c.* 1610–*c.* 1615

Signed: *Elias Allen fecit*; undated

Brass, gilded; diameter 60, thickness closed 18

Comprising from the top:

a) nocturnal

b) table of latitudes of 30 towns

c) equinoctial sundial

d) magnetic compass

e) tide computer

85 a.

a) The nocturnal is unusual in form, and is analogous to the pictorial type (**83**, **86**). At the edge is a 24-hour circle, divided in quarter hours, and labelled in arabic numerals. There follows a large volvelle, with its rim divided in months, which are indicated by their full names, subdivided in 5-day intervals, not numbered. From the centre 12 radial lines are cut to intersect each month. By each line is the name of a star. This list resembles the 12 navigation stars (seven are identical) of Willliam Bourne published in his *Almanacke* (1571); reproduced in Taylor (1963, pp. 99–101).

Around the centre rotates an index, which could be of help if one of the possible stars were occluded. When one of the named stars is on the meridian, the volvelle is turned to match, when the day of the month gives the time.

b) The latitude table of 30 towns is on the inside of the lid, with six columns and five rows listing the latitudes in degrees and minutes.

c) When fully opened, the chapter ring in its cradle is erected, and the appropriate latitude set by a quadrant that turns to the vertical. The quadrant arc is divided in degrees, 0–90°, numbered every ten. The gnomon is attached to the quadrant, and erects with it. The upper, summer, side of the ring is numbered in hours, IV–VIII, and the lower, winter, VI–VI; the signature is on this rim. The inside of the ring has the divisions to half an hour.

d) The compass card is type A-2, and has a diameter of 57 including the circle of degrees, twice 0–90°–0. This is divided to 2°, and labelled in tens. The pointers for the 32 wind directions are coloured in blue, faded red, pale

Luc ♈	α Ari	Ocu ♉	α Tau	Seg ori:	ζ Ori
Can min	α CMi	Cor ♌	α Leo	Cau ♌	β Leo
Spi ♍	α Vir	Lanx bor	β Lib	Cap oph	α Oph
vultur	α Aql	Os peg	ε Peg	Ext ala	γ Peg

258

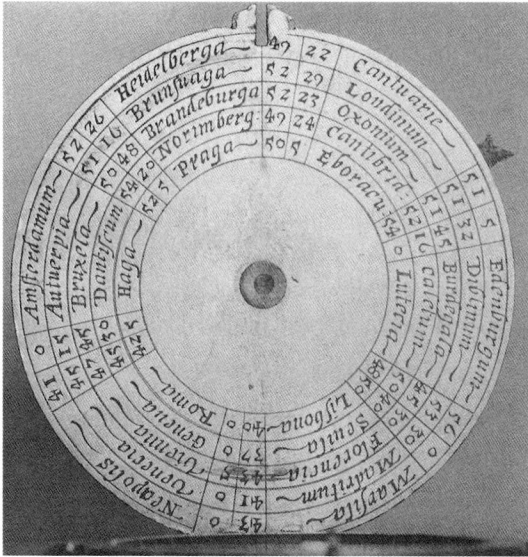

85 b.

hours in roman numerals. The innermost circle is a wind rose with 16 of the points labelled with initials of the wind directions. Rotating at the centre is a volvelle with a long index and a scale of the Moon's age; divided in 29½ days and labelled every 2 days. Finally, there is a lunar volvelle with a long index, phase aperture, and planetary aspects.

Provenance

Presented to the British Museum by A.W. Franks, 1863.

Literature

Ward (1981, p. 126), no. 362.

Location

The British Museum, London (1863, 9–29.2).

green, white. The magnetic needle is straight at both ends, and is a replacement. The glass cover contains air bubbles.

e) The base is a tide computer. The form and sequence of information is the same as on other Allen compendia, and the Charles Whitwell 1600 compendium, **58**. The outer circle gives the months, with the days in units of two, numbered in tens, followed by the Zodiac, each Sign shown by its sigil, divisions in 2° intervals, numbered in tens. Then comes a 24-hour scale divided in half hours, the

86 Compendium *c.* 1610–*c.* 1615

Signed: *Elias Allen fecit Londini*; undated

Silver; diameter 53, thickness 19

Comprising from the top:

a) nocturnal

b) table of latitudes of 42 towns

c) equinoctial sundial

d) magnetic compass

e) moonlight to solar time computer

Cantuarie	51 5	Edenburgum	56 0	Marsila	43 0
Londinium	51 32	Dublinum	53 30	Madritum	41 0
Oxonium	51 45	Burdegala	45 30	Florencia	43 5
Cantibrid:	52 16	Caletum	50 40	Seuila	37 0
Eboracu:	54 0	Lutecia	48 50	Lisbona	40 0
Neapolis	41 0	Amsterdamum	52 26	Heidelberga	49 22
Venecia	45 15	Antuerpia	51 16	Brunsuaga	52 29
Vienna	47 45	Bruxela	50 48	Brandenburga	52 23
Geneua	45 30	Dantiscum	54 20	Norimberg:	49 24
Roma	42 5	Haga	52 5	Praga	50 5

86

a) The nocturnal is of the pictorial type, and at the edge is a 24-hour circle, labelled in roman, divided in quarter hours. There follows a volvelle, with its rim divided in months, which are shown by their initial letters, and subdivided in days, numbered in tens (28, 31 as appropriate). From the centre, radial lines are cut to the beginning of each month. Disposed over the central part are pictorial representations of six constellations, with a simplified star pattern on each. The time is found by matching the orientation of a readily

observable constellation on the instrument with that seen in the sky, and then reading off the time from the date. See also **83**.

b) The inside of the lid has a circular table of six columns and seven rows listing the latitudes in degrees and minutes of 42 towns. In the central space is a coat-of-arms for John Wood (d. 1640) of Faversham, Kent: *[Argent] on a mount [Vert] in base an oak tree with acorns proper*, surmounted by a helm, mantled.

c) When fully opened, the chapter ring, in its floriated semicircular bracket, is erected, and the appropriate latitude set by a quadrant that turns to the vertical. The quadrant arc is divided in degrees, 0–90°, numbered every 10°, the latitude of use set against a notch. The gnomon is attached to the quadrant, and erects with it. The upper side of the chapter ring is numbered in hours, twice 12, marked in roman numerals, the lower side (with the signature), V-VI. The inside of the ring has the divisions to half an hour.

d) The compass has a glass cover, and the original magnetic needle, of the arrow head and

Cantarberi	51 10	Newcastell	54 30	Midelburg	51 30
London	51 32	Barwick	55	Antwerp	51 12
Oxford	51 45	Edenburg	56	Amsterdam	52 26
Cambridg	52 16	Aberdien	57 20	Bruxells	50 45
Notingham	53	Glosgow	57	Dunkerke	51
Lincoln	53 30	Dublin	53 30	Colona	50 56
Yorke	54	Paris	48 50	Wesel	51 30
Hedeburg	49 36	Ratisbona	49 9	Constantinopel	43
Ingolstat	48 42	Brandenburg	52 30	Ierusalem	32 10
Norenburg	49 26	Luxenburg	49 50	Neapolis	40 42
Hamburg	53 43	Lunenburg	53 36	Madrid	40 45
Magdenburg	51 15	Franckford	50	Lisbona	38 30
Lubeca	53 58	Vienna	48 22	Venecia	45 15
Brema	53 8	Praga	50 6	Roma	42 2

86 a, e.

sea blue, orange red, apple green, and white; all three rings are gold.

e) The bottom of the instrument is a simplification of the tide computer, where the small size of the instrument has caused the omission of the Zodiac and calendar scales. The outer edge has a 24-hour scale in half hours, followed by 16 points of the compass, identified by capital initials. The solar volvelle has an index at the new moon position, and the rim is marked out with the Moon's age in days, $0-29\frac{1}{2}$, labelled every two. The lunar volvelle has an index opposite the aspectarium that shows the phases. The central part has a diagram of the aspects: opposition, trine, quartile, sextile. Above the catch is an oval brass pendent loop.

Provenance

Purchased in 1921 by Lewis Evans from P. Webster.

Location

Museum of the History of Science, Oxford (LE 223; 36259).

cross variety, has a thin brass rod set at right angles to steady the movement of the needle. The underside of the glass is ground away in a circle above the boss of the needle. The card is type A-2, with a diameter of only 48 since the degree circle has been cut off because of limited space. All the points are labelled and coloured,

87 Compendium 1617

Signed and dated: *Elias Allen Fecit / 1617*

Brass, gilded; diameter 64, thickness when closed 25

Comprising from the top:

a) the Royal Arms

b) the names and latitudes of 30 English and Scottish towns

c) equinoctial sundial

d) magnetic compass

e) tide computer

87

87 a; Royal Arms of James I of England.

f) table connecting Prime Numbers with Epacts

g) nocturnal

a) The lid is engraved with the Stuart arms of James I, now with Scotland and Ireland in the second and third quarters, and the Welsh dragon replaced by the unicorn as a supporter.

b) The names of the towns are arranged in five rings divided into six columns. The central space is occupied by:

THE LATITVD OF CITIES AND TOWNES IN INGLAND AND SCOTLAND

c) The equinoctial sundial is the same as those fitted by Charles Whitwell in 1602 (see **59**) but without any decoration.

d) The magnetic compass is supplied with a type C wind rose, identical to those found on

two Whitwell instruments dated 1604 (**60**) and 1610 (**62**). The colours are blue, green, and gold. The degree band is present, divided in single degrees, 0–90°–0, twice. The blued-steel needle has a cross at one end and a plain spike at the other. The cover is glass.

e) The tide computer, for the establishment of the port, is underneath the compass. It is a full version with calendar and Zodiac, and is exactly the same as that on **83**.

f) The table connecting prime numbers with Epacts is in two bands, and begins:

The prime 3 4 5 ... / *and Epact* 3 14 25 ... / *beginning* 1617.

g) The nocturnal is of the slit form, with the disc held in position by six lugs with short pegs. The rim is marked out with the names of the months, then a division into days in 2-day units marked in tens (28 or 31). The solar volvelle follows with a 24-hour scale in half hours, the numbering in roman. The volvelle with the slit completes this side. The pattern is the same as on the Whitwell 1610 compendium (**62**), but the sequence of months is reversed. The meridian line runs between 27 October (hinge) and 24 April (catch).

By the North pointer on the compass card are the initials R G in black. They are not part of the engraved copperplate from which the card was printed. The significance is not known; however, another Elias Allen compendium is said to have the initials R.G. on the compass card. This was sold at Christie, Manson & Woods, King Street, London, sale 30 March 1977, lot 104. What may be more than a coincidence, however, is the existence of the initials RG engraved on the silver chapter ring of a diptych sundial (**99**), attributed to Elias Allen.

Provenance

Presented by William Henry Smyth, RN, FRS, FSA, in 1839, to the Museum of the Royal United Services Institution in the old Banqueting Hall of Whitehall Palace, London. Acquired by the Victoria and Albert Museum in 1963.

Literature

Blair (1964).

Location

Victoria and Albert Museum, London (M 51–1963).

88 **Theodolite** early 17th Century

Signed: *Elias Allen fecit*; undated

Brass; diameter of circle 315; radius of quadrant 170; length of sighting bar 315

The azimuth circle is divided to 15 minutes of arc from the South, 0–360°. Four quadrants are cut away, leaving cross-bars, at the centre of which is a large compass box (diameter 100) above, and below, a cylindrical socket to fit on a post. Rotating round the compass box is an alidade (length 345) with sights at either end that attach by dovetail slides. The vanes each have slit and wire sighting apertures. Over the compass box spans the base of a peractor, which attaches by wing-nuts to the alidade.

London	51 32	Newarke	53 10	Newcastle	54 57
Royston	51 56	Dancaster	53 28	Barwicke	55 20
Huntigton	52 6	Yorke	53 58	Edenburgh	55 36
Grantham	52 50	Rippon	54 12	Glasguo	55 20
Lincolne	53	Durham	54 46	S Andres	56 41
Carleele	54 55	Leester	52 48	Cantarbury	51 12
Lancaster	53 45	Warwick	52 43	Chechester	51 0
Darby	52 55	Northhampton	52 36	Southampton	51 5
Notingam	52 58	Cambridg	52 10	Bristol	51 30
Chester	52 54	Hartford	51 45	Oxforde	51 45

88

The paper compass card is identical to the one illustrated in Arthur Hopton, *Speculum Topographicum* (1611), pp. 17, 147, sig. Ee. The diameter of this published version is 91 mm. Within a double scale, labelled in fives to 120, is a magnetic azimuth-dial. The time is found by lining the sights to the Sun, when the time is read by the magnetic needle. The card is positioned East of North to allow for the magnetic deviation. There is a table of the day of the month when the Sun enters into the 12 Signs of the Zodiac, similar to Hopton's illustration.

For measuring altitudes, this instrument is provided with a peractor, the invention of which was claimed by Aaron Rathborne in his book *The Surveyor*, published in 1616. It has a geometrical quadrant, engraved and labelled in the same way as the one described under **76**. The sights on the vanes are of the hole and peg type.

Provenance

The Landau Collection, Paris.

Literature

Guye and Michel (1971, p. 277), plate 282.

Location

Musée du Louvre, Paris (OA.10760).

89 Compendium 1577[?]

Unsigned; tentatively attributed to Humfrey Cole; inscribed with the date 1577

Silver; diameter 61

The only evidence for the existence of this instrument lies in a printed auction catalogue. The description, published in 1957, proposes Cole as maker by comparison with a 1575 compendium at the British Museum. The initials I.C. are claimed to be present, but in a curious position. Without a photograph of this part it is not possible to check the accuracy of this claim. The compendium is included here on the supposition that it might be genuine. If so, then Cole could be a candidate, through the description of its parts, and the evidence of the towns and their latitudes that are visible in the printed photograph. The construction with eight operative parts is matched by other Cole compendia, of which there are two earlier (1575) and two later (1579). From the latitude table one sees that the engraving is consistent with that of Cole. Of 40 towns, 30 are frequently used by Cole (perhaps five more that cannot be read); another five are not known as used by Cole (but they are by other English makers of compendia). These are all ports, and could be present at the request of the owner.

This compendium is 25 per cent greater in area than the four others produced by Cole in

FIG. 89 Reproduced from a Sotheby catalogue.

the 1570s. This allows some larger features to be included: celestial orbs, world map, armillary sphere (for which Cole had a fondness). The use of silver is matched by the azimuth dial, **14**, also dating from the middle of the 1570s.

From the photograph in the catalogue of the auction sale, this compendium appears to follow the design of most of those made in Elizabethan England. The only existing record of this instrument is the account and photograph published in *Catalogue of Objects of Vertu Scientific instruments Watches, miniatures, coins and Works of Art by Carl Faberge*, Sotheby & Co., New Bond Street, London, sale 16 May 1957, lot 125. The following description and gloss is copied from the sale catalogue.

The Property of G.E.C. Ashford, Esq.

125 A VERY RARE ELIZABETHAN SILVER POCKET COMPENDIUM in the form of a watch and *dated* 1577, the circular case chased with broad chevron and narrower ropework bands, the pendant with stiff leaves. The outside of the hinged cover is engraved with a calendar of Holy Days and dominical letters surrounding an armillary sphere, the back with a calendar giving the dates and names of the main ecclesiastical festivals, surrounding a map of the world. On the inside of the lid is engraved another calendar indicating, among other things, when to let blood, and to bath, in the centre a delineation of the four elements; fitting loosely into the cover is a silver table of the tides for British and Channel Ports, facing this is a volvelle with two rotating circles, the inner engraved with the signs of the aspects, the concentric triangle, square and pentagon and showing the phases of the moon, the outer with the month divided into $29\frac{1}{2}$ days, the pointer engraved with the sun and the

plate engraved with the signs of the zodiac and the month divided into days. This central portion hinges further to disclose a table of latitudes of English and European towns surrounding a representation of the solar system, facing it a compass dial (the bare needle missing), and between them is hinged (the hinge broken) an equinoctial sundial adjustable for various latitudes and *dated* 1577 on the gnomon, 2³/₈ *in. diam.*

★ It is tempting to regard this superbly engraved instrument as the work of Humphrey Cole, the most famous Elizabethan maker of scientific instruments, whose fully signed works date from 1568–1586. The only possible signature to be found on this compendium are the initials I.C. which also act as the nose and mouth of the sun face engraved on the volvelle's pointer, but these may well be the owner's and not maker's initials.

Two gilt-brass dials of similar type by Humphrey Cole, one dated 1575, are recorded by Gunther in 'The Great Astrolabe ad other Scientific Instruments of Humphrey Cole,' *Society of Antiquaries*, 1927, figs 33–5. One of these is in the British Museum and the other in the Museum of Antiquities, Edinburgh.

There are later examples of universal dials, but without the tables and calendar, of similar type by Charles Whitwell (fl. c. 1590–1606) at the National Maritime Museum, Greenwich. See E.G.R. Taylor, *The Mathematical Instrument Makers of Tudor and Stuart England*, 1954, pl. IV, and also biographies nos. 88 and 112.

This instrument has many similarities to the celebrated V.C. compendium of 1554 in the Bodleian and likewise the tables it contains correspond exactly with those in Leonard Digge's [sic] *Prognostication*, 1555.

[*See* ILLUSTRATION]

It is thought that the person most likely to have written this account was the late Philip Coole (1916–69), who was employed as a watchmaker by Malcolm Gardner in Holborn; from 1960 he looked after the new Horological Students' Room of the British Museum. Coole

also described horological items for Sotheby, which he could be expected to do fully and accurately. Elements of the instrument are familiar, and identifiable now that this piece can be compared with over 20 other Elizabethan compendia.

Turning now to the printed photograph: the screen used to prepare the plate makes it difficult to decipher the words. However, the table of latitudes of various towns can be sufficiently established, and is quite familiar. It is difficult to make out the lines on the compass card. The equinoctial dial is also familiar, and it still has its plumb-line for levelling; on no other extant compendium of the period has this item survived. It is the style of the decoration that is clear on the photograph, and this is of considerable interest. As the description says, the edges of the case have chevron and ropework bands. Ropework yes, but chevrons are not known. Usually there are acanthus leaves, the motif constantly found on Elizabethan and later instruments. The semicircular cradle for the equinoctial dial is chased with a pattern not seen so far on any English instrument. This consists of vine trails and double spirals. Enough has been said to make clear that this silver compendium is a curiosity. It has not surfaced during the past 40 or more years, and until it does, unanswered questions are bound to remain.

It seems that the various parts of the compendium function as follows:

Top of lid   Easter calendar; at centre an armillary sphere.

Inside lid   The Moon's place in the Zodiac for blooding, purging, and bathing; at centre the Four Elements.

Table of latitudes

Copostell	?	?	?	Cracouia	51 0	Oxforde	51 50
Lisbona	39 38	Louania	50 58	Dantisc	55 0	Bristowe	51 42
Parisius	47 55	Bruges	51 30	Emden	53 42	Canterbe	51 23
Marsili	43 6	Tolesa	43 30	Colonia	51 0	London	51 34
Roma	42 0	Praga	50 4	Venitia	44 50	Exiter	51 0
T?	?	?	?	N.hampton	52 50	Carlyl?	55? 2
Plimmuth	51 2	Harford	52 2	Notingame	53 0	N.castle	55 0
Sādwich	51 27	Norwich	52 10	Lincoln	53 6	Barwick	56 23
S.hāpton	51 12	Licester	52 50	Chester	53?	Edenbo	57 0
Douer	51 0	Rye	51 5	Hull	?	Yorke	54 0

Loose disc   Tide table. (Fold-out discs are - extras with the compendia **26, 27**.)

Central leaf   (a) Lunar volvelle tide computer. (b) Table of latitudes of 40 towns; at centre the Ptolemaic world system.

Middle   Equinoctial sundial.

Inside base   Magnetic compass.

Base   Table of fixed feast days of the Church; at centre a map of the world.

The five rings with names of towns leave a large circle, which is usually blank on Cole's other known compendia, but which is here filled with a Ptolemaic diagram of the Celestial Orbs. Below the ring of towns is a narrow band shaded to match the columns of degrees, and then the celestial band, which bears asterisms (the Eighth Sphere). There follow more bands, each marked with a planetary symbol: Saturn, Jupiter, Mars, Sun, Venus, Mercury, Moon, at the centre the Earth. Such a diagram was published by Thomas Digges in his edition of Leonard Digges, *A Prognostication* (1576). This same book also gives models for the Easter calendar, the Moon's place for purging, etc., the ebb and flow of tides, and the fixed feasts.

90   Case of architectural drawing instruments *c.* 1585

Unsigned and undated; inscribed in a cartouche on underside of base: *Barthelmewe Newsum*

Brass, gilt; case 193 × 83 square

All four sides of the case are finely engraved with classical figures representing Peace, War, Poverty, and Abundance. Above each, in a cartouche, is engraved a motto in Latin. The base is supported by cherubs. There are 20 gilded instruments remaining (a few have been lost), and these plug into shaped holes in a wooden block. The instruments include adjustable dividers, 3:1 proportional dividers, compasses, beam compasses, folding rule, folding set-square, scissors, knives, pens, and scribers; a complete list is in Stone (1906), where the case and all the instruments are fully illustrated in two plates and two figures.

It is doubtful whether this set of drawing instruments was made by Newsam: he was an expert maker of portable clocks. The allegorical figures on the case, and its fine finish do not recall English work of the period, nor does the elaborate arabesque decoration on each instrument; on the contrary, Flemish or French work is more likely. Three of the rules are engraved with small numbers. They are carefully executed, but do not correspond to the work of English mathematical instrument engravers of the period, such as Humfrey Cole or Augustine Ryther. This elegant case may well have been a gift to Newsam.

Bartholomew Newsam, as spelled in the *DNB*, was a clock-maker in the Strand near Somerset House by 1568, and obtained a Royal Appointment by 1582 as clock-keeper to Queen Elizabeth, and as clock-maker in 1590; he died in 1593. His Will (1586) exists, and has been drawn on by some writers on clocks, for example Britten (1982, pp. 315–16). Because the Will mentions 'a sonne dyall of copper gylte', and another 'sonne dyall to stande upon a post', some have presumed he made dials and was, therefore, an instrument maker; see Gunther (1920–23, pp. 173–4), and Taylor (1954, pp. 176–7). Such a conclusion is unjustified; the Will includes a vice, hammers, and so on, which would not be considered to have been made by Newsam. This case of drawing instruments is included here to challenge the attribution as maker to Newsam.

Provenance

Acquired by the grandfather of Dr English of Sleights, Yorkshire, who was the owner when it was noted by Stone (1906) in *Archaeologia*. Bequeathed by Max Rosenheim in 1912.

Literature

Stone (1906); Ward (1981, p. 86), no. 239; Hambly (1988, p. 152), fig. 145 (instruments in case); G. Turner (1996, p. 242), fig. 81 (instruments displayed).

Location

The British Museum, London (1912, 2–8.1).

91 Horizontal sundial *c*. 1590

Unsigned and undated

Bronze, 144 × 142; thickness 2

The rectangular plate, now dull with age, has the hour circle contained between bands; the outer edge has three concentric circles, and the inner two. At the four corners are the fixing holes. The hours are numbered IIII–VIII, and are divided to quarter-hour intervals. The gnomon angle is about 54°, and the hour angles engraved on the plate show that they were calculated for an angle of about 55°. The gnomon is thin, with a bevelled edge, and the meridian hour line makes no allowance for its thickness. At the point of the gnomon is a seven-petalled rose. At the North is a cross *pattée*; at the South is an eight-pointed star.

The gnomon (vertical height 78) has lugs that fit into a pair of slots in the plate, which is rough finished underneath. The northern side of the gnomon is decorated by two ogee curves, resembling **79**.

Location

Private collection.

92 Horizontal sundial 1591

Unsigned [?]; dated 1591

Bronze or brass; diameter *c*. 300

By good fortune this dial has remained, since it was made, at its present location at Haddon Hall, Bakewell, Derbyshire. The dial is now in a museum showcase, and could not be examined closely. The surface is blackened with age, so parts of the engraving could not be distinguished clearly; consequently it is not known whether there is a maker's signature or mark. A sketch was made on my visit, and this

92 Author's drawing of dial at Haddon Hall.

is correct for the date, and roman numerals marking the hours. The back edge of the gnomon is in the form of a double ogee curve; the approximate length of the leading edge is 190. The angle could not be measured; it is probably 53° to suit the location (modern value 53° 11′).

Fortunately, the original site in the corner of a retaining wall in the upper garden remains, and three stone steps allow one to view the dial plate. The top of the stone pillar in the corner is 320 mm square; a recess was cut in the top to receive the circular plate, and a groove cut to take the lugs holding the gnomon. The four fixing points are clearly visible, and they correspond to the screw holes in the plate.

92 Author's photograph of the dial's fixing position.

The terracing, in a corner of which the sundial was placed, was inspired by Sir John Manners (d. 1611) during the late sixteenth century.

Location

Haddon Hall, Derbyshire, England.

93 Compendium [fragment]
1596

Unsigned; dated 1596

Brass; diameter 78

This disc will have been the lid of a compendium, and is made to be a nocturnal. The outer edge has a calendar with months and single days numbered in tens. The next scale is the Zodiac, the Signs identified by name and sigil, divided to 1° and numbered in tens. The third circle is a wind rose, with 32 triangular points, eight named with the direction by their initials. The last circle before the vol-

93 Nocturnal

velles divides the full day into 24 hours, labelled in arabic numbers.

The first of the volvelles has an hour scale from 4 am to 8 pm, each hour with a notch similar to a ratchet wheel. A long index extends from the 12th hour. Engraved below the hours is a circle divided to $29\frac{1}{2}$, so giving the age of the Moon. The second volvelle is a plain disc with a long arm extending over the edge of the instrument. This is for aligning to the Guards of the Little Bear, when sighting the Pole Star through a small hole at the centre of the disc. The meridian line of the nocturnal passes through 10 April and 12 October.

At opposite sides are the catch and the small cover to the hinge. This is engraved with a rising Sun.

The reverse has in an outer ring a brief Church calendar, as follows:

Ian: 1 Circumci: 6 Epiphany
Feb: 2 Purif: of mary · 25 Mathi
Mar: 25 annunti: of Marye
Apr: 23 Georg 25 Mark
May 1 Phillip and Iacob
Iun 24 Iohn bap 29 Peter
Iuly 25 Iames
Aug: 24 Bartholmew
Sep: 21 Mathew 29 michael
Oct 18 Luk 28 simon and iud
Nou: 1 all saints 30 andrew
Dec: 21 tho ·25· nat· 26 ste

Subsequently, over four bands, is a perpetual calendar. The function is explained in the central circle:

This table begin=eth at 1596 and seru=eth for euer.

It lists the Prime Number, Epact, Dominical Letter, and leap years. It is similar in form to a table on a complete Kynvyn compendium dated 1593 (**66**).

There are some errors. In the second Epact column, a *2* is over a *1*, and in the next box an unwanted *1* has been rubbed down slightly. Below Epact *14*, the Dominical Letter *g* is over *f*, and under Epact *18*, the *g* is over *a*. In Aquarius, the *3* is over *2*. It has not been possible to attribute the engraving. It seems too regular for Kynvyn, and although the quality is closer to Whitwell, it is not his hand, and there are too many errors for him.

Location

Harvard University Collection of Historical Scientific Instruments, Cambridge, Mass (7320).

94 **Horizontal sundial** 1596

Unsigned; dated 1596

Brass; 111 square; thickness 2

The rectangular plate has the hour circle contained between bands; the outer edge has two concentric circles, and the inner three. At the four corners are the fixing holes. The hours are numbered IIII–VIII, and are roughly divided to

94

half-hour intervals by punched stars, a motif used all over the dial in two sizes. The gnomon is a replacement, with an angle of 47°. The hour angles engraved on the plate show that they were calculated for a latitude of about 52°. The meridian hour line makes no allowance for the thickness of the gnomon. The point of the gnomon touches the circumference of a double circle, the diameter of which is on the meridian. On either side of the gnomon, and within the circle, is a punched head of an animal, probably a boar or ox. The gnomon has lugs that fit into a pair of slots in the plate, which is rough finished underneath.

In the space at the southern side of the dial plate are engraved, within the hour circle, the initials T.H., while below is the dedication:

Ex dono Lislei Caue 1 Ianuarii Anno dño 1596

A possible explanation for this dedication is that it was a new year's gift to Thomas Hood from one of the students at his course on the mathematics of navigation held in the city of London. Leslie Cave has not been identified.

The most detailed study on the career of Thomas Hood (*c.* 1577–1620) is that published by Stephen Johnston (1991). After being the City of London's first public lecturer on mathematical subjects from 1588 to 1592, he returned to Cambridge to study medicine. He was in London to give private lectures for the two years between 1595 and 1597.

Provenance

Sotheby's New Bond Street London, sale 25 September 1984, lot 280.

Literature

Archinard (1987), illustrates the dial plate.

Location

Private collection.

95 Ptolemaic armillary sphere
after 1593

Unsigned and undated

Brass; overall height 998; diameter of horizon ring 610; radius of base 235

The solid cast brass base is engraved on each foot with the following motifs: a crescent inside the garter, inscribed HONI SOIT QVI MALY PENSE; cap of maintenance surmounted by a lion; the arms of the Earl of Northumberland. The horizon ring is supported by three cast brass lions with, above, grotesques on the brackets. At the centre of the base is a socket for a magnetic compass, now lost. The sphere has all its parts labelled and numbered by punches. An Earth ball is held by an iron rod.

The armillary was probably made in the Low Countries, and was fitted with its base in England. Certainly, the arms and badges on the base appear to be of English workmanship. The armillary sphere itself is not English, but is of a high quality, especially in the vignettes in each Sign of the Zodiac.

This instrument belonged to Henry Percy, 9th Earl of Northumberland (1564–1632). Through his interest in scientific matters, he was the patron of Thomas Harriot among others; he was known as the Wizard Earl. He served under Robert Dudley (?1532–88), Earl of Leicester, in the Low Countries campaign of 1585–86, against the Spanish Armada in 1588, and at Ostend in 1600.

Henry Percy succeeded to the earldom in 1585. On 23 April 1593 he was installed as a Knight of the Garter, and this honour is recorded by the engravings on the base. He may have disposed of the sphere in 1600 when raising money for the Flanders expedition. Thus the date of the base must be between

95 Reproduced from Gunther (1923), plate 60.

1593 and 1601. The armillary sphere itself is likely to be earlier than 1593.

The sphere was presented to the University of Oxford in 1601 by Sir Josiah Bodley, the brother of Thomas, who founded the Bodleian Library. It arrived at the library in 1613, and was lent in 1951 to the Museum of the History of Science, Oxford.

Literature
Longstaffe (1860), on the Percies; Gunther (1923, pp. 148–50), no. 60.

Location
Museum of the History of Science, Oxford (51–55; 70229).

96 Quadrant, nocturnal, and Regiment of the Pole Star

4/4 16th Century

Unsigned and undated; probably English, late 16th century

Brass; diameter 132

A simple disc of brass, with a handle stub, has a quadrant with a pair of sights on one side, and on the other a volvelle with Moon phase aperture and Regiment diagram.

The quadrant

One side is marked out with a small shadow square with below a horological quadrant. Both are used with a pair of pin-hole sights; the hole for the plumb-line is at the apex, but no line is present. The shadow square is divided along each edge in units to 6 at the 45° position. The arc of the quadrant is divided in single degrees that are labelled in tens to 90°. Immediately above the degree scale is the solar altitude scale divided into 5-day intervals from 10 December to 10 June, that is, from the Winter Solstice to the Summer Solstice. These dates mean that the Julian, or Old Style, calendar was being used. The placing of this altitude scale lies between $14\frac{1}{2}°$ and $61\frac{1}{2}°$, which means the quadrant is for use at the latitude of 52°.

The hour lines have become faint through polishing, done, most likely, during the past century. The lines give the equal hours from 4 to 12 noon, and then to 8 in the evening; their form is that of Stöffler (1524). Just visible is the wide curve of the ecliptic line, marked at either end and in two places along its length

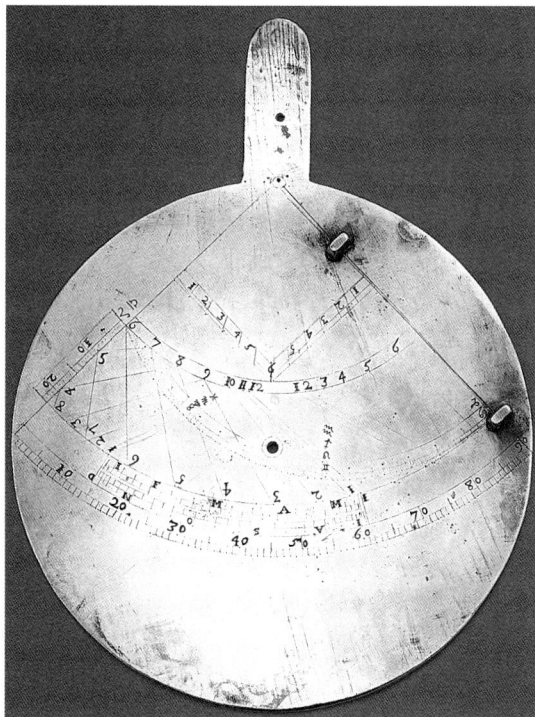

96 Side 1, quadrant.

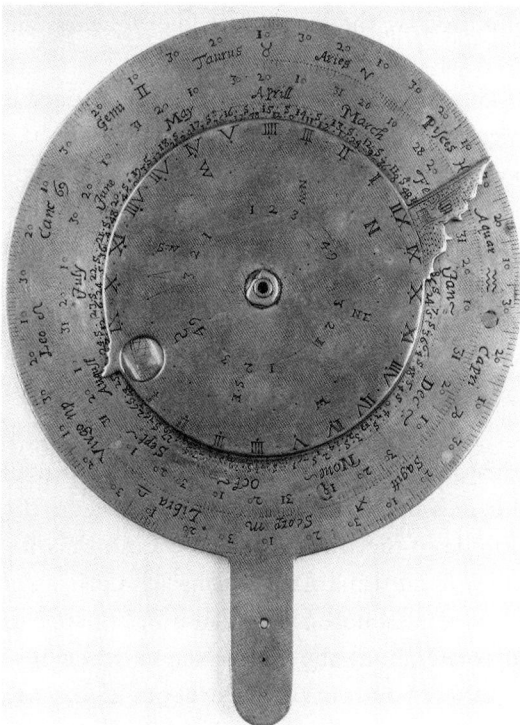

96 Side 2, nocturnal.

The nocturnal and Regiment of the Pole Star

On the other side, the outer edge is divided into the 12 Signs of the Zodiac, each with 30° units numbered in tens, and labelled with the names and sigils. There follows the calendar, in units of 1 day numbered in tens (or to 28 and 31), with names in English. The third band provides the age of the Moon, 0–29½ days, beginning and ending at the handle. This position is also marked 12S, meaning the Moon is due South at 12 o'clock. At each following day there is an increment of 48 minutes in the time the Moon is South. The phase of the Moon is simulated by the aperture in the volvelle that rotates about the centre of the disc.

The volvelle has two indices, the longer is engraved with a left hand, and points to the position of the Sun in the Zodiac. On the opposite side an index points to the age of the Moon. The edge of the volvelle is marked out in twice 12 hours, numbered in roman.

The central part is engraved with eight points of the compass and a band of degrees in two sequences, from 0 to 4° 9′ to 0. Across the band is a set of parallel lines connecting pairs of like numbers. It is this part that gives the measure of the difference between the altitude of the true North Pole from the altitude, as measured, of *Polaris* (α UMi). An adjustment is necessary because *Polaris* appears to rotate about the true Pole, so when it is North or South of the Pole, about 4° have to be subtracted or added to the altitude as measured by a quadrant; when East or West, no correction need be made to the value obtained for the altitude. Originally, a curved arm would have been attached to a boss at the centre. This arm, known as 'the Horn', represented the constellation Ursa Minor, with *Polaris* at one end and the 'guard' *Kochab* (β UMi) at the other. It is possible to see where this curved arm was originally attached. The star *Kochab* also serves as an indicator when the instrument is used as a nocturnal. When the handle is held vertically, the meridian line passes through 9° Scorpio (22 October), which is correct for the period.

The maker of this instrument is unknown, and not expert at engraving. The letters and numerals, for example the shape of 1 and 8, with a reversed S for South on the volvelle, do not match the work of any known craftsman. The mathematical lines are scratched rather than engraved in the normal manner. Nevertheless, this is an item of considerable interest. Not only is the Regiment of the North Star provided, but it has an early, and erroneous, Polar Distance of 4° 9′. The customary value became 3½°. It is worth quoting the explanation of David Waters (1958, pp. 45–6), which makes the position clear:

> …the earliest detailed description of [such an instrument] in print appears to be contained in Martin Cortes's *Arte de Navegar*, written in 1545, published in 1551. This … consisted of a disc or volvelle marked with the four cardinal points termed also 'The Head', 'The Foot, 'The Right Arm', and 'The Left Arm', with an inner circle drawn on it marked with the degrees of correction to be applied to the Pole Star to find the true pole. The navigator rotated a pointer in the form of a trumpet, marked with the seven stars in the constellation of Ursa Minor, until it coincided with the position of Ursa Minor, holding the instrument up meanwhile and sighting the Pole Star through a hole in the centre…

. Unfortunately he used the astronomer Werner's erroneous (1541) Polar Distance of 4° 9′ instead of the seamen's more accurate 3° 30′ of that time, which, however, he included in the text, so that he unwittingly condemned his navigators to faulty observations.

It seems clear from its features that this instrument was made in England, perhaps by a navigator himself. The English translation of Cortés book, *The Arte of Navigation* (1st edition 1561, 10th edition 1630), became well known, and was drawn upon by William Bourne (*c.* 1535–82) for his *A Regiment for the Sea*, which enjoyed 11 editions between 1574 and 1631 (Adams and Waters 1995; Taylor 1963). The Polar Distance of 4° 9′ suggests that the illustration in the English translation of Cortes's book was used as the model, and not the text.

Provenance
Christie's South Kensington, sale 17 December 1998, lot 149.

Location
Private collection.

97 Gunner's level
4/4 16th Century

Unsigned and undated; probably English

Brass; overall width 172, overall height 73; radius of outer arc 122, of inner 96; thickness of arc 1.1–1.5

It is possible that the 90° arc was cut from a larger curve. The back is roughly finished, and shows casting fissures. The cutting of two apertures, and the clipping of the outer ends of the arc, have removed the defining dots on the outer edge of the scale. Their presence is seen where the base joins the arc in three places.

The ends of the base are strengthened by pieces of brass plate riveted in position. The central part of the base is curved to avoid instability when the instrument is placed on a surface.

The arc is divided in single degrees by lines defined at each end by punched dots. Transversals are cut crossing five concentric curves to give six parts in each degree, therefore a reading to 10′. Labelling is in 5° in both directions from the central position to 45°. This labelling is in two bands, fives below and tens above. The numbers are roughly scratched, and not engraved in the normal way.

Three holes (diameter 2.5) have been drilled in the arc from the front, because the lines adjacent to the holes have been pushed aside. At the back the holes have burst through, leaving a raised edge. The two holes either side

97 Front.

97 Close-up.

of the centre line are not well positioned, while the third at the bottom has no clear purpose. The holes are threaded, and two screws were found in place. One has a raised flattened head for hand turning, the other has a high curved head with a slit. The upper pair of holes was probably for attaching some superstructure from which was suspended a plumb-line.

It is not easy to date this incomplete instrument, but the poor quality of the cast brass suggests the late sixteenth century. It is impossible to say with certainty that it is English-made; this is a case where a general attribution is based on the design, choice of scales, and the whole construction of the instrument. Transversals are found on Elizabethan instruments; for example, the Hood-type sector attributed to Kynvyn, **69**. The shape suggests a use by a gunner, for whom the maximum range is achieved at an elevation of 45°. There are many later gunner's levels with arcs of this form, also on bases with a two-point prop (Bennett and Johnston 1996, pp. 35–49).

Location

Private collection.

98 Compendium 1602

Signed and dated: ✳ T ✳ W ✳ *Fecit* ✳ 1602 ✳

Brass, gilt, and silver; overall diameter 48; thickness 26

Comprising:

a) lunar volvelle tide computer

b) table of latitudes for 13 towns, plus 1 on the gnomon

c) equinoctial sundial

d) magnetic compass

e) nocturnal

98 a.

a) The outer edge of the top is engraved with twice 12 hours, marked in Roman numerals. There follow the 32 wind directions, eight labelled N, NE, etc., the rest marked with triangles. Over these scales runs a fiducial edge attached to a rotating disc on which are engraved the days of the Moon's age from 0 to 30 (not to $29\frac{1}{2}$), the days indicated by punched dots and labelled every 2 days. The central volvelle has a protruding pointer and a lunar phase aperture; the centre is decorated with a double Tudor rose.

b) The underside of the top has around the edge the names of some English ports and the Moon's compass direction to establish the port (Waters 1958, pp. 31–2).

98

1602 · *Plimouth· E·* ; *Bristowe· W· b· S·* ; *Grauesend· S· S· W·* ; *London· S· W·* ; *Portsmouth· S· b· E·* ; *Poole· S· E·*.

The latitude table is in the central circle.

The hinged gnomon has engraved on one side the single town *Barnestabell 51*.

c) Opening the lid allows a semicircular bracket to rise, which supports the chapter ring of the equinoctial dial. The rim is divided in hours, twice 12 with roman numbers; on the underside (for use in winter) only the hours VI–VI are marked, followed by the initials and date. The gnomon swivels into position, with a small quadrant to set the latitude, which is divided in 2° intervals, marked from 20° to 9[0]°. The plummet is missing.

d) The printed compass card is plainer than with most other Elizabethan examples (see Chapter 4, and Plate 14h). The copperplate from which it was printed was cut by TW, as is clear from the E, with serifs typical of those used elsewhere on the instrument. There are 32 divisions, alternately shaded in olive green, with three subsequent rings of narrow triangles: 8, 4, 4; arranged to point in 16 directions, and variously shaded in olive green, pale yellow, and red. A fleur-de-lis is the North point, E, S, W, complete the cardinal points. The magnetic declination is marked on the card one point East of North with a drawn needle, which has a 'drop' head at the South, and † at the North. The magnetic needle (length 32) of blued iron has a plain South

London 51	York 54	Barweek 56	Cambridge 52
Roma 42	Venes 45	Vlma 48	Compostella 45
Buda 47	Rochill 47	Calliz 37	Antwarpe 51
Plimouth 51			

98 e.

engraved meridian. The silver volvelle that follows is cut with prominent teeth at each hour from 8 pm to 4 am with an extra long marker at 12 o'clock. Finally, the index arm on the central volvelle has an extension arm that swivels to pack away in a case (not present); there is a peg stop to position the extension arm. A Tudor rose is the final decoration.

Being on the underside of the magnetic compass, this nocturnal cannot have a central hole for viewing the Pole Star. Even without this facility, it is possible to align an index roughly with one of the Guards in the Little Bear. But as pointed out above, this particular nocturnal does not work with *Kochab*, so there is the possibility that the maker had a different star in mind for the pointer to *Polaris*.

The identity of TW is not known. He has tried to make a pretty, small instrument, but with a pointless choice of Continental towns and only five English, and with a flawed nocturnal. The engraving is not expert, guide lines are prominent, letters and numbers are deeply cut. There are very pronounced serifs on numbers, and, oddly, on risers of certain minuscules (notice *b*, *d*, *k*, *l*). The minuscule *l* often has a flick to the left at the middle (*Calliz*, *Poole*). The effect is to produce a script that is more like pen-writing than that used by instrument or map engravers.

point, and at the North a ⊤. The glass cover appears to be original, clear of colour, with some air bubbles as inclusions; two are long and narrow, about 2.5 mm in length.

e) The outer edge of the bottom of the instrument is marked out in the days of the months, each month with an initial letter, and the divisions into 2-day intervals (except for months with 31 days). The next band is the Zodiac, with the sigils, and 2° divisions, labelled at 20° and 30°. The meridian line cuts 10 March and the First Point of Aries; also 14 September and the entry of Libra, but this is not correct for a normal nocturnal. For the Julian calendar, as here, the date should be about 20 October for the transit of *Kochab* (β UMi), but this point is some 35° off the

Provenance

Bought by Lewis Evans from P. Webster for £50 early in the twentieth century.

Location

Museum of the History of Science, Oxford (LE 222; 54727).

99 Diptych sundial *c.* 1600

Inscribed: *R* ★ *G*; undated

Ivory and silver; length 68, width 53, thickness 15

The instrument is made from ivory, with the chapter ring in silver, and the hinge and catches in a copper alloy, probably brass. The form is octagonal, with the base leaf holding a magnetic compass, and the underside of the top leaf having inset a circular map of England and Wales (see Plate 11). Both the chapter ring and the map are surrounded by a double circle inscribed in the ivory, and on each leaf a pair of semicircles fills vacant spaces. The top, bottom, and side surfaces are quite plain.

A string gnomon (missing) attaches to a small hole in a tracery extension from the chapter ring, and to a hole above the map. The hours are divided into quarters, and are numbered in roman numerals. The hour angles are suitable for the latitude of 52°.

The edge of the compass rose (type B-1) is divided into the 32 wind directions, or points of the compass, and 16 are emphasized by triangles divided by central lines, and parti-coloured. The colours used are blue–yellow, red–buff, and red–green (see Plate 14e. This rose appears to be from the same copperplate as that with a similar diptych dial, **56**. The magnetic needle has a large triangular head and a triangular tail, resembling a crossbow bolt; it is a replacement.

The map of England and Wales has the counties identified by their initial letters, and the boundaries are coloured, in red, green, or yellow. The diameter of the map is about 48 mm (under 2 inches), and such miniature country maps are very rare. This particular one is not recorded; however, a somewhat similar one, but nearly twice the size (diameter 84), was printed in Arthur Hopton, *Speculum Topographicum* (1611), title page and pp. 2, 86. For a note on the Hopton map, see G. King (1996, p. 80). The other ivory diptych dial (**56**) also has a miniature map of England and Wales (diameter 56), but the engraving is of a much higher quality, and it is attributed to Charles Whitwell.

The significance of the initials R.G. on the chapter ring is not known; they are found on two other instruments; see **87**).

99

Literature

Holbrook (1974), no. 31, plate 18; Holbrook (1992, p. 39) fig. 36.

Location

Horniman Museum, Forest Hill, London (31–187 A).

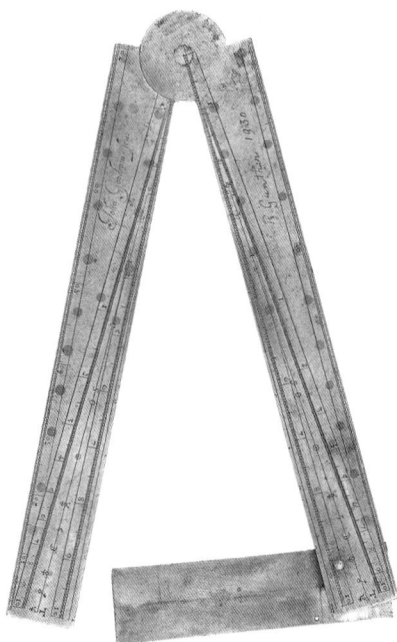

100 Side 1.

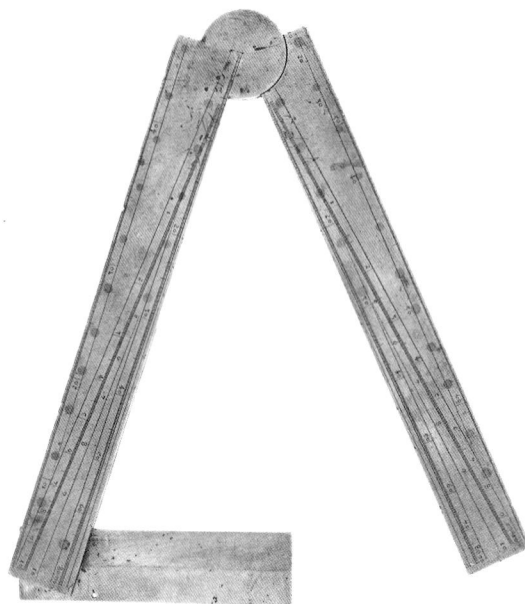

100 Side 2.

100 **Sector** ¹/4 17th Century

Inscribed: *John Goodwin · ſcu:*; undated

Brass; length of an arm 178 (7-inch); width 22; in modern slip case with red cloth cover

Provenance

Acquired in 1930 from R.T. Gunther, Curator of the Museum.

Location

Museum of the History of Science, Oxford (30–22; 29026).

101 **Sector** ¹/4 17th Century

Inscribed: *John Goodwin scu:*; undated

Brass; length of an arm 178 (7 inches); width 22

Location

Private collection

102 **Sector** ¹/4 17th Century

Inscribed: *John Goodwin scu:*; undated

Brass; length of an arm 257 (10 inches); width 28; thickness 6

Provenance

Part of the Gabb Collection acquired in 1937 by the Museum.

Location

National Maritime Museum, Greenwich (NAV0152; old number CI/S.9. 37–8).

★ ★ ★

The (**100, 101, 102**) sectors are attributed to Elias Allen. All three have the same scales shown in the illustration at the front of Edmund Gunter, *The Description and vse of the Sector* (1624), printed by William Jones, and sold by Edmund Weaver.

Side 1	*Side 2*
Line of meridians	Line of tangents
Solids	Secants
Equated bodies / Metals	Superficies [area]
Line of equal parts	Segments
Inscribed bodies	Sines
	Quadrature

The sectors are included here because they bear the name of a man who was an Elizabethan, though their execution indicates very clearly that they were made by Elias Allen. He, of course, later signed other examples of Gunter's sector, and was advertised as their prime producer. The significance of the 'scu:' after Goodwin's name is not clear. A possible explanation is that, since Goodwin died before 1616, and Allen became free of a livery company only in 1612, these early examples of the Gunter sector were made at the end of Allen's long apprenticeship to Charles Whitwell, were commissioned by Goodwin, and therefore carried his name. For an outline of the early Gunter sector, and more on Goodwin, see Chapter 4, 'The sector'.

103 Compendium
1/4 17th Century

Signed: *R Grinkin Fecit*; undated

Brass, gilded; oval 61 × 52, thickness when closed 25

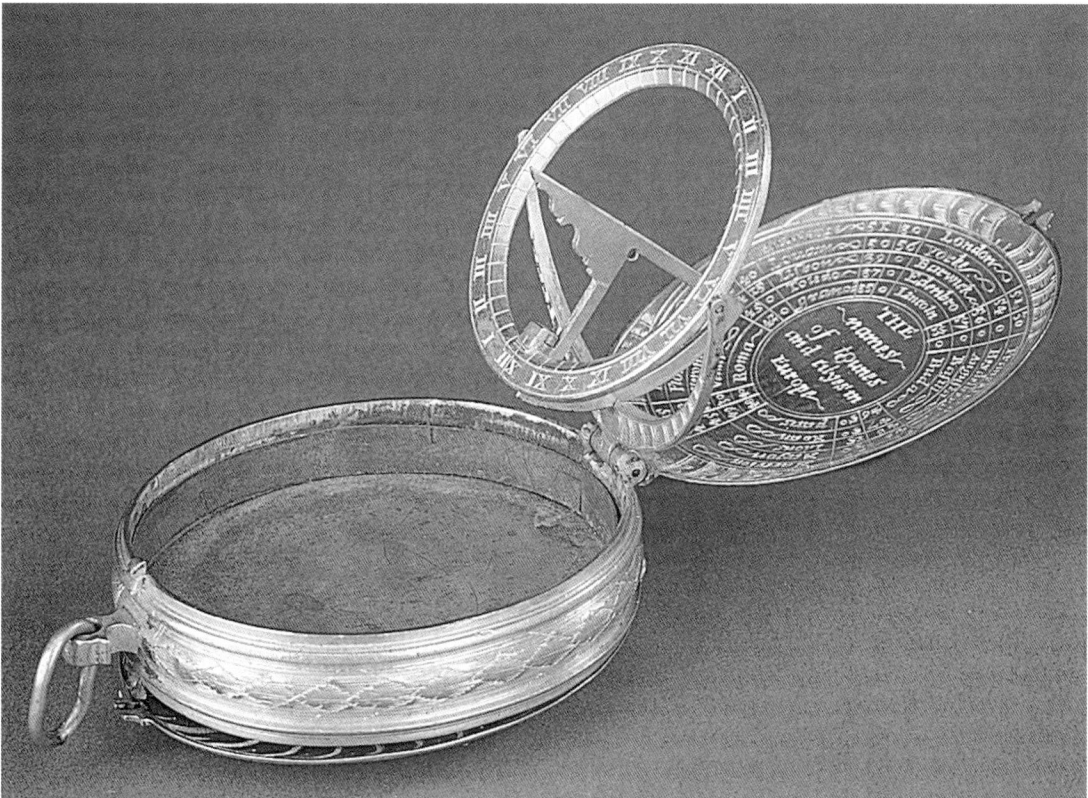

103

London	51 30	Franckford	50 10	Barcelon	43 15
Yorke	54 0	Heydelberg	49 30	Bayon	44 30
Barwick	56 0	Augusta	48 15	Lions	45 15
Edenbro	57 0	Basill	47 50	Roan	49 30
Lincoln	53 0	Buda	46 50	paris	48 40
Neaples	40 40	Antwerpe	51 30		
Florence	42 40	Louain	50 56		
Verona	44 15	Lisbon	39 0		
Venice	45 0	Toledo	37 0		
Roma	42 0	granad	35 0		

Comprising from the top:

a) wind rose

b) table of latitudes of 25 towns

c) equinoctial sundial

d) compartment to hold a compass (missing)

e) shallow compartment

f) underside of nocturnal

g) nocturnal

h) rim

a) The wind rose is engraved on the outer domed surface, where the gilding has rubbed away through use. The 32 winds are named on ribbons, the pattern being the A-2 type, which is first noticed on a chart engraved by Ryther of 1592; it appears again on a compass dial by Whitwell of *c.* 1600, on a compendium by Whitwell of 1606, and on three compendia by Allen of *c.* 1610. At the centre of the rose is a small hole (diameter 0.8 mm) that is most likely to have been to insert a wind vane.

b) Inside the top plate is a table of five columns and five rows listing the latitudes of 25 towns in degrees and minutes. This table is titled in a central circle: THE *names of tounes and cityes in Europe.* The values of latitudes do not match exactly those found on other com-

pendia described in this book. It seems that the maker rounded the values for minutes of arc, generally to tens, with 15 in four cases, but 56 in one. The gold plating is perfect.

c) The equinoctial dial is the same as those made by Whitwell and Allen. The gnomon is set by a quadrant divided from 0 to 90° in 2° intervals, labelled every 10°. The plummet is missing. The chapter ring on the upper (summer) side is divided in half hours, I–XII, twice. The under (winter) side is divided in half hours alternately shaded (unlike the upper side), VI–VI. The signature is engraved on this side. The act of lifting the top plate automatically raises the equinoctial dial to the vertical, by means of a lug protruding from the central part of the hinge.

d) An oval compartment, 9 mm deep, has a plain non-gilded surface, with a bronze strip round the inside to grip a compass unit, now missing.

e) An oval compartment, 3 mm deep, with a plain non-gilded surface, is also empty.

f) The gilded underside of the nocturnal shows the four attachment grips on the sighting volvelle. The surrounding surface is engraved with a scoop decoration.

103 Signature and quadrant.

g) Much of the surface gilding of the nocturnal has worn away. The outermost circle is engraved with the days of the months in 2-day intervals, labelled in tens (or 28, 31). There follow the names of the months: *Ianuari, Februari, March, Aprill, Maye, Iune, Iulie, August, September, October, Nouembe, Decembe.* There follow two volvelles, the outer with a 24-hour circle, marked out as twice 12, the hours divided in halves, alternately shaded. The edge is cut with a 'saw' tooth at each hour, with a longer index at one 12 o'clock position. The inner volvelle is cut with a slit from the centre to the edge for viewing the Pole Star and the Lesser Bear. The earliest known use of this form of noc-turnal is on the compendium dated 1590 by Humfrey Cole (**33**). At the edge in line with the slit is an index for pointing to the time on the previous volvelle. At either end, the oval surface is engraved with a foliate decoration.

h) The rim is inscribed with an acanthus motif. The extension holding the suspension ring is quite small. Closure of the top and bottom plates is by snap catches.

Robert Grinkin senior, a Member of the Blacksmiths' Company, died in 1626, and some six watches of his are known. He was succeeded as a watchmaker by his son of the same name. Grinkin senior is known to have taken as apprentices Edmund Bull, *c.* 1598, free 1607, and John Willow, *c.* 1608, free 1617 (Loomes 1981, p. 269). Assuming he was 25 when he took his first apprentice, and assuming this was Bull, then Grinkin's date of birth could be about 1573, and if so he could have been 53 when he died. His Will is extant (1626 87 Hele. PROB 11/149/87), and is dated 26 May 1626: he bequeathed his tools and instruments to his son Robert.

The engraved letters and numerals on this compendium continue the form that was practised by Ryther, Whitwell, and Allen. Although competent, Grinkin's hand is not equally well controlled. Some small points of difference can be seen in the bottom curves of *3* and *5*, which are full and not foreshortened. His *k* has a loop at the top right. One is making the assumption that the engraving actually is by Grinkin; it does not resemble that of any maker known at present.

Note. This compendium (in a private collection) was brought to my attention after this book was completed. It was therefore necessary to include it at the end of the numbered sequence. Grinkin senior was obviously an Elizabethan craftsman who combined instrument making with watch making.

Provenance

Christie's South Kesington, 13 April 2000, lot 28.

Location

The British Museum.

THE ATTRIBUTION OF UNSIGNED INSTRUMENTS

By my life this is my lady's hand! these be her very c's, her u's and her t's; and thus makes she her great P's. It is, in contempt of question, her hand.

[Malvolio, thinking Maria's hand was that of Olivia; Shakespeare, *Twelfth Night*.]

Any archaeologist, anyone who studies the applied arts, knows how much can be learnt from the really detailed examination of three-dimensional objects. Curiously, however, the crafting of scientific instruments has not been considered with anything like the attention given to their function in relation to the development of scientific thought. Even when faced with the obvious challenge of dating a mathematical instrument, attempts have been made to find a rule of thumb that would make it possible to establish a date from the astronomical information on it (a process fraught with logical and historical problems; see G. Turner 2000), rather than from examination of the engraving style, the decoration, and other distinguishing craft features. This is not to suggest that it is simple to postulate either a maker or a date for unsigned, undated instruments. But what may be called the forensic approach is always rewarding, and can frequently result in reasonably confident attributions.

Before the invention of printing with movable type, in the mid-fifteenth century, information was recorded by carving into stone, metal, or wood, or by writing with a quill pen on vellum or paper. The skills of writing and reading were initially practised by ecclesiastics, until, with the Renaissance, the demands of commerce, and the increasing complexity of society, required improved means of communication. Professional scribes were needed, and they developed their own individual styles of writing that could reflect national or regional characteristics, or the requirements of a particular profession, such as the law, or an individual trade. With the development of various styles of writing, and a growing need for instruction in the art, writing-books were in demand, giving specimens of various styles.

One of the most influential of these manuals, and indeed by far the most important in relation to the engraving on maps and mathematical instruments, was the treatise on italic script published in 1540 by Gerard Mercator (1512–94), with the title, *Literarum latinarum, quas Italicas, cursoriasque vocant, scribendarum ratio* (reproduced, with translation, in Osley 1969, Pt 2). The book was solely concerned with his new italic hand, ignoring the range of other contemporary scripts. Mercator, already a maker of globes and maps when he wrote the treatise, realized that italic was the ideal script for maps, being compact, legible, and potentially decorative. He believed that maps should contain as much information as possible about the regions portrayed, and knew from experience that providing it might involve on occasion using extremely small letters and numbers. Variants of Mercator's italic hand are used by all the instrument makers whose work is described in the present book, because nearly all the Tudor makers of instruments were also map makers. Indeed, Mercator had direct links with Thomas Gemini, whose arrival in London initiated the instrument-making trade in England.

It is important to recognize the interrelation between calligraphy, engraving, type design, and instrument making. The London makers of the sixteenth century were engravers, who used their skill with the burin to engrave in reverse on the copper plates used to print maps and globe gores, and also to engrave directly on to brass instruments. The techniques of printing, however, also included type cutting, which involves cutting metal away to produce a letter or number in relief, while the engraver incises metal, producing an intaglio letter or number. Instrument makers in Flanders, and following them, in London, used engraving. In Germany and Italy (Turner 1995), however, the general practice was to use punches; it was rather like the difference between handwriting and using a typewriter. Interestingly, in the same way in which the type-face of an individual typewriter can be recognized forensically, letter and number punches used by a particular instrument maker can be identified through such features as size, wear, and damage to a particular punch.

The chief tool of the engraver is the burin or graver. This is a small steel rod 4–5 inches in length, with a cross-section that is square or diamond-shaped, with the face (cutting point) filed to an angle of about 45°. The handle is usually mushroom-shaped, frequently with the lower side removed so that the burin can be held at a low angle in the palm of the hand. This tool is the equivalent to the writer's pen, and each engraver develops his own characteristic hand, just as a penman does. This is the basic technique. Mathematical lines were usually incised with a wing-compass with metal-cutting points, or dividers. It was normal for the engraver to place the work on a leather cushion set on a rotating table, which is turned so that the direction of the hand working the burin is always the same while forming the letter or numeral. Some of these tools can be seen in Holbein's 1528 portrait of Nicolas Kratzer in the Musée du Louvre. Kratzer served as astronomer for a period at the court of Henry VIII.

In order to make reliable attributions, what has to be studied is how each craftsman performed the task of making an instrument. The hardness of the brass was sometimes tested by small cuts with the burin on parts that would not be seen, even with trials of letters, a fine example being that of James Kynvyn preparing his theodolite of 1597 (70). Some can cut a zero that looks perfect, such as Mercator, others form the numeral by a series of short jabs. Then there may be a pair of guide lines, especially in broad spaces, since within degree or month scales the circular dividing lines provide the requisite control. The kinds of error may give a clue to a particular maker.

With Humfrey Cole errors are rare, but when numbering scales on long rules a numeral may be cut out of place, and so had to be rubbed down and over-engraved. It seems that with his early instruments Kynvyn had some difficulty in concentration when numbering lengthy scales. Those numerals started in the wrong place are abandoned, and one may see which part of the figure was cut first, e.g., the diagonal line in a *2*. With those makers who were commonly concerned with map engraving, where all the lettering has to be in reverse, one sometimes finds a reversed numeral on a mathematical instrument. The alternate shading of degree divisions, and the use of dots or stars of 3, 4, 5, or 6 points can also be of assistance in establishing an individual worker's practice. The calendar of Church feasts and the Easter perpetual calendar, as well as tables of the latitude of towns, may all point to a period and/or to a particular maker. Then there is the ingenious way in which a mechanical problem may be overcome; for examples, see the section on Humfrey Cole in Chapter 2. Finally, a monogram or a coat of arms can reveal the patron and thus the approximate date of manufacture.

It is the considerations outlined above, amongst others, that have to be brought to bear before an analysis of the forms of letters and numerals can be successful. Planetary and Zodiac sigils also can be highly personal. The complete process resembles closely the mode of attack by forensic document examiners, who assemble a concatenation of data before making an attribution. Those who examine the handwriting on documents are in the same position as those who scrutinize engravings. David Ellen has explained the difficulties, and the way to achieve a high probability of success in attributions.

> Handwriting comparison does not permit exactly reproducible measurements, therefore precise calculations of the probability that two writings came from one source cannot be made. This means that some degree of subjectivity must be present; without a technique which automatically produces the exact result any analytical method must depend on the experience and ability of the analyst. In every analytical method the limitations must be appreciated. It is erroneous to express any conclusion with a certainty which does not recognise the limitations of the method and of the accuracy of the observations on which it is based. However, although these limitations exist, conclusions can be properly drawn if they are recognised; the danger of wrong results occurs when they are not. (Ellen 1997, p. 3)

The best way to examine engraving is to employ a microscope, preferably a stereo microscope with a camera port at the top. This gives the clearest and largest view of how the work was built up, and provides a permanent record. However, this is not always possible, and recourse to photographs of the instrument or to parts of it has to suffice. A macro lens is essential, but again it is not always possible to order such photographs from some institutions. For this present study, an Olympus SZ-STS stereo zoom microscope was used where possible, and otherwise an Olympus OM-2N camera with a Vivitar 55 mm macro lens or a Tamron 80–210 mm tele macro lens. This photographic equipment was used at the Museo di Storia della Scienza, Florence, where the 20 instruments of Sir Robert Dudley are preserved, and also at the Museum of the History of Science, Oxford, where 15 instruments have been studied.

Photographs of the calligraphy and of the sigils on both signed and unsigned instruments were examined by scanning and storing the images in a data base. This enabled like elements to be brought to a common size and compared side by side. This is described below.

An unreliable method of dating that has been favoured by some researchers is a chemical analysis by X-ray fluorescence. This method is poor at identifying alloying elements with concentrations between 2 and 5 per cent. Another disadvantage comes from the shallow surface penetration of the X-rays. With old brass one must expect diffusion of elements within the brass, and obvious surface effects, where the copper will attract oxygen to form cuprous oxide, which has a reddish tone. Some have declared that this reddishness shows that some parts of an instrument are of copper, while other parts that appear more yellow are brass. Such effects depend on the composition of the material, which may not be uniform on the same instrument; on the way the metal was fabricated, e.g., cast, hammered, annealed; and how it has been kept over some 400 years. Spectroscopic chemical analysis is a very much better and more reliable technique, but it has been used only once on a mathematical instrument. No analysis is of value unless it is part of an extensive, well-planned research programme that tests hundreds of similar items. Such studies have been carried out with, for example, memorial brasses and candlesticks. Two scientists formerly at the Research Laboratory for Archaeology and the History of Art, Oxford, formed these opinions when endeavouring to test whether a brass instrument was truly old or was an imitation:

> all that can be done is to compare its composition with those of known genuine pieces … It is also possible that perfectly genuine instruments will have analyses outside the range expected for a particular country and date (Pollard 1984).

> Although some composition patterns are emerging in certain workshops and countries, it still seems unlikely that judgements on authenticity can be made purely on the basis of major element chemical analysis. (Mortimer 1987)

To the non-scientist, the idea of a quick probe by a fine beam of X-rays seems to be a simple way of dating, even if only to 50 or 100 years. It is but a vain hope, appearing to offer an escape from the laborious effort necessary to examine objects with the thoroughness required for reliable attribution.

It is instructive to consider what happens about authentication, attribution, and dating in the field of fine art. The Rembrandt Research Project has been the most ambitious study ever undertaken to investigate a single artist's work. Three volumes of the *Corpus of Rembrandt Paintings* have been published, and a fourth is due in 2000. The latest volume, however, will represent a shift in treatment, for, instead of categorizing the paintings as genuine, doubtful, or not by the artist, the evidence and arguments will simply be presented, with the judgement left to the reader. This is because faith in scientific studies has faded, and more reliance has had to be placed on looking for characteristic signs of the artist's style, method, and handiwork—in fact, what is sometimes called 'eye-balling'. This is exactly the process that, in my experience, has given the greatest success in attributing unsigned mathematical instruments to a particular maker.

DESCRIPTION OF A DIGITAL CAPTURE TECHNIQUE FOR ENGRAVED CHARACTERS ON BRASS

Mark St J. Turner

Introduction

The purpose of the technique is to capture digitally the individual characters, symbols, and numbers engraved on brass instruments in order to characterize the style of the engraver. Subsequent comparative analysis may be used to assign to an engraver an instrument that has no signature.

The technique uses the latest computer hardware and software to produce efficient, high quality images suitable for publication. Once digitally captured, the character images may be presented in any way that will aid later comparative analysis.

Computer tools used

Hardware

Intel 350 mHz Pentium II processor; 9.1 GB hard disk; 128 MB RAM; 4 MB SGRAM video card; 18 inch monitor; A4 flatbed scanner; desktop laser printer.

Software

Adobe Photoshop v 5.0; Adobe PageMaker v 6.5; Microsoft Windows 95.

Preparation of the Character Frame

To present the individual, digitally captured, character in a consistent manner that aids comparative analysis, a Character Frame (Fig. A1)

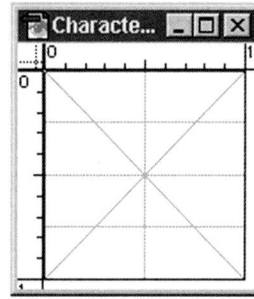

FIG. A1

template was constructed. This matrix of horizontal, vertical, and diagonal lines forms the background of every digitally captured character. Each character is sized to fill the Character Frame, so that comparisons of engraved characters of different sizes can be easily made.

Description of technique

Step 1: Scan the original

Place original image (photograph, book illustration, actual instrument) on the scanner. Start Photoshop and initiate the image import software (third-party software from the scanner manufacturer). Make sure the settings are for a colour scan. This gives better definition than that obtained for a black and white scan. The resolution of the scan should be 600 dpi (dots per inch) or greater. A scanner resolution of 600 dpi gives a digital image that is of publishable quality (Fig. A2). Bear in mind that increasing the resolution will dramatically increase the file size of the scanned image. Initiate the scan.

Step 2: Improve the scanned image

When the scan has finished, the image will be presented as a window in Photoshop. This scanned image will almost certainly require some improvement prior to capture of individ-

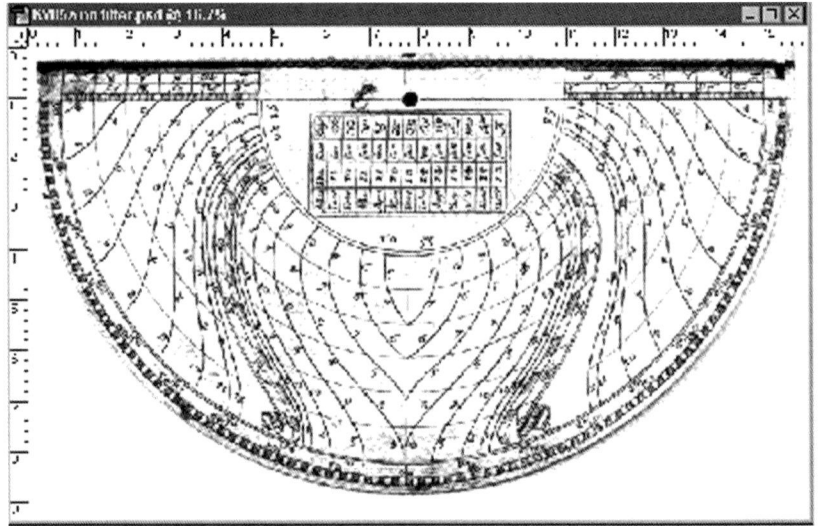

FIG. A2

ual characters. First of all, the scanned image should be turned into a *greyscale* image. The scanned image will contain colour information, since the scanner was set for a colour scan. However, since colour is not essential in this technique, and was only used to optimize definition, the colour information can be converted into a *greyscale*. By discarding the colour information the file size of the image will be reduced to approximately one-third of its original size. This is a definite advantage as these images can occupy many megabytes of disk space. Next, the image must be sharpened to enhance the contrast between the characters (which should be as dark as possible) and their background. The best technique for this is the *Unsharp Mask* filter in Photoshop. *Unsharp Masking* is the traditional film compositing technique used to sharpen edges in an image. The *Unsharp Mask* filter corrects blurring introduced during scanning. This filter locates pixels that differ from surrounding pixels by the threshold specified and increases the pixels' contrast by the amount specified. In addition, the radius of the region to which each pixel is compared can be specified. Fig. A3 shows the *Unsharp Mask* filter window in Photoshop.

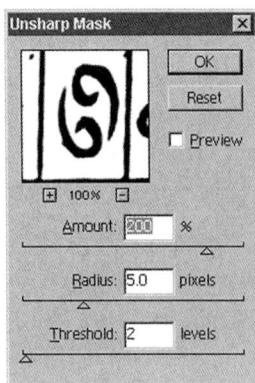

FIG. A3

The best settings for the *Unsharp Mask* filter are: Threshold = 2 levels; Amount = 200%; Radius = 5 pixels. Figure A4 shows the effect that the *Unsharp Mask* filter has on the appearance of the engraved characters. The illustration on the left is before the application of the *Unsharp Mask* filter. The illustration on the right is after the application of the *Unsharp Mask* filter.

290

 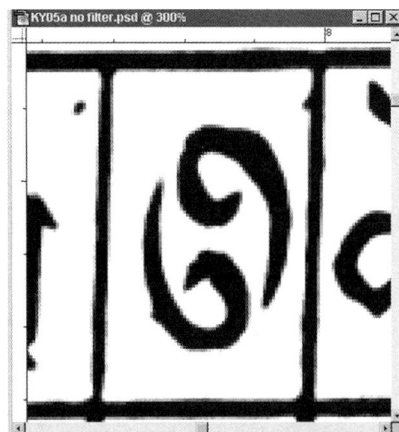

FIG. A4

After completing these improvements, *save* the image to hard disk.

Step 3: Capture individual characters

Locate the character that is to be captured and *zoom* onto this character so that it fills the window. Using the *Polygon Lasso* tool in Photoshop, draw an outline around the character (Fig. A5).

Copy this portion of the image to the *Clipboard*. *Open* the file containing the

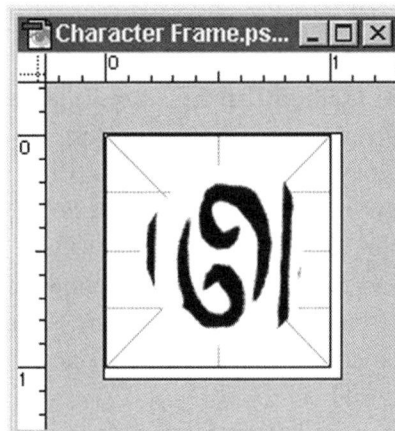

FIG. A6

FIG. A5

Character Frame and then *Paste* the character into it (Fig. A6).

The pasted character now needs to be centred within the Character Frame. Select *Free Transform* from the *Edit* menu in Photoshop. This enables the character image to be translated, rotated and sized to fit within the Character Frame (Fig. A7).

Once this has been achieved, the character must be isolated from the remains of its back-

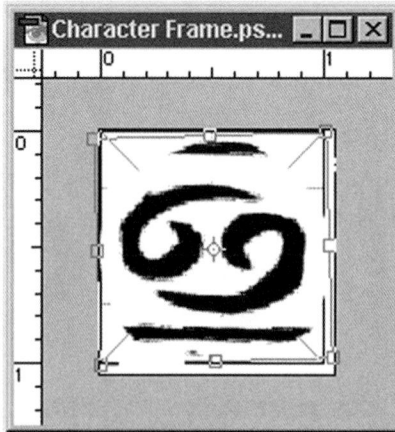

FIG. A7

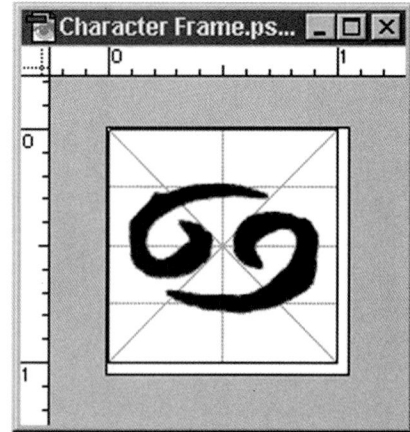

FIG. A9

ground. This is done using the *Magic Wand* tool in Photoshop. This tool allows a consistently coloured area to be automatically selected, without having to trace its outline manually. The sensitivity of this tool is set by specifying a *Tolerance* in pixels. A value between 0 and 255 may be entered. Enter a low value to select colours very similar to the pixel selected by the tool, or a higher value to select a broader range of colours. Using the tool, click on the area of the character whose

colour best represents the whole of the character. All adjacent pixels within the tolerance range are automatically selected (Fig. A8).

Once the *Magic Wand* tool has selected the correct outline of the character, *Inverse* the selection so that it is the background that is outlined. Then select *Cut* from the *Edit* menu finally to get rid of the background (Fig. A9).

The image may now be saved to the hard disk.

Step 4: Presentation

The captured images of characters may be presented in different ways depending on the aim of the analysis. Using the desktop publishing software package PageMaker, images and text may be laid out in a very precise manner. Images may be inserted and then resized and positioned to a very high level of accuracy. In this way tables of character classes (e.g. numbers, upper and lower case letters, Signs of the Zodiac) allow comparison of the similarities and differences between different engravers. Similarities of character form that become apparent in these tables can help to assign an engraver to an unsigned instrument.

FIG. A8

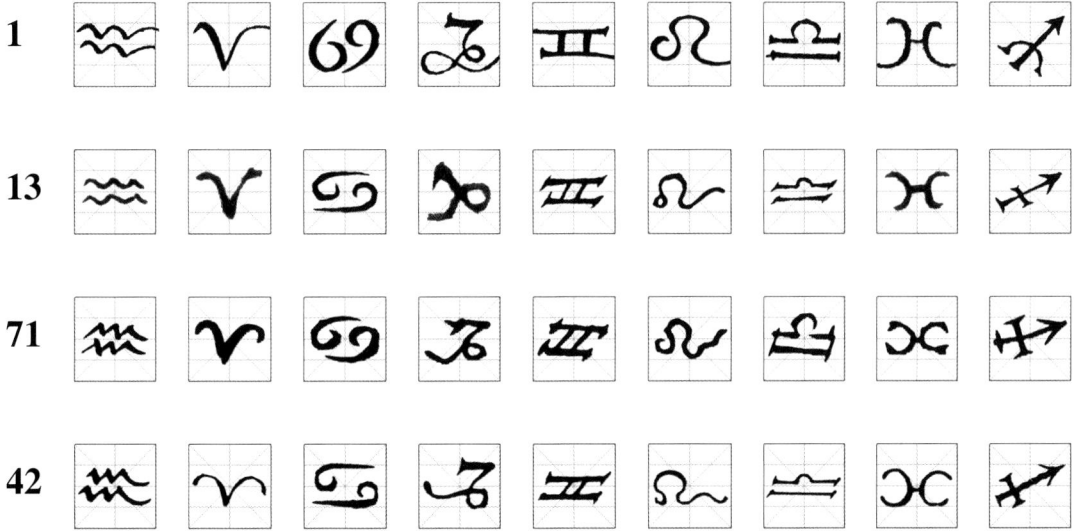

FIG. A10 Zodiac sigils used by four Elizabethan makers. **1** Gemini; **13** Cole; **71** Kynvyn; **42** Whitwell.

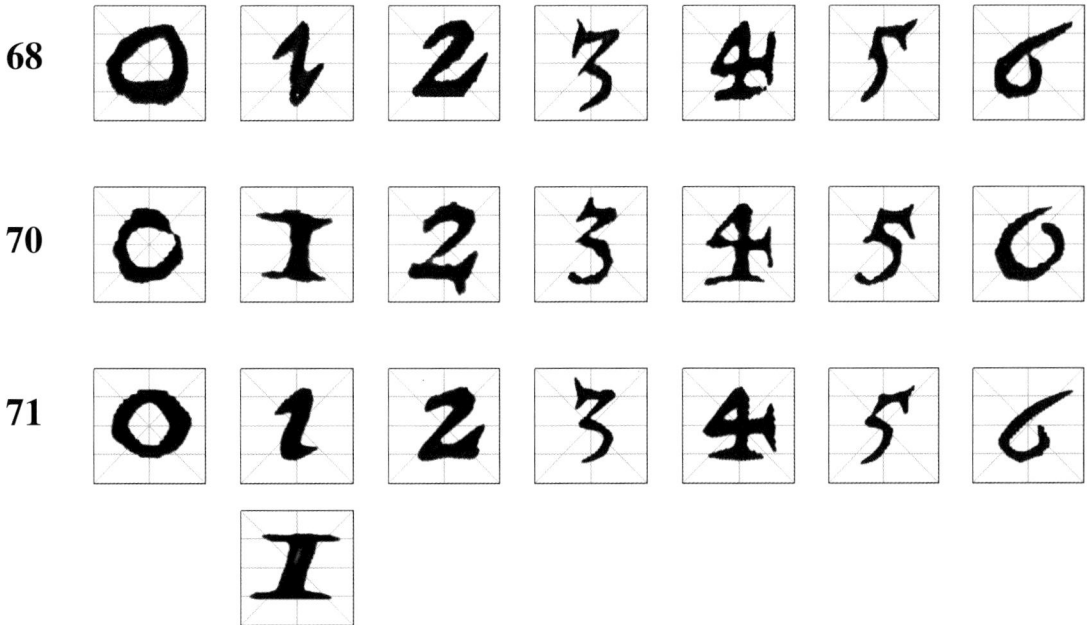

FIG. A11 Comparison of the number forms on two signed and dated instruments by James Kynvyn, and one attributed to Kynvyn.

BIBLIOGRAPHY

Ackermann, S. (ed.) (1998). *Humphrey Cole: Mint, Measurement and Maps in Elizabethan England,* British Museum Occasional Paper, No. 126. British Museum, London.

Ackermann, S. and Cherry, J. (1999). Richard II, John Holland, and Three Medieval Quadrants. *Annals of Science,* **56**, 3–24.

Adams, S. (ed.) (1995). *Household Accounts and Disbursement Books of Robert Dudley, Earl of Leicester, 1558–1561, 1584–1586,* Camden Fifth Series, Vol. 6. Cambridge University Press.

Adams, T.A. and Waters, D.W. (1995). *English Maritime Books Printed before 1801 ....* John Carter Brown Library, Providence, RI, and National Maritime Museum, Greenwich.

Allen, H.S. (1928). James Gregory, John Collins, and Some Early Scientific Instruments. *Nature,* **121** (24 March), 456.

Apian, Peter (1524). *Cosmographicus liber Petri Apiani.* I. Weissenberger, Landshut.

Apian, Peter (1534). *Instrvmentvm primi mobilis.* Petreius, Nuremberg.

Apian, Peter (1541). *Instrvmentvm sinvvm sev primi mobilis.* Petreius, Nuremberg.

Arber, E. (1875). *A List, based on the Registers of the Stationers' Company, of 837 London Publishers ... between 1553 and 1640.* Privately printed, London.

Archinard, M. (1987). A Note on Horizontal Sundials. *Bulletin of the Scientific Instrument Society,* No. 14, 6–7.

Armada (1988). *Armada 1588–1988.* Catalogue of an Exhibition at the National Maritime Museum 20 April–4 September 1988. Penguin Books, London.

Bakich, M.E. (1995). *The Cambridge Guide to the Constellations.* Cambridge University Press.

Barlow, William (1597). *The Navigators Svpply. Conteining many things of principall importance to Nauigation, ...* G. Bishop, London.

Beauchesne, J. de, and Baildon, J. (1570). *A Booke containing divers sortes of hands...* T. Vautrollier, London.

Bendall, S. (1997). *Dictionary of Land Surveyors and Local Map-Makers of Great Britain and Ireland 1530–1850,* 2 vols (2nd edn). British Library, London.

Benham, W.G. (1924). *Benham's Book of Quotations, Proverbs and Household Words.* London.

Bennett, J.A. (1987). *The Divided Circle: A History of Instruments for Astronomy, Navigation and Surveying.* Phaidon, Oxford.

Bennett, J.A. and Johnston, S. (1996). *The Geometry of War 1500–1750.* Museum of the History of Science, Oxford.

Bennett, J.A. and Meli, D.B. (1994). *Sphaera Mundi: Astronomy Books in the Whipple Museum 1478–1600.* Whipple Museum, Cambridge.

Blackmore, H.L. (1976). *The Armouries of the Tower of London: I Ordnance.* HMSO, London.

Blagrave, John (1585). *The Mathematical Jewel ....* Walter Venge, London.

Blair, C. (1964). A Royal Compass-Dial. *The Connoisseur,* December 1964, 246–8.

Bond, J.J. (1875). *Handy-Book of Rules and Tables for Verifying Dates with the Christian Era.* G. Bell & Sons, London.

Bourne, William (1567). *An Almanacke and Prognostication for three yeares ... 1571. and 1572. & 1573.* Thomas Purfoote, London. Transcribed in Taylor (1963).

Bourne, William (1574). *A Regiment for the Sea: Conteyning most profitable Rules, Mathematical Experiences, and perfect knowledge of Nauigation ....* Thomas Hacket, London.

Bourne, William (1578). *The Arte of Shooting in great Ordnance.* H. Bynnyman, London.

Bourne, William (1578). *Inuentions or Deuises.* T. Woodcock, London.

Boxmeer, H. Van (1996). *Instruments anciens de L'Observatoire Royal de Belgique*. Brussels.

Britten, F.J. (1982). *Britten's Old Clocks and Watches and their Makers* (9th edn) (ed. G.H. Baillie, C. Ilbert, and C. Clutton). Bloomsbury Books, London.

Broecke, S. Vanden (2001). Dee, Mercator, and Louvain instrument-making: An undescribed astrological disc by Gerard Mercator (1551). *Annals of Sciences*, **58**, in press.

Brown, J. (1979). *Mathematical Instrument-Makers in the Grocers' Company 1688–1800*. Science Museum, London.

Brown, O. (1982). *Catalogue 1: Surveying*. Whipple Museum, Cambridge.

Bruce, John (1866). Description of a Pocket-Dial made in 1593 for Robert Devereux, Earl of Essex … in a Letter addressed to the possessor of the Dial, Edward Dalton, Esq. LL.D. F.S.A. *Archaeologia*, **40**, 343–56.

Bryden, D.J. (1988). *Catalogue 6: Sundials and Related Instruments*. Whipple Museum, Cambridge.

Bryden, D.J. (1992). Evidence from Advertising for Mathematical Instrument Making in London, 1556–1714. *Annals of Science*, **49**, 301–36.

Burnett, John (1969). *A History of the Cost of Living*. Pelican Books, Harmondsworth.

Capp, B. (1979). *Astrology and the Popular Press: English Almanacs 1500–1800*. Faber & Faber, London.

Catalogo (1954). *Catalogo degli Strumenti del Museo di Storia della Scienza*. Florence.

Challis, C.E. (1975). Mint Officials & Moneyers of the Tudor Period. *British Numismatic Journal*, **45**, 51–76.

Challis, C.E. (1989). The Introduction of Coinage Machinery into the Royal Mint by Eloy Mestrell. *British Numismatic Journal*, **59**, 256–62.

Challis, C.E. (ed.) (1992). *A New History of the Royal Mint*. Cambridge University Press.

Cheney, C.R. (ed.) (1961). *Handbook of Dates for Students of English History*. Royal Historical Society, London.

Ciano, Cesare (1980). *I primi Medici e il mare: Note sulla politica marinara roscana da Cosimo I a Ferdinando I*, Biblioteca del Bollettino storico Pisano, 22. Pacini Editore, Pisa.

Clagett, Marshall (1995). *Ancient Egyptian Science: Vol. 2 Calendars, Clocks, and Astronomy*, Memoirs of the American Philosophical Society, Vol. 214. Philadelphia.

Clark, J.W. (ed.) (1905). *Cantabrigia illustrata*. Macmillan and Bowes, Cambridge.

Clark, J.W. and Gray, A. (1921). *Old Plans of Cambridge 1574–1798: Pt II, Plans*. Bowes and Bowes, Cambridge.

Cleempoel, K. Van (1998). Aspects of Scientific Instrument Production in Louvain between 1550 and 1600. Unpublished PhD thesis. University of London.

Clifton, Gloria C. (1995). *Directory of British Scientific Instrument Makers 1550–1851*. Zwemmer, London.

Coignet, Michiel (1580). *Nieuwe Onderwijsinghe op de principaelste Puncten der Zeeuaert*. Antwerp.

Coignet, Michiel (1581). *Instructions nouvelle des poincts plus excellents & necessaires touchant l'art de nauiger*. Antwerp.

Colvin, Sidney (1905). *Early Engraving and Engravers in England 1544–1695*. London.

Connor, R.D. (1987). *The Weights and Measures of England*. HMSO, London.

Cortés, Martin (1551). *Breve compendio de la esfera y de la arte de navegar*. Seville.

Cortés, Martin (1556). *Breue compendio de la sphera y de la arte de nauegar, con nueuos instrumentos y reglas*. Seville.

Cortés, Martin (1561). *The Art of Nauigation … Written in the Spanysh Tongue by Martin Curtes* (trans. R. Eden). R. Jugge, London.

Crawforth, M.A. (1987). Instrument Makers in the London Guilds. *Annals of Science*, **44**, 319–77.

Cuningham, William (1559). *The Cosmographical Glasse, conteinyng the pleasant principles of cosmographie, geographie, hydrographie, or navigation*. J. Day, London.

Cushing, H. (1962). *A Bio-Bibliography of Andreas Vesalius* (2nd edn). Hamden, CT.

Davis, J. (1595). *The Seamans secrets* …. T. Dawson, London.

Dee, John (1570). *A very fruitfull Præface made by M.I. Dee*. In *The Elements of geometrie of the most auncient Philosopher Evclide of Megare* (trans. H. Billingsley). J. Daye, London.

Dekker, Elly (1995). An Unrecorded Medieval Astrolabe Quadrant from *c*.1300. *Annals of Science*, **52**, 1–47.

Dekker, Elly (1999). *Globes at Greenwich: A Catalogue of the Globes and Armillary Spheres in the National Maritime Museum*. Oxford University Press.

Dekker, Elly and Turner, G. (1997). An Unusual Elizabethan silver Globe by Charles Whitwell. *The Antiquaries Journal*, **77**, 393–401.

Digges, Leonard (1555). *A Prognostication of right good effect.* T. Gemini, London.

Digges, Leonard (1556). *A Prognostication euerlasting of ryght good effecte* (2nd edn). T. Gemini, London.

Digges, Leonard (1556). *A boke named Tectonicon briefelye shewynge the exacte measurynge … all maner lande, squared tymber, stone ….* J. Daye for T. Gemini, London.

Digges, Leonard (1564). *A Prognostication euerlasting of ryght good effecte.* T. Marshe, London.

Digges, Leonard (1571). *A Geometrical Practise, named Pantometria, finished by T. Digges.* Henrie Bynneman, London.

Digges, Leonard (1576). *A Prognostication euerlastinge of righte good effecte.* [With] *A Perfit Description of the Cælestiall Orbes,* by Thomas Digges. T. Marshe, London.

Digges, Thomas (1591). *A Geometrical Practical Treatize named Pantometria.* A. Jeffes, London.

DNB. Dictionary of National Biography. 1885–1900 and later supplements. Oxford University Press.

Dodgson, C. (1917). Note on Thomas Geminus's 'Morysse and Damashin'. *Proceedings of the Society of Antiquaries,* 2nd series, **29**, 210–14.

Donald, M.B. (1955). *Elizabethan Copper: The History of the Company of Mines Royal 1568–1605.* Pergamon Press, London.

Donald, M.B. (1961). *Elizabethan Monopolies: The History of the Company of Mineral and Battery Works from 1565 to 1604.* Oliver and Boyd, Edinburgh.

Doursther, Horace (1840). *Dictionaire universel des poids et mesures anciens et modernes, contenant des Tables des monnaies de tous les Pays.* Brussels; reprinted Meridian Publishing Co., Amsterdam, 1965.

Drake, S. (1978). *Galileo Galilei: Operations of the Geometric and Military Compass.* Smithsonian Institution Press, Washington, DC.

DSB. Dictionary of Scientific Biography. 1970–80 (ed. C.C. Gillispie). Charles Scribner's Sons, New York.

Dudley, Robert (1646). *Dell'Arcano del Mare,* Vol. 1, Books I, II; Vol. 2, Books III, IV. F. Onofri, Florence.

Dudley, Robert (1647). *Dell'Arcano del Mare,* Vol. 3, Books V, VI. F. Onofri, Florence.

Dunn, Richard (1994). The True Place of Astrology among the Mathematical Arts of Late Tudor England. *Annals of Science,* **51**, 151–63.

Eade, J.C. (1984). *The Forgotten Sky: A Guide to Astrology in English Literature.* Oxford University Press.

Edmunds, F. (1872). *Traces of History in the Names of Places; with a vocabulary of the roots out of which names of places in England and Wales are formed.* London.

Ellen, D. (1997). *The Scientific Examination of Documents: Methods and Techniques.* Taylor & Francis, London.

Estácio dos Reis, A. (1997). *Medir Estrelas (Measuring Stars),* text in Portuguese and English. CCT Correios, Lisbon.

Feingold, M. (1984). *The Mathematicians' Apprenticeship: Science, Universities and Society in England 1560–1640.* Cambridge University Press.

Field, J.V. and Wright, M.T. (1985). Gears from the Byzantines: A Portable Sundial with Calendrical Gearing. *Annals of Science,* **42**, 87–138.

Fitzherbert, John (1523). *Here begynneth a ryght frutefull mater: and hath to name the boke of surueyeng and improumentes.* R. Pynson, London.

Gabb, G.H. (1937). The Astrological Astrolabe of Queen Elizabeth. *Archaeologia,* **86**, 101–3.

Gatty, M. (1872). *The Book of Sun-Dials: Collected by Mrs. Alfred Gatty.* London.

Gatty, M. (1889). *The Book of Sun-Dials: Collected by Mrs Alfred Gatty* (2nd edn) (ed. H.K.F. Gatty and E. Lloyd). London.

Gatty, M. (1900). *The Book of Sun-Dials: Originally Compiled by the late Mrs. Alfred Gatty* (4th edn) (ed. H.K.F. Eden and E. Lloyd). London.

Gemini, Thomas (1545). *Compendiosa totius anatomie delineatio, æra exarata.* J. Herford, London. For later editions, see Chapter 3.

Gemma Frisius (ed.) (1529). *Cosmographicus liber Petri Apiani.* R. Bollaert, Antwerp.

Gemma Frisius (1556). *De Astrolabio catholico et usu ejusdem.* Antwerp.

Gemma Frisius (ed.) (1564). *Cosmographia Petri Apiani, per Gemmam Frisivm … Additis eiusdem argumenti libellis ipsus Gemmæ Frisij.* A. Birkmann, Antwerp.

Gemma Frisius (ed.) (1584). *Cosmographia, siue Descriptio vniuersi Orbis, Petri Apiani & Gemmæ Frisij.* Ioan. Bellerum, Antwerp.

Gettings, F. (1990). *The Arkana Dictionary of Astrology* (2nd edn). Arkana (Penguin Group), London.

Gibbs, S.L., Henderson, J., and Price, D. de Solla (1973). *A Computerized Checklist of Astrolabes.* Department of History of Science, Yale University, New Haven, CT.

Gouk, P. (1988). *The Ivory Sundials of Nuremberg 1500–1700.* Whipple Museum, Cambridge.

Gunter, Edmund (1623). *The Description and vse of the Sector. The Crosse-staffe and other instruments*. Printed by William Jones, sold by John Tomson, London.

Gunter, Edmund (1624). *The Description and vse of the Sector. The Crosse-staffe and other instruments*. Printed by William Jones, sold by Edmund Weaver, London.

Gunther, R.T. (1920–23). *Early Science in Oxford*, Vol. 1. Privately published, Oxford.

Gunther, R.T. (1923). *Early Science in Oxford*, Vol. 2. Privately published, Oxford.

Gunther, R.T. (1926). The Astrolabes and Dials of Humphrey Cole of London. *Illustrated London News*, 14 August 1926, 292–4.

Gunther, R.T. (1927). The Great Astrolabe and other Scientific Instruments of Humphrey Cole. *Archaeologia*, **76**, 273–317.

Gunther, R.T. (1929). The Uranical Astrolabe and other Inventions of John Blagrave of Reading. *Archaeologia*, **79**, 55–72.

Gunther, R.T. (1932). *The Astrolabes of the World*. Oxford University Press.

Gunther, R.T. (1937). The Astrolabe of Queen Elizabeth. *Archaeologia*, **86**, 65–72.

Guy, John (1988). *Tudor England*. Oxford University Press.

Guye, Samuel and Michel, Henri (1971). *Time & Space: Measuring Instruments from the 15th to the 19th Century* (trans. D. Dolan). Pall Mall Press, London.

Hall, A.R. (1952). *Ballistics in the Seventeenth Century*. Cambridge University Press.

Hambly, Maya (1988). *Drawing Instruments 1580–1980*. Sotheby's Publications, London.

Hamilton, H. (1967). *The English Brass and Copper Industries to 1800* (2nd edn). Frank Cass, London.

Harley, J.B. and Woodward, D. (1987). *Cartography in Prehistoric, Ancient, and Medieval Europe and the Mediterranean*, The History of Cartography, Vol. 1. University of Chicago Press.

Harvey, Gabriel (1593). *Pierces Supererogation, or, A new prayse of the old asse, a preparatiue to certaine larger discourses, intituled Nashes fame*. J. Wolfe, London.

Harvey, Paul (ed.) (1967). *The Oxford Companion to English Literature* (4th edn). Oxford University Press.

Hausmann, Tjark (1979). *Alte Uhren*, Kataloge des Kunstgewerbemuseums Berlin, Bd. VIII. Kunstgewerbemuseum, Berlin.

Heawood, E. (1932). *English County Maps in the Collection of the Royal Geographical Society*. R.G.S., London.

Higgins, K. (1950). An Elizabethan Quadrant-Dial in Silver by Humphrey Cole. *The Connoisseur*, **125**, 118–19.

Higton, H.K. (1996). Elias Allen and the Role of Instruments in Shaping the Mathematical Culture of Seventeenth-Century England. Unpublished PhD thesis, Cambridge University.

Hind, A.M. (1938). *Early Italian Engraving: A Critical Catalogue ....* Quaritch, London.

Hind, A.M. (1939). An Elizabethan pack of Playing-cards. *British Museum Quarterly*, **13**, 2–4, plates 2, 3.

Hind, A.M. (1952). *Engraving in England in the Sixteenth and Seventeenth Centuries: A Descriptive Catalogue with Introduction*, Part I: The Tudor Period. Cambridge University Press.

Holbrook, Mary (1974). *A Girdle Round About the Earth: Astronomical and Geographical Discovery, 1490–1630*. Holburne of Menstrie Museum, Bath.

Holbrook, Mary (1992). *Science Preserved: A Directory of Scientific Instruments in Collections in the United Kingdom and Eire*. HMSO, London.

Holinshed, R. (1577). *The Chronicles of England, Scotlande, and Irelande*. G. Bishop, London.

Holinshed, R. (1587). *Chronicles, comprising ... the description and historie of England ... Ireland ... and ... Scotland* (ed. W. Harrison *et al.*). I. Harison, London.

Hood, Thomas (1598). *The Making and vse of the Geometricall Instrument, called a Sector*. John Windet, London.

Hopton, A. (1611). *Speculum Topographicum; or The Topographical Glasse*. N.O. for Simon Waterson, London.

Hopton, A. (1612). *A Concordancy of Yeares*. Company of Stationers, London.

Jagger, Cedric (1988). *The Artistry of the English Watch*. David & Charles, Newton Abbot.

Johnston, S. (1991). Mathematical Practitioners and Instruments in Elizabethan England. *Annals of Science*, **48**, 319–44.

Karrow, Jr, R.W. (1993). *Mapmakers of the Sixteenth Century and Their Maps*. Speculum Orbis Press, Chicago.

Keynes, G. (1959). The *Anatomy* of Thomas Geminus: A Notable Acquisition for the Library. *Annals of the Royal College of Surgeons of England*, **25**, 171–5.

Kiely, E.R. (1947). *Surveying Instruments: Their History and Classroom Use*. Columbia University, NY; reprinted Columbus, OH, 1979.

King, D.A. (1975). al-Khalīlī's *Qibla* Table. *Journal of Near Eastern Studies*, **34**, 81–122.

King, D.A. (1987). *Islamic Astronomical Instruments.* Variorum Reprints, London.

King, D.A. and Turner, G.L'E. (1994). The Astrolabe Dedicated to Cardinal Bessarion in 1462, Attributable to Regiomontanus. *Nuncius: Annali di Storia della Scienza*, **9**, fasc. 1, 165–206.

King, G.L. (1996). *Miniature Antique Maps.* Map Collector Publications, Tring.

Koonings, W. (1964). *Negentiende Eeuws Zilver.* Antwerp.

Leader, J. Temple (1895). *Life of Sir Robert Dudley, Earl of Warwick and Duke of Northumberland.* Florence; reprinted Amsterdam, 1977.

Lee, A.G. (1964). *The Son of Leicester: The Story of Sir Robert Dudley.* Gollancz, London.

Leigh, Valentine (1577). *The Moste profitable and commendable science, of Surueying of Landes, Tenementes, and Hereditamentes ....* Andrew Maunsell, London.

Lemay, R. (1978). Gerard Cremona. In *Dictionary of Scientific Biography*, Vol. 15, pp. 173–92. Scribner's, New York.

Linschoten, Jan Huygen van (1598). *Discours of Voyages into yᵉ Easte & West Indies. Deuided into Foure Bookes* (trans. William Phillip). John Wolfe, London.

Lloyd, S.A. (1992). *Ivory Diptych Sundials 1570–1750.* Harvard University Press, Cambridge, MA.

Longstaffe, W.D.H. (1860). The Old Heraldry of the Percies. *Archaeologia Æliana: or, Miscellaneous Tracts Relating to Antiquity*, New Series, **4**, 157–228.

Loomes, B. (1981). *The Early Clockmakers of Great Britain.* N.A.G. Press, London.

Lucar, Cyprian (1590). *A treatise named Lucarsolace.* R. Field for J. Harrison, London.

Macdonald, A.R. and Morrison-Low, A.D. (1994). *A Heavenly Library: Treasures from the Royal Observatory's Crawford Collection.* Royal Observatory & National Museums of Scotland, Edinburgh.

Maddison, F. (1966). Hugo Helt and the Rojas Astrolabe Projection. *Rivista di Faculdade de Ciências*, **39** (Agrupamento de Estudos de Cartografia Antiga, XII, secção de Coimbra). Coimbra.

Madrid (1997). [Exhibition catalogue] *Instrumentos cientificos del siglo XVI: La Corte española y la Escuela de Lovaina.* Fundación Carlos de Amberes, Madrid. [English text] *Scientific Instruments in the Sixteenth Century: The Spanish Court and the Louvain School.* Idem.

Malin, S.R.C. and Bullard, E. (1981). The Direction of the Earth's Magnetic Field at London, 1570–1975. *Philosophical Transactions of the Royal Society*, **299**, 357–423.

Meskens, A. (1992). Michiel Coignet's Nautical *Instruction. Mariner's Mirror*, **78**, 257–76.

Meskens, A. (1997). Michiel Coignet's Contribution to the Development of the Sector. *Annals of Science*, **54**, 143–60.

Meskens, A. (1998). *Familia universalis: Coignet.* Koninklijk Museum voor Schone Kunsten, Antwerp.

Metcalfe, W.C. (1885). *A Book of Knights Banneret, Knights of the Bath, and Knights Bachelor.* Mitchell & Hughes, London.

Michel, Henri (1947). *Traité de l'Astrolabe.* Gauthier-Villars, Paris; reprinted 1976.

Michel, Henri (1961). Thomas Lambert ou Lambrechts, dit Gemini. In *Biographie Nationale ... de Belgique*, Vol. 31, cols 386–95. Émile Bruylant, Brussels.

Michel, Henri (1966). *Les Cadrans Solaires de Max Elskamp.* Editions du Musée Wallon, Liège.

Michel, Henri (1974). *Catalogue des Cadrans Solaires du Musée de la Vie Wallonne* (2nd edn). Editions du Musée Wallon, Liège.

Millburn, J.R. (1995). William Deane and his Ordnance Bills. *Bulletin of the Scientific Instrument Society*, No. 45, 12–14.

Minadoi, G.T. (1595). *History of the Warres between the Turkes and Persians* (trans. A. Hartwell). London.

Mitchell, G.A.G. (1953). Who was Geminus? *British Medical Journal*, 27 June, 1449–50.

Moreland, C. and Bannister, D. (1989). *Antique Maps: A Collector's Guide* (3rd edn). Phaidon–Christie's, London.

Mortimer, C. (1987). X-ray Fluorescence Analysis of Early Scientific Instruments. *Proceedings of 1987 Archaeometry Symposium*, 311–17.

NMM (National Maritime Museum) (1976). *The Planispheric Astrolabe.* National Maritime Museum, Greenwich.

Norman, Robert (1581). *The New attractive; Containyng a short discourse of the magnes or lodstone.* J. Kyngston, London.

O'Malley, C.D. (1959). Introduction to a facsimile of the 1553 edition of Gemini, *Anatomy.* Dawson, London.

Ortelius, Abraham (1570). *Theatrum orbis terrarum.* Egidius Coppens Diesth, Antwerp.

Ortroy, F. Van (1920). *Bio-Bibliographie de Gemma Frisius.* Académie Royale de Belgique, Memoires, 2nd series, Vol. 11, fascicule 2. Brussels.

Osley, A.S. (1969). *Mercator: A Monograph on the Lettering of Maps, etc in the 16th century Netherlands.* Faber and Faber, London.

Osley, A.S. (1980). *Scribes and Sources: Handbook of the Chancery Hand in the Sixteenth Century.* Faber and Faber, London.

Oughtred, William (1632). *Circles of Proportion and the Horizontall Instrument, Both inuented and the vses of both Written in Latine by Mr. W.O.* A. Mathewes for Elias Allen, London.

Oughtred, William (1633). *Circles of Proportion and the Horizontall Instrument, Both inuented and the vses of both Written in Latine by Mr. W.O.* [With] *An Apologeticall Epistle.* A. Mathewes, London.

Patten, J. (1978). *English Towns 1500–1700.* Dawson, London.

Pollard, A.M. (1984). Authenticity of Brass Objects by Major Element Analysis. Unpublished proceedings of the 1984 Archaeometry Symposium. Oxford.

Price, D.J. (1955). An International Checklist of Astrolabes. *Archives Internationales d'Histoire des Sciences,* **8**, 243–63, 363–81.

Ræder, H., Strömgren, E., and Strömgren, B. (eds) (1946). *Tycho Brahe's Description of his Instruments and Scientific Work.* Det Kongelige Danske Videnskabernes Selskab, Copenhagen.

Rasquin, V. (1984). Les Instruments non mecaniques. In *La Mesure du Temps dans les Collections belges,* Exposition organisée par la Société Générale de Banque, 26–1/7–4–1984, pp. 27–143. Brussels.

Rathborne, Aaron (1616). *The Surveyor in Foure Bookes.* W. Stansby, London.

Recorde, Robert (1543). *The Ground of Artes Teachyng the Worke and practise of Arithmetike.* R. Wolfe, London.

Reisch, Gregorius (1512). *Margarita Philosophica.* Freiburg.

Richeson, A.W. (1966). *English Land Measuring to 1800: Instruments and Practices.* MIT Press, Cambridge, MA.

Richmond, B. (1956). *Time Measurement and Calendar Construction.* E.J. Brill, Leiden.

Röttel, Karl (ed.) (1995). *Peter Apian: Astronomie, Kosmographie und Mathematik am Beginn der Neuzeit.* Polygon, Eichstätt.

Saunders, H.N. (1984). *All the Astrolabes.* Senecio, Oxford.

Saxton, Christopher (1579). *Atlas of the Counties of England and Wales.* London.

Schilder, Günter (1987). *Monumenta cartographica Neerlandica, II.* Canaletto, Alphen aan den Rijn.

Schneider, I. (1970). *Der Proporzionalzirkle, ein universelles Analogrecheninstrument der Vergangenheit,* Abhandlungen und Berichte des Deutschen Museums, Heft 2. Munich.

Shaw, W.M. (1906). *The Knights of England.* London.

Skelton, R.A. (1965). *Decorative Printed Maps of the 15th to 18th Centuries* (2nd edn). Spring Books, London.

Skelton, R.A. (1970). *County Atlases of the British Isles 1579–1850.* Carta Press, London.

Smith, Lucy T. (ed.) (1906). *The Itinerary of John Leland in or about the years 1535–1543.* London.

Stern, V.F. (1979). *Gabriel Harvey, his Life, Marginalia and Library.* Oxford University Press.

Stimson, Alan (1988). *The Mariner's Astrolabe: A survey of Known, Surviving Sea Astrolabes.* HES, Utrecht.

Stöffler, Johann (1524). *Elucidatio fabricae ususque astrolabii.* J. Köbel, Oppenheim.

Stone, P.G. (1906). A Sixteenth Century Mathematical Instrument Case. *Archaeologia,* **50**, 531–3, plates 42, 43.

Stott, Carole (1991). *Celestial Charts: Antique Maps of the Heavens.* Studio Editions, London.

Strong, Roy (1987). *Gloriana: The Portraits of Queen Elizabeth I.* Thames and Hudson, London.

Tartaglia, Niccolì (1546). *Quesiti et inventioni diverse.* Venice.

Taylor, E.G.R. (1930). *Tudor Geography 1485–1583.* Methuen, London.

Taylor, E.G.R. (1934). *Late Tudor and Early Stuart Geography 1583–1650.* Methuen, London.

Taylor, E.G.R. (1951a). The Oldest Mediterranean Pilot. *Journal of the Institute of Navigation,* **4**, No. 1.

Taylor, E.G.R. (1951b). Early Sea Charts and the Origin of the Compass Rose. *Journal of the Institute of Navigation,* **4**, No. 4.

Taylor, E.G.R. (1954). *The Mathematical Practitioners of Tudor & Stuart England.* Cambridge University Press.

Taylor, E.G.R. (ed.) (1963). *A Regiment for the Sea and other writings on navigation by William Bourne of Gravesend, a gunner,* Hakluyt Society, 2nd series, Vol. 121. Cambridge University Press.

Taylor, E.G.R. (1971). *The Haven-Finding Art: A History of Navigation from Odysseus to Captain Cook* (new edn). Hollis & Carter, London.

Thompson, David (1987). A Verge Watch with Alarm and passing strike by Isaac Simmes, London. *Antiquarian Horology*, **16**, 499–506.

Thompson, H.Y. (ed.) (1919). *Lord Howard of Effingham and the Spanish Armada, with exact facsimiles of the 'Tables of Augustine Ryther', A.D. 1590 ....* Roxburghe Club, London.

Turnbull, H.W. (ed.) (1939). *James Gregory Tercentenary Memorial Volume.* Bell, London.

Turner, A.J. (1981). William Oughtred, Richard Delamain and the Horizontal Instrument in seventeenth-century England. *Annali dell'Istituto e Museo di Storia della Scienza di Firenze*, **6**, fasc. 2, 99–125.

Turner, G.L'E. (1983). Mathematical Instrument-Making in London in the Sixteeth Century. In Tyacke, S. (ed.) (1983). Pp. 93–106, plates 47–52.

Turner, G.L'E. (1985). Charles Whitwell's Addition, *c.* 1595, to a Fourteenth-Century Quadrant. *The Antiquaries Journal*, **65**, Pt. 2, 454–5, 476, plates 99–101.

Turner, G.L'E. (1990). *Scientific Instruments and Experimental Philosophy 1550–1850.* Variorum, Aldershot.

Turner, G.L'E. (1991). *Gli Strumenti. Storia della Scienza*, Vol. 1 (ed. P. Galluzzi). Einaudi, Turin.

Turner, G.L'E. (1994). The Three Astrolabes of Gerard Mercator. *Annals of Science*, **51**, 329–53.

Turner, G.L'E. (1995). The Florentine Workshop of Giovan Battista Giusti, 1556–*c.* 1575. *Nuncius: Annali di Storia della Scienza*, **10**, fasc. 1, 131–72.

Turner, G.L'E. (1996). Later Medieval and Renaissance Instruments. In *Astronomy Before the Telescope* (ed. C. Walker), pp. 231–44. British Museum Press, London.

Turner, G.L'E. (2000). A Critique of the Use of the First Point of Aries in Dating Astrolabes. In *Sic itur ad astra: Studien zur Geschichte der Mathematik und Naturwissenschaften. Festschrift für den Arabisten Paul Kunitzsch zum 70. Geburtstag,* (ed M. Folkerts and R. Lorch), pp. 548–54. Harrassowitz, Wiesbaden.

Turner, G.L'E. and Dekker, E. (1993). An Astrolabe attributed to Gerard Mercator, *c.* 1570. *Annals of Science*, **50**, 403–43.

Turner, G.L'E. and Cleempoel, K. Van (2000). A Tudor Astrolabe by Thomas Gemini and its Relationship to an Astrological Disc by Gerard Mercator of 1551. *The Antiquaries Journal*, **80**, in press.

Tyacke, S. (ed.) (1983). *English Map-Making 1500–1650.* British Library, London.

Tyacke, S. and Huddy, J. (1980). *Christopher Saxton and Tudor map-making.* British Library, London.

Ubaldini, P. (1590). *A Discovrse concerninge the Spanishe fleete inuadinge Englande in the year 1588 ... and translated for A Ryther: ....* A. Ryther, London.

Underwood, E.A. (1953). Medicine and the Crown. *British Medical Journal*, 30 May, 1185–90.

Usherwood, S. and Usherwood, E. (1983). *The Counter-Armada, 1596: The Journall of the 'Mary Rose'.* Bodley Head, London.

Vaughan, D. (1993). A very Artificial Workman: The Altitude Sundials of Humphrey Cole. In *Making Instruments Count: Essays on Historical Scientific Instruments presented to Gerard L'Estrange Turner* (ed. R.G.W. Anderson, J.A. Bennett, and W.F. Ryan), pp. 191–200. Variorum, Aldershot.

Vesalius, Andreas (1543). *De Humani corporis fabrica.* Basel.

Wagenaer, Lucas Janz. [1588]. *The Mariners Mirrovr...* (trans. with additions by A. Ashley). J. Charlewood, London.

Wallis, R.V and Wallis, P.J. (1993). *Index of British Mathematicians. Part 3: 1701–1800.* Project for Historical Biobibliography, Newcastle upon Tyne.

Ward, F.A.B. (1981). *A Catalogue of European Scientific Instruments in the Department of Medieval and Later Antiquities of the British Museum.* British Museum, London.

Warner, G.F. (ed.) (1899). *The voyage of Robert Dudley ... to the West Indies, 1594–1595,* Hakluyt Society, 2nd series, Vol. 3. Hakluyt Society, London.

Watelet, Marcel (ed.) (1994). *Gérard Mercator, cosmographe.* Fonds Mercator Paribas, Antwerp.

Waters, D.W. (1958). *The Art of Navigation in England in Elizabethan and Early Stuart Times.* Hollis & Carter, London.

Waters, D.W. (1966). *The Sea- or Mariner's Astrolabe,* Agrupamento de Estudos de Cartografia Antiga, XV, secção de Coimbra. University of Coimbra.

Webster, R. and Webster, M. (1998). *Western Astrolabes,* Historic Scientific Instruments of the Adler Planetarium & Astronomy Museum, Vol. 1. Adler Planetarium, Chicago.

Williams, A. and de Reuck, A. (1995). *The Royal Armoury at Greenwich 1515–1649: A History of its Technology.* Royal Armouries Monograph 4. London.

Winkler, W. (ed.) (1988). *A Spectacle of Spectacles*. Edition Leipzig, Leipzig.

Worsop, Edward (1582). *A Discoverie of Sundrie Errours and Faults Daily Committeed by Landemeaters Ignorant of Arithmeticke and Geometrie*. H. Middleton, London.

Wray, E.M. (1984). *Historical Scientific Instruments from the Collection of the Department of Physics, University of St. Andrews*. University of St Andrews.

Wright, Edward (1599). *Certaine Errors in Navigation, arising either of the ordinarie erroneous making or using of the sea chart ...*. V. Sims, London.

Wright, Edward (1610). *Certaine Errors in Navigation, detected and corrected ... with many additions that were not in the former edition*. F. Kingston, London.

Wright, J.K. (1923). Notes on the knowledge of latitudes and longitudes in the middle ages. *Isis*, **5**, 75–98.

Wynter, H. and Turner, A. (1975). *Scientific Instruments*. Studio Vista, London.

Zinner, E. (1990). *Regiomontanus: His Life and Work* (trans. E. Brown). North-Holland, Amsterdam.

Zupko, R.E. (1981). *Italian Weights and Measures from the Middle Ages to the Nineteenth Century*. American Philosophical Society, Philadelphia.

INDEX

Numbers in normal type are page numbers; numbers in bold type are catalogue numbers